Industrial Safety, Health Environment and Security

Industrial Safety, Health Environment and Security

By

Basudev Panda

O.A.S. (Retd.) M.A. in Political Science, L.L.B., M.B.A. (HRM),
Diploma in HSE and Security Handled Industrial Safety,
Occupational Health, Environment and Security Functions as
Head of Government Offices, Public Sector Undertaking
and Multinational Companies, Balsore
Orissa

UNIVERSITY SCIENCE PRESS

(An Imprint of Laxmi Publications Pvt. Ltd.)

BANGALORE • CHENNAI • COCHIN • GUWAHATI • HYDERABAD
JALANDHAR • KOLKATA • LUCKNOW • MUMBAI • PATNA
RANCHI • NEW DELHI

Published by :

UNIVERSITY SCIENCE PRESS

(An Imprint of Laxmi Publications Pvt. Ltd.)

113, Golden House, Daryaganj,
New Delhi-110002

Phone : 011-43 53 25 00
Fax : 011-43 53 25 28

www.laxmipublications.com
info@laxmipublications.com

First Edition : 2011

OFFICES

✆ **Bangalore**	080-26 75 69 30	✆ **Chennai**	044-24 34 47 26
✆ **Cochin**	0484-237 70 04, 405 13 03	✆ **Guwahati**	0361-251 36 69, 251 38 81
✆ **Hyderabad**	040-24 65 23 33	✆ **Jalandhar**	0181-222 12 72
✆ **Kolkata**	033-22 27 43 84	✆ **Lucknow**	0522-220 99 16
✆ **Mumbai**	022-24 91 54 15, 24 92 78 69	✆ **Patna**	0612-2300097
✆ **Ranchi**	0651-221 47 64		

UIS-9619-245-INDUS SAFE HEAL ENV SEC-PAN C—

Typesetted at : Shubham Composer, New Delhi. Printed at : Repro Knowledgecast Limited, Thane

Dedicated to

Bapi (Piyush)

My Engineer (Mech.) Son, abroad

With Love for his better appreciation and guidance

CONTENTS

PREFACE

William Shakes Peare in his famous Drama, Henry IV, has said "*Out of this nettle, danger, we pluck this flower, safety*". In God's creation, danger and humanfalibility do exist in all spheres. Industry too has it's share. Since independence, India has tried to take wide strides by embarking upon a planned economy with a view to generate all-round industrial development. In recent years, industrial development has resulted in advent of large enterprises with huge labour force. Obviously, the larger the labour force, the greater are the potential problems, particularly, their safety. Besides, the recent trend indicates that the employee can no longer be viewed as a commodity one must heed the writing on the wall. The state must move the alert manager to a recognition of the importance of the safety approach. The labour force today is better organized through the increasing strength of trade unions and their mighty movements.

Worldwide, every year, around 60,000 people die and huge numbers are disabled for life in accidents. The I.L.O. findings portray an alarming picture on this score. Compared to other countries, there is precious little authentic data in respect of accident rates, causes and preventive measure taken by the Indian industries. No agency till date has been assigned the responsiblility to compile such records and no voluntary efforts have been made in this regard. However, as per one report at an all India level, 165 per one thousand workers get injured during construction activities. This is obviously very high, compared to the rates in the developed countries. Findings of I.L.O. reveal that the accident rate among industrial workers is highest in India, touching 4 per 1000 and a major share of it is accounted for by the building and construction sector.

The chief cause of the majority of safety related losses is the absence of systematic process to identify and mitigate work place hazards and total unsafe work practices. It is the product of failure of supervision and management to effectively communicate the importance the safety has, on the continuing economic viability of the organization employing them.

The connotation of safety is far more than a worker, wearing a hat in the workshop of a factor on construction sites. It is rather a philosophy that identifies and eliminates job site hazards throughout life cycle of a project. It is a philosophy that discourages work practices that place individuals at risk of injury and death. It is the integration of safety in to the daily work process. It is also the promotion of an environment, where each person in the hierarchy, has a role and responsibility for safety. It is an established fact that out of one hundred accidents, eighty eight account for human error and only twelve relate to machine failures. There are numerous international codes, laws, and legislations in developed countries regulating the whole gamut of industrial activities starting from the drawing and design state up to completion and regular operation, upkeep and maintenance thereafter. In India, there are numbers of codes framed, basing on The Factory Act 1948 for protection of health and environment at work places. As regards construction industry.

National Building Code and other dozens of codes have been framed to regulate unsafe practices of late, Government of India (Ministry of Labour and Employment) framed "The Building and other workers (Regulation of Employment, Construction of Service Act 1996C Act 27 of 1996) on 9th August 1996, and Occupational Health and Safety Act 2004, which insulates the workers against hazards, exploitation and largely guarantees safety.

With regard to protection of environment, urban heat island is considered as one of the major problems in the 21st century, posed to human being due to rapid urbanization and industrialization. The large amount of heat generated from urban structures (Greenhouse gases) as they consume and re-radiate solar radiations and from the anthropogenic heat sources are the main causes of urban heat island. The two heat sources increase the temperatures of an urban area as compared to it's surroundings, which is known as U.H.I. The Problem is even worse in cities and metropolises with large population and extensive economic activities. The estimated three billion people living in the urban areas in the world are directly exposed to these problems. Environment friendly construction (Green Building) carries lots of messages for the planners and engineers to develop and adopt, for shaking the conventional system.

Coming to the security system in our country, the threat perception from within and without has also posed a gigantic problem. The Maoists, Naxalites, the Zehadi forces and dozens of disruptive groups are having their operations. The state force is found to be inadequate to tackle this problem. Therefore, the institution of private security has largely entered in to this arena. Both the wings *i.e.*, public police and private security should work with active partnership mean while, cogitation and articulation be continued to pave the way for a meaningful and healthy relationship between law enforcement and private security.

This book is thus an attempt to help entrepreneurs, managers and general public, who are required to perform the safety functions. It is aimed at aiding such functionaries, develop the appropriate procedures and attitudes, whereby they would succeed in encouraging initiative and foster enthusiasm among their colleagues, subordinates and immediate stake holders. Undoubtedly, this book is likely to help the policy makers in framing stringent laws in order to enforce the existing safety system and encourage liberalization safer adoption of public, private partnership policy is concerned.

The author hopes that this book would be found beneficial by the students of safety management and fire engineering in India and abroad. The author would heartily welcome suggestions from all concerned to enable him make the subsequent editions even more useful.

I would conclude in quoting Mr. Peter Drucker. "*The first duty of business is to survive and the guiding principle of business economics is not the maximization of profit, but the avoidance of loss*".

—Author

1

INDUSTRIAL SAFETY

INTRODUCTION

The world at present is passing through a very critical stage and the very existence of the entire animal and plant life is in danger. Man has strived to go beyond and behind nature to seek the meaning and purpose of existence. Is nature's realm only a veil behind which lies concealed an invisible inscrutable ineffable reality? Are natural objects and effects but a flickering procession of shadows of things beyond the reach of human sense as Plato speculated in his haunting parable of the cave. Are they merely the illusory, ephemeral appearances of an absolute that lies beyond speech, thought, invention, progress and prosperity? The man is madly in search of the absolute. Continual changes and development all over the world are accompanied with changes in machinery, technologies, materials used and changes of working means and labour relations. These changes no doubt require adequate adjustments in labour protection. In many areas, this regulation has caused changes in previous approaches and methods and also in the philosophy of labour protection, increasing moral, social, economical and legal obligations. One of the key principle of labour protection policy is the obligation to assess risk at work. This principle has developed into comprehensive theories from a new scientific field—industrial hazard and prevention studies.

Industrial hazard effect to Health and Safety of Humans. Execution of adverse measures according to safety regulations and standards are not enough to ensure safety at work. It is necessary over the scope of legal requirements to assess risks. The border line of the acceptable risk is not fixed, it changes according to the degree of technological and cultural standards. Employees, users and other persons concerned have to be informed, on residual risks *i.e.*, on the possibilities of accident occurrence, prevention and in worst cases, minimal causations. Methods of the risk assessment set procedures and systematically those things, which may be dangerous for people and on the other hand to focus on the most serious faults and risks. The risk has to be coped with, by those who create it *i.e.*, the designer in his design, producer in his product, employer in the work, he sets for accomplishment. An overall understanding these principles and their application in practice as well as acquiring adequate knowledge on methods of risk assessment in real life, confronts with certain difficulties, particularly in small and medium enterprises, where experts in this field are not available. Even if they are available in small measures, the entrepreneurs do not engage them on the plea of cost effectiveness.

This is unarguably a suicidal policy, they adopt. Scientific analysis of costing a product which includes the exact cost of production (Fixed and variable cost), cost per unit decreases, when production increases. To keep the usual tempo, tone and tenor of production, every thing should move as per set norms keeping an eye on the target. A commodity produced in large numbers, reduces the cost per unit, although profit margin remains the same. An accident results less production. In order to recoup, the company has to put its extra efforts to keep hours of work, paying overtime, engaging more hours and replacement etc., payment of compensation, 'insured sum', satutory claims, fighting litigations etc. In any case, the variable cost price goes up, and the company fails to compete with the other of his peers in the market, which undesirably results in increase of post price and drastic fall of profit. These concepts naturally are applicable to each and every interruption in the work process, which in turn is a result of some wasted activity or resource.

Therefore without a doubt, every business interruption must effect profits. In fact, one invariably finds a definite co-relation between profits and the disabling injury frequency rates of firms. One of the reasons for this is that those elements which assist in running a business efficiency, also contribute to reducing accidents. Examples include-selection, training, placement, use of correct tools, work methods and communication.

The effect on the profit margin is tremendous. However, safety and security professionals must ensure that the cost of downgrading incidents are effectively one communicated to executive management. If done, the management will be motivated and certainly committed to the loss control programme.

Therefore, it is necessary to identify and establish control for all that could result in maior or catastrophic damage of people, property and the profit.

WHAT IS SAFETY?

H.W. Heinoich is considered as father of safety. According to him it means. *"Freedom from Danger/ Hazards or freedom of person from injuries, diseases and of property from damage."*

Hazard: In simple term, it is a method to identify hazards at various stages. It is a technique for systematic examination of process, engineering, intentions to assess potential of all conceiveable maloperation, malfunction and their consequences.

Accident: An accident is an unplanned and uncontrolled event in which the action or reaction of an object, substance, person or radiation, results in personal injury or probability there of.

History of safety movement: Safety movement is as old as man's struggle for existence and supremacy world story is replete with internecine wars among various clans, tribes, groups to take over possession of fiefdoms, territories and fertile lands for habitation, cultivation, trade and commerce. It is as old as Ramayan, the epic, that vividly described the wars between Aryas and Asuras. The fierce war between Ram and Ravan reveals, how both the fighting commanders adopted safety strategies to minimise loss of life and property in their side. Mahabharat is replete with such strategies like "Vyuhas" adopted by Kauravas and Pandavas. In each side, the war mongers, the commanders had the gumption to innovate new strategies for minimal loss prevention. American war of Independence in the year 1776 exhibited various safety and security strategies adopted by both the parties, *i.e.*, Rulers and the revolutionaries. However, safety movement become an integral part of the society with the advent of industrial revolution in the year 1793. It took shape and formally adopted by the administration in the year 1833, when industrial revolution started in

Europe. In U.K., the textile factories came into existence and various machineries were invented to accelerate production. The factory owners and employers adopted policy of laisserz-faire and were in a dominating situation. They dictated both the wages and conditions of service of the workers resulting in many industrial and social ills, such as low wages, unduly long hours of work, poor working conditions and persecution of trade union activity. The plight of workers was miserable. With the invention of new machines, the working condition in the factories became horrible. The safety aspect was totally absent and there was absolutely no law to curb the exploitation of workers and their protection from accident occurrences, improvement of working conditions and payment of compensation to the accident victims. A new awakening among the working class was brought about and the 1st, one such law "Factory Act" came in to existence. This historical Act curbed the exploitation of workers and various safety measures in the dingy factory premises were introduced. In the year 1866, first national Board of Fire under writers came in to force in U.S.A. Similarly, workmen compensation laws were Implemented in U.S.A. in the year 1911 to compensate accident victims in Industries. The 1st, world war too brought some new changes. After the war, workers began to take retaliatory action to help themselves get a new deal. They began to strike against safety abuse and apathy of the owners. In their turn, the Employers retaliated by declaring lock-outs and thus Industrial war started. In the beginning, the Government did not intervene but later forced to frame enactments to protect the workers and employees of industries and Establishments, including its own departments. In 1929, H.W. Heinrich brought about his pioneering work "Foundation of a major injury" which created safety awareness tremendously among the people, largely involved with various industries. In the year 1931, his "Domino Theory" highlighted the science of accident investigation.

He evolved five link chain:

(*a*) Ancestry and social environment

(*b*) Fault of person

(*c*) Unsafe condition and unsafe act

(*d*) Accident

(*e*) Injury.

According to him, the easiest link to be effectively dealt is the unsafe act and unsafe condition. Again, the global havoc, that was created during the 2nd world war, the life and property destroyed and the plight of the wounded and disabled, created awareness among the waring nations and as a result, Army Services Forces Safety Programme was introduced in U.S.A. in the year 1942. Coming to the Railway Safety, George Stephenson, suggested use of steam whistle on locomotives in the year 1833. Likewise, safety of life at sea convention was held in London in the year 1914.

The Wanton destruction and injury that occurred during Cremean War between England and France, brought about a sea change in International Law of War and Peace and Safety in Health and Environment. Florence Nightingale, reverentially called as "The Lady with the Lamp", a french nurse, set a new trend and ideology in treatment of war wounded soldiers. Her struggles and sufferings changed the psychology of the statesmen all over the globe and safe management of heath of wounded and battered soldiers was given prime importance.

In it's gradual development, safety promotion and propaganda was in it's peak starting from 1935 to 1960. Post 1960, saw massive efforts made in managing the accidents.

Most countries now have wide sweeping legislation on the following areas imposing heavy duties and responsibilities on the shoulder of the employers.

(*a*) To provide safe place of work.

(*b*) To provide safe system of work.

(*c*) To provide safe access and egress.

(*d*) To provide adequate information.

(*e*) To provide safe plant, equipment and facilities.

(*f*) To provide adequate supervision.

A vast amount of detail regarding the physical state of the premises or mine, plant and equipments together with the Personal Protective Equipments (PPEs.), that must be supplied and worn, is elaborately spelt out in various Acts, statutes and information circulars.

Responsibility of the Management

Almost all the Industry owners have fallen in to the trap of thinking that the worker's compensation will meet all the costs involved in case of accident victim. This certainly is misconceived and largely fallacious. Assuming that the workman is covered even if the person is off work, he is entitled to receive compensation in the form of percentage of his salary. This, otherwise mean that in order to meet his financial commitments (various payments), he may well have to face hardships. Most countries have legislation of this nature, but there are limitations to the compensation which can be paid. Therefore management has the responsibility of organising it's work in such a manner that workers are not injured.

HAZARD AND OPERABILITY STUDIES (HAZOP)

One of the key principles of labour protection policy is the obligation to assess risk at work. This principle has developed into comprehensive theories of safety and accident prevention, and today these theories form a new scientific field (HAZOP).

Safety means acceptance of certain degree of risk. Hazard is anything capable of producing adverse effect to the health and safety of Humans. Execution measures according to safety regulations and standards are not enough to secure safety at work. It is necessary over the scope of legal requirements to assess risks. Employees are to be informed on risk factors. The risk has to be coped with by those who create it *i.e.*, designer in his design, producer in his product, employer in the work, he gives.

The technique of Hazard and Operability Studies (HAZOP) has been used and developed over approximately four decades for identifying potential hazards and operability problems caused by deviation from the design intent of both new and existing process plants. HAZOPS were initially invented by Mond Division of ICI in 1960 in U.K. but the technique only started to be more widely used within the chemical process industry after the flixborough disaster, in which a chemical plant explosion killed 28 people. Bhopal tragedy in India in 1984 is an glaring example of man made disaster. Initially, the system was adopted by Petroleum Industry, which has similar potential for major hazards. This was followed by Food and Steel Industries.

HAZOP study is done to identify hidden hazards internal to a process, to Identify accident potentials and to derive remedial measures, to either eliminate hazard or minimise the risk involved.

What is HAZOP?

In simple term, it is a method to identify hazards at various stages. It is a technique for systematic examination of process/Engineering intentions to assess the hazard potential of all conceivable mal-operation/mal-function and their consequences. It defines hazards at various stages. It is an examination procedure and the purpose is to identify all possible deviations. Once identified, an assessment is made as to whether such deviations and their consequences can have a negative effect upon the safe and efficient operation of the plant. Then action is taken to remedy the situation.

Costs of HAZOP

Can you afford, the cost of process failure? Are you prepared to risk injury to your employees and the consequent costs resulting from injuries and their investigation. If the answer is no, then the cost of HAZOP is well spent.

Why HAZOP study?

The following are the reasons:

1. A potential dangerous plant has to be properly designed.
2. Even if the current safety performances is good, a desire to improve upon it is always there.
3. Current safety measures are inadequate,
4. The last is to satisfy Statutory Authorities to avoid punishments.

Objective of HAZOP

1. To identify hidden hazards,
2. To identify accident potential,
3. To derieve remedial measures to elimate or minimise the risks involved.

HAZOP study approach

There are four approaches:

1. Evolutionary approach
2. Educational approach
3. Do it yourself approach
4. The crash programme.

Explain in details

1. An introductory study is conducted on trial basis to determine whether the technique can be applied and convince the management that it is worth trying. Once it is adopted and benefits accrued, it could be extended to other parts too and becomes part of the standard procedure.
2. Necessary training is imparted to cover more and more employees.
3. A suitable person is appointed as study leader who, studies HAZOP. Selects team, train the team members and arrange trial study.

HAZOP Study Steps

The steps followed in the study are the assessment of:

1. Intention
2. Deviation
3. Causes
4. Consequences
5. Safeguards
6. Remedial measures.

1. **Intention:** It defines how the part is expected to operate. It can be diagrammatic or descriptive.
2. **Deviation:** These are departures from intention which are discovered by systematically applying guide words.
3. **Causes:** Potential causes which would result in the deviation occurring.
4. **Consequences:** The consequences which would arise both from the effect of the deviation and if appropriate, from the cause itself. When assessing the consequences, one should not take any credit for protective systems or instruments, which are already included in the design.
5. **Safeguards:** Any existing protective device, which either prevent the cause or safeguard against the adverse consequences would be recorded in this column. Safeguards need not be restricted to hardware-where appropriate. Credit can be taken for procedural aspects such as regular plant inspections, if any one is sure that they will actually be carried out.

   Where credible cause results in a negative consequence, it must be decided whether some action should be taken. It is at this stage that consequences and associated safeguards are considered. If it is deemed that the protective measures are adequate then no action would be taken and recorded in the action column. Action to contain hazards are of 4 kinds.

   (*a*) A change in the process

   (*b*) A change in the process conditions.

   (*c*) An alternation to physical design.

   (*d*) A change in the operative method.

Action Falls into Two Groups

1. Action that removes the cause.
2. Action that mitigate or eliminate the consequences.

The risk assessment theory and practice are opening a new point of view on labour protection and safety of Technical equipment. It requires changes in the process and forms of preventive measures and changes in health and safety protection management system. Zero level of risk does not exist, nothing is absolutely safe. It is critical to perform risk assessment for dangers that may cause an accident or working disease.

Safety and reliability in the design of plant intially relies upon the application of various codes of practice or design codes and standards. These represent accumulation of knowledge and experience of both individual experts and the industry as a whole. Such application is usually backed by the experience of the engineers involved, who might well have been previously concerned with the design, commissioning or operation of similar plant. HAZOP is a better measurement or performance and is an independent assessment. It results in reduced injury, ill health, reduced property damage, reduced direct and indirect costs of injury and health, improved productivity and profitability from better designed work places and methods, higher morale amongst employees, improved industrial relations and the satisfaction of legal obligations on management and Employees.

FUNDAMENTALS OF INDUSTRIAL SAFETY, TYPES AND CATEGORISATION

Why Injuries take Place?

H.W. Heinrich in his book "Industrial Accident Prevention" published in the year 1920 developed certain axioms, which today form the very foundation of accident prevention. He attempted to analyse and categorise the various causes and results due to errors. Later, Frank E Bind jr. took Heinrien's axioms step further by relating them to the work done by Louis A.Allen in the field of management. This was developed to a further stage by Safety Projects International through the introduction of the management by objectives methodology, the concept of recognition in the form of star grading rounding off the system.

ACCIDENT

What is an Accident?

An accident is an unplanned and uncontrolled event in which the action or reaction of an object, substance, person or radiation results in personal injury or probability there of.

How it is Caused?

(*a*) Lack of management control results on failure to maintain work performance standards for selection, training, tooling, processing, communication and inspection.

(*b*) Lack of management control permits existence of certain contributing causes of accidents, that downgrade the business operation. They are personal factors, consisting of lack of knowledge or skill, improper motivation and physical and mental problems. Job factors such as inadequate work standards, inadequate design and maintenance, inadequate purchasing standards, normal wear & tear, abnormal usage etc., all downgrade the operation and result in avoidable losses.

An accident occurs due to the following causes

(*a*) Unsafe acts of persons.

(*b*) Exposure to unsafe mechanical conditions.

(*c*) Exposure to bad evironmental conditions.

Unsafe actions and conditions are caused by fault of persons. Fault of persons are caused by Environment or acquired by inheritance. Almost 88% of all accidents are attributed chiefly to unsafe practices of working and only 12% attributed to machines.

Unsafe acts by people and unsafe conditions at the work place are to be controlled. For example,

(*i*) operating without authority, failure to warn and secure.

(*ii*) operating at improper speed.

(*iii*) making safety devices in operable.

(*iv*) unsafe conditions on the other hand are inadequate guards and protection.

(*v*) use of defective tools, equipments and substances.
(*vi*) congestion.
(*vii*) poor house keeping.
(*viii*) working on moving or dangerous equipments.
(*ix*) unsafe loading, placing, mixing, combining, directing, teasing, abusing, startling.
(*x*) failure to use safe attire or personal protective equipments.

Unsafe mechanical or physical conditions

(*i*) Inadequate Guanded, Guards of improper height, strength and mass.
(*ii*) Unguarded, absence of required guards.
(*iii*) Use of defective, rough, sharp, slippery, decoyed, cracked machines
(*iv*) Unsafely designed machines and tools etc.
(*v*) Unsafely arranged, entrance and exits blocked.

Environmental conditions

(*i*) Inadequately lighted
(*ii*) Sources of glare etc.
(*iii*) Unsafely clothed, working without P.P.Es, and dangerously dressed
(*iv*) Unsafe processes (mechanical, chemical, electrical and nuclear etc.)
(*v*) Affected by the fumes of solvent vapours which may lead to alteration of judgement by a worker and may move close to the cutting blade.
(*vi*) Biological factors like, infected by diseases, doing a. hazardous job.
(*vii*) Situational factors like substituted equipments like improper ladder, used in construction industry, improper storage and handling of hazardous materials.

Example: Oxygen cylinders stored in an unstable manner and ready to be toppled over with the slightest impact. Inadequately manning, layout and design.

Example: Welding shed located near combustible materials.

To sum up: Industrial injury is caused by accidents. Accident is caused by unsafe acts of persons. Exposure to unsafe mechanical conditions. Unsafe actions and conditions are causes by fault of persons and fault of persons are caused by environment or acquired by inheritance.

Management of Accidents

Objective of accident management are minimise risk, business performance, legal requirement and social obligation.

(*a*) **Identification:** The 1st step is to identify the areas of the work situation, which require immediate control inorder to achieve the desired objective. Some of the key areas of management would be those areas, which are required by statute. Identification of work required to achieve desired loss control objectives. Work could involve the following factors:

1. Loss Investigation
2. Inspection Procedure
3. Group Communication
4. Personal Communication
5. Job Analysis Procedures
6. Rules and Regulations

7. Job Observations
8. Physical Protection
9. Supervisory Training
10. General Promotion
11. Proper Reparative Action
12. Emergency Action Plan
13. Salvage
14. Design, Maintenance-Purchasing etc.
15. Skills Training
16. Research.

(*b*) **Standard established:** The next step in the control function is the activity of establishing a standard for the work performance, a standard of accountability and a valid understandable and practical measurement of the work performed.

(*c*) **Measuring:** The performance must be measured against the established standard *i.e.*, measured in terms of output.

(*d*) **Evaluating:** The work performance must be evaluated to determine what and whether action should be taken in order to rectify the matter.

(*e*) **Corelating:** There is need in corecting the deficiency in output or standards set or Identification of any key areas to be controlled.

(*f*) **Commending:** The last step is to commend the good performance. People respond well, when their efforts are recognised and will continue to work hard to wards excellence.

Management by Objective (MBO)

Prevention of all loss producing events is neither economically feasible nor practicable. But it has been proved beyond doubt that with an unlimited budget, a near perfect system is possible. The manager must strive to optimise loss control effectiveness within the constraints of the budget and available resources. The program must be carefully planned to evolve at the fastest possible rate over period of time, subject to the some business considerations involved in productions quality and cost improvement, one of the most successful methods of carrying out this task is by introducing 5 Star Health and Safety System, which is based on the principle of management by objectives.

Management consultants like Louis A. Allen and John Humble make it quite clear that effective management is always management by objectives. Establishing objectives means, logical action is spelt out in such a manner that effectiveness is improved. The concept of management by objectives, which is defined as a process, where by the supervisor and the subordinate of an organisation, jointly identify its common goals, define each individual's major area of responsibility in terms of the results expected of him and uses these measures as guide lines for operating the unit and assess the contribution of each of its members. The management principles, summarised by Louis A. Allen, underlying the Safety Projects International 5 Star Health and Safety System, include the following.

1. *The principle of critical few*: This relate to the causes and costs of loss, anatomical location of injury, occupational and job involvement. It's application is wide and its value so great in terms of time and effort conservation that, it is considered one of the most important management principles. The most useful applications of this principle is, its use in arriving at the objectives, which contribute to making a successful accident prevention programme.
2. *Principle of recognition*: Motivation to accomplish results tends to increase, as people are given recognition for their contribution. Application of personal and behaviourable recognition serves as powerful motivating force in reinforcing management and employee efforts. It needs the individual's strong desire to feel, he is contributing, regardless of his position in the organisation.

3. *The principle of participation*: Participation involves making systematic provision for consultation with those involved with the programme. This is applicable when determining the priorities of the objectives, which should be tackled first in establishing the programme.
4. *Measuring technique*: In measuring against the standard, one should attempt to eliminate as much subjectivity as possible.
5. *Safety audit*: The safety programme is an ongoing system of activities so structured to fit into the day to day management functions. For the system work, certain records and registers are to be maintained. The safety auditor needs to check the details and look for the concrete or tangible items and from these he is to determine how well the abstract activity, the controlling function was carried out.

To sum up, more can be expected of management to do considering the men, material, methods and money available.

2

ACCIDENT PREVENTION

WHAT IS ACCIDENT PREVENTION?

The concept of prevention and predictive maintenance of accident and loss prevention is being introduced because techniques are available, if implemented adequately, the impact of hazard predictive, accident prevention and loss prevention can be envisaged on a process plant, which is large in scale and equipped with intricate production machines.

Mainly, there are three reasons why accident prevention becomes more imperative for discussion.

1. The legal requirement based on statutory compulsions of national and international authorities.
2. The humanitarian reasons for accident prevention.
3. The most important for any organisation is the exorbitant cost, which the organisation is compelled to pay directly or indirectly.

Broadly, accident prevention hinges on these following three things:

(*a*) Control of man's performance
(*b*) Control of machine performance
(*c*) Control of physical environment.

What are the Approaches for Accident Prevention?

(*a*) Immediate approach
(*b*) Larger range approach.

Examples

(*a*) To immediately remove the Environmental deficiencies such as lighting the area, place fire extinguishers in the ventage points etc.
(*b*) Revision of mechanical process, design, alternation of policies, programmes, installation of new sophisticated machines, fault detectors, computer remote controller, change of personnel, persuasion, appeal, safety training, public relation on safety etc.

How accident prevention is achieved ?

There are three basic principles.

(*a*) Creation of maintenance interest
(*b*) Fact finding
(*c*) Action based on facts.

Five steps are to be taken for prevention of accidents

1. Application of remedy, Selection of remedy, Analysis
2. Fact finding organisation
3. Supervision, Education
4. Personnel, Adjustment, Placement, Instruction and Persuasion
5. Engineering revision.

Detail Elaboration

Analyse the frequency of happenings, its severity, location, occupation, direct causes, subcauses, underlying major causes, tools and equipments used and obstacles faced.

To find facts by supervising, inspecting, observing, reviewing of records on injury, judgement and investigation.

Examples on Organisation

Safety committee, Safety officers' role, Safety direction, Support and following systematic proceedings.

(*a*) Arranging safety training programmes, creating and maintaining interest on accident prevention.
(*b*) Adjusting the existing personnel, new placement, instruction, appeal, persuasion and if need be punish the erring employees.
(*c*) Engineering revision, both major or minor.

For control and prevention accidents, the following steps are to be adopted. It is called three 'E's.

(*a*) Engineering
(*b*) Education
(*c*) Enforcement.

Examples

(*a*) Mechanical guarding and correction of mechanical and physical hazards are first requirement of a complete safety programme.
(*b*) Unsafe acts of persons like operating the machine without clearance, failure to secure warn, operating in unsafe speed, making safety devices 'in operative', using unsafe equipments, unsafe loading, placing, mixing, combining, taking unsafe positions, distracting, teasing, abusing the operator, failure to use safe attire, and personal protective equipments.
(*c*) Inadequate or improper guarding, unguarded and unsafely designed machines, tools, poor house keeping, inadequate lighting, ventilation, impure air sources, unsafely clothed, working without P.P.E. unsafe processes etc. The popular slogan in Accident Prevention is "Full proof the task mechanically when possible and then proceed with control of personal unsafe conditions."

Managerial and Supervisory Faults

Lack of organised safety procedure.

Example: Safety inspector, safety officer, safety committees, accident investigation, safety analysis, etc., lack of executive direction, partition, failure to provide guarding, fire prevention, fire fighting, first aid etc. Lack of suitable personnel for the job, prevalence of poor morale of Employees, lack of training, adoption of safe practice rules and failure to place responsibility for accident and taking remedial measures.

Accident prevention can be done through:

1. Safety management organisation
2. Comprehensive and well communicated safety policy
3. Improved safety management system and procedure
4. In time identification, evaluation and control of risks
5. Regular safety training for new and old employees
6. Safety work permit system and implementation
7. Plant or equipment maintenance system and procedure.

Hazard prediction through safety inspection, job safety analysis, direct contact with workers,

Incident recall technique, surveys, Hazard Operability Studies and Hazard Analysis (HAZOP and HAZAN).

- Hazard assessment and evaluation can be, carried, out by enlisting hazards, allocating hazard weightage and standardising with some numerical scales and prioritisation, Hazard control needs examining priority list, taking short term and long term measures.
- And above all implementation of preventative and predictive maintenance system and procedure specially for critical machines and equipments which have higher potential for accidents.

MAINTENANCE AND INSPECTION

Maintenance includes care to ensure long term life of company assets.

Routine care to maintain uninterrupted service and appearance:

- Repair work required to restore, improved service and appearance.
- Maintenance and inspection are defined in BS3811:1969 as follows: Maintenance work is undertaken in order to keep or restore every facility *i.e.*, every part of site, building and contents, to an acceptable standard.
- Inspection is the process of ensuring by assessment that a facility reaches the necessary standard of quality or performance and that the level is maintained. Maintenance thus clearly implies some measures of inspection as part of its function to determine, What maintenance is needed and when, and whether the results are acceptable.

Forms of Maintenance

1. **Unplanned and Ad hoc. maintenance:** Cheapest and most flexible.
2. **Planned maintenance:** Some loss of flexibility and some outlay in making the plan and implementing it.

Planned Maintenance

The following are some of the planned maintenance ways.

1. Replacement and scrapping of worn items,
2. Replacement and sale of worn items.
3. Breakdown maintenance
4. Preventative maintenance and
5. Predictive maintenance.

As indicated in the below figure, analysis of history as well as knowledge of current production requirement and long range forecasts are needed to establish courses of action. Logically, a maintenance problem can be eliminated if service line of units or components can be extended and corrective actions considered. Or, if it is evident that a problem will have to be tolerated because, the cost of modifications cannot be justified. Preventative and predictive maintenance may be considered to minimise the effects on the availability of the manufacturing or supporting equipment.

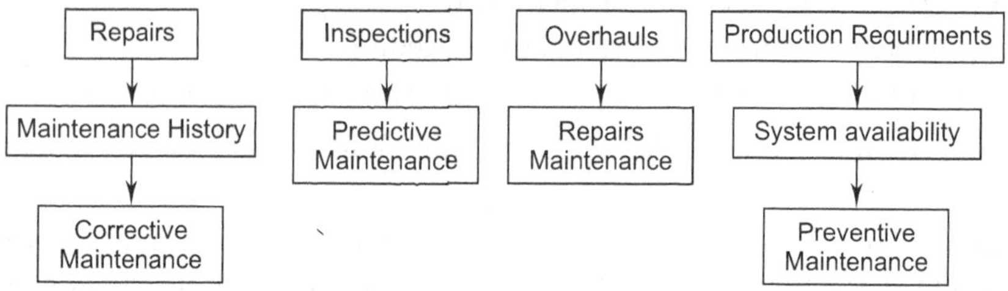

Preventative Maintenance

It is usually justified economically on continuous and/or highly automated processes, where the cost of lost production is high and where the failure of one piece of equipment brings the entire process to a halt on safety grounds, preventative maintenance is necessary on other types of equipment, *e.g.*, boilers and pressure vessels, whose failure may become dangerous. It also requires a thorough analysis of the hazards that may arise:

1. Through failing to carry it out at the right time;
2. In the maintenance operations themselves;
3. Through faulty maintenance which leaves the equipment in a more dangerous condition than it was originally.

   This latter necessity is recognised in the Factories Act, 1948 for types of plant and machinery, so that a limited amount of preventative maintenance has become virtually obligatory.

Predictive Maintenance

The role of predictive maintenance in accident prevention and loss prevention is becoming pivotal day-to-day. Due to rapid rise in production targets, the plant and equipment sizes has gone very big. Automation has replaced batching process and subsequently the process parameters have developed enormous safety hazards. Every component of machine and equipment has its own service life after which the component breakdown occurs and consequences may be catastrophic. Predictive maintenance is a system which provides complete condition monitoring of various parameters. The system is entirely different from both breakdown and fixed time maintenance. Breakdown maintenance is just to wait for it to happen then try it and fix it. Preventative maintenance, which is normally based on statistical failure data of a similar equipment or service. In preventative maintenance, replacement of spares is not carried out before a fixed time. Obviously, there is no indication of built in hazard being developed inside the machine. If there is an indiction of deterioration of any machine component the maintenance can be scheduled in a planned shutdowns and equipment is not left to run to a catastrophic failure.

MONITORING TECHNIQUES

Predictive maintenance can be classified into two major areas of condition monitoring techniques:

1. On-Load Monitoring Techniques — Monitoring without interrupting the operation.
2. Off-Load Monitoring Techniques — Monitoring which requires shutdown.

Both the categories have their own advantages and limitations. The range of methods used in normal practice is very wide. Application of a particular method varies with the nature of equipment, process and industry.

Safety Tips for Employees

There is a general feeling that offices are safe. But this is not always true. These can be safe only if you make safety a part of your business. Many office accidents do happen. Employees suffer injuries. As in other place, mishaps are costly interms of pain, discomfort, disruption in work, lost wages, medical expenses and insurance cost etc. The main causes for mishap are negative attitude and bad habits. Some of them are:

(*a*) *Over confidence*: "It could never happen to me."
(*b*) *Stubbornness*: "I will do it in my own way."
(*c*) *Sarcasm*: "What Safety ? Safety is kid stuff."
(*d*) *Carelessness*: "Why should I bother about it."
(*e*) *Impatience*: "It will take me a lot of time to do it the safe way."
(*f*) *Ignorance*: "I was not aware, it is dangerous to do it that way."
(*g*) *Laziness*: "It will take a lot of effort and trouble to do it the right way."
(*h*) *Showing off*: "No problem, see how I do it."

If you avoid the above, you can help in reducing threats to Safety and Health.

What is Probable Accident?

In the definition of accident, it has been laid down earlier that probability of any injury, underlying in any operation is also termed as an accident. It is other wise called near mis accident. It is an occurrence such as major emission of fire or explosion resulting from uncontrollable development in course of an industrial activity, leading to a serious danger to a man or may not involve personal injuries or disablements immediate or delayed inside or outside establishment and to environment and involving one or more dangerous substance.

Examples: Falling of material or equipments from a great height, collapse of crane, roof, bridge, channel etc.

Preventative Maintenance and Inspection required under The Factories Act. 1948 and other regulations

Facility concerned	Act or Regulation	Section No.	Max. Interval between attentions	Attention required	Details of facility and attention required
Factory cleanliness	The Factories Act 1948	Sec. 11	Daily	Cleaning	Accumulation of dirt or other nuisance
		- do -	Weekly	Cleaning	Floors of work rooms to be washed or cleaned by other approved method if effective.
		- do -	6 months	Cleaning	Inside walls and celling painted with washable water paint, surface to be washed or cleaned by other approved method.
		- do -	14 months	Cleaning	Where they are painted or varnished or where they have smooth impervious surfaces.
		- do -	14 months	Whitewash or colour wash	Inside walls and cellings it not painted or varnis-hed.
Hoists and lifts	The Factories Act 1948	Sec. 29	6 months	Examination	Competent person to thoroughly examine
Chains, ropes and lifting tackle	The Factories Act 1948	Sec. 29	12 months	Examination	Competent person to thoroughly examine
Cranes and other lifting machines	• The Factories Act 1948 • Factories Model Rule	Sec. 29 Model Rule 62	12 months	Examination	Competent person to thoroughly examine, the capability of (*i*) Carrying the stated load with adequate safety (*ii*) Operating properly and (*iii*) Possessing adequate stability
Pressure vessels	• The Factories Act 1948 • The Boiler Act 1923 • Static and Mobile Pressure Vessels (Unfired) Rules 1981	Sec. 31	6 months	Inservice test and examina-tion	Pressure vessel testing: externally
			12 months	- do -	Internally
		Rule 63	2 years	Hydraulic test	
			4 years	Hydraulic test	If above cannot be carried out because of contin-uous process, test at 1.5 P or 1.25 design pressure, whichever is less.

Hazard Identification and Risk Management

Hazard is an existing or potential condition in the work place, which by itself or by inclusion with other variables can result in the unwanted effects of death, injury, property damage and other losses.

Advantage of Methods of Safety Analysis

1. It maps out all the details of an operational so that they can be studied and restudied.
2. It is quick, simple, factual and objective, so that its results can be easily understood with confidence.
3. It permits a ready comparison of the undesirable method and the proposed new method, demonstrating the type of hazards and number of steps involved in each.
4. It presents a picture of the effect on production, which a safety improvement will have.
5. It is an aid showing management when necessary, so that a properly designed safety feature does not hinder productive output and therefore can assist in convincing management of the benefit to be gained by improving the safety method.
6. It permits the Safety Officer to discern and improve ways and means for increasing productivity of an operation.
7. It facilities the analysis of the Safety potential of a job before an accident has occurred.
8. It assists in through investigation of those methods, which have involved accidents but in which the causes are obscure.
9. Perhaps, the greatest importance is the evidence it provides to support the safety officers' too often disputed contention that "Improving a Shop's Safety will increase it's production efficiency."
10. Operation integrity management system (OIMS) is designed help avoid accidents. Should an incident occur, O.I.M.S. procedures aid determining why it occurred and what can be done to prevent it or a similar occurrence from happening again.

Methods of Safety Analysis-By Whom?

(*a*) *Operation analyst*: He sees the operational aspects. It's efficient running. He is the Engineer or a technical person in charge of the Mechanical workshop.

(*b*) Safety officer sees the hazardous steps elimination.

Job Operation

These are three job operation units.

(*a*) **Man:** The operator, supervisor, foreman

(*b*) **Method:** Whether working procedure is adopted

(*c*) **Machine:** The machine, tools, whether running correctly.

Accident Analysis

(*a*) Master Sheet

(*b*) Work Sheet

These sheets are prepared by the Safety Engineer/Officer, on reports by the Supervisor/Inspector. In large organisation/establishment, the punch card method of recording is feasible, thus providing ready means of drawing of totals by kind. In small establishments, annual procedure is practicable. See the formats of both master sheet and worksheet as given follows.

Sample of Master Sheet for use by Safety Officer

Name of the firm:

Location,............ From............ To ....

Branch............ Section............ Division............

By whom prepared : Mr............ Safety Officer

Date when Submitted............

1	2	3	4	5	6	7	8	9
Name of injured Person	Date of Accident	Department (Foreman)	Nature of Injury	Kind of Accident	Nature of unsafe Act	Reason of Accident	Agency Part	Mechanical Hazard

Sample of Work Sheet of Safety Officer. (Work Sheet No. 1)

Name of the firm:

Location, ... From............ To ....

Branch............ Section............ Division............

By whom prepared : Mr............ Safety Officer

Date when Submitted............

Worksheet No.1

1	2	3	4	5	6	7	8	9	10	11
Kind of Accident Agency	Striking Against	Struck by	Caught in between	Fall on same level	Fall to different level	Bodily reaction	Contact Temp. extreme etc.	Rubbing Abrassing etc.	Contact with Elctrical current	Contact wtih corrosive Poisonous gas

Worksheet No.2

1	2	3	4	5	6	7	8	9	10	11	12	13
Unsafe acts Agency	Operating without authority not warning others	Operating at unsafe speed	Making safe devices in operative	Using unsafe equipments	Unsafe loading	Taking unsafe position	Working on moving or dangerous equipments	Distracting Abusing Teasing etc.	Failure to use safe attire P.P.E	Miscellaneous.	Total	Remarks

Accident summery sheet

C	A	U	SE		IDENTIFICATION						
1	2	3	4	1	2	3	4	5	6	7	8
Deptt.	Kind of injury	Part of the body	Kind of Accident	Month of Accident	Day of Accident	Hour of Accident	Mechanical Hazard	Unsafe Act	Reason for unsafe Act	Part played by agency	No. of Accidents

Plant emergency organisation (Fire Brigade)

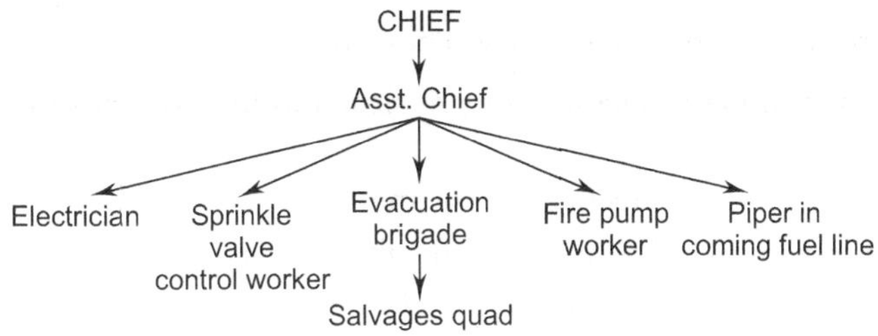

Merits of accident analysis

1. It leads to discovery of personal and mechanical hazards that predominate and should be given first attention.
2. The firms, who do not make use of tabulating machines will often find it advantageous to prepare an accident summery sheet on which complete accident data may be recorded.
3. Selecting predominant causes for attack. From the accident summery, causes of accident are easily selected and duly attacked in taking remedial measures.

Remedial Measures

"Remember, unsafe actions occur several hundred times before injury occurs." This is the age old adage which is considered as a warning for the personnel working in an organisation. There are four basic remedies.

1. *Engineering revision*: This includes guarding of machines and tools, isolation of hazards, revision of procedures and process, illumination, ventilation, color and color contrast, provision of personal protective devices, substitution of safer tools, replacement and repair etc.
2. Instruction, persuasion, appeal, training, instruction, reinstruction, persuasion and appeal through motivation characteristics of persons (shop psychology), visual as well as oral approaches, safety education and safety organisation with its activities.
3. *Personal adjustment*: This included selection and placement with regard to the requirement of the job and the physical and mental suitability of the worker, medical treatment and advice.
4. *Discipline*: Mild admonishment, expression of disappointment, fair insistence statement of past record, transfer to other work and penalties.
5. Creating and Maintaining interest.

Making use of known incentives as under:

(*a*) Self preservation
(*b*) Personal and Material gain
(*c*) Loyalty
(*d*) Responsibility (Recognition of obligation)
(*e*) Pride (Self satisfaction, desire for praise)
(*f*) Conformity (Fear of being thought different from others)
(*g*) Rivalry (Desire to compete)
(*h*) Leadership (Desire to be outstanding)
(*i*) Logic (Special ability to reason)
(*j*) Humanity (Desire to serve others)

Selection of Remedy

There are five procedures in selection of remedy. They are:

(*a*) Guarding and elimination of machines or physical and material hazards without the use of cause analysis.

(*b*) Selection should be based on practical cause analysis..

(*c*) It should be based on knowledge of personal sub-clauses.

(*d*) It should be based on analysis of underlying clauses.

(*e*) It should be based on psychological analysis.

Reasons of unsafe acts and recommended remedy

	Reasons	Recommended remedy
1.	Willful disregard, reckless, lazy, disloyal, co-operative, fearful, oversensitive, egoistical, jealous, impatient, absentminded, excitable, obsession, phobia, inconsiderate, intolerant, mentally unsuited in general.	Personal adjustment, placement, medical attention including psychological, disciplining as a last resort and engineering revision etc.
2.	Lack of knowledge or skill, insufficiently informed, misunderstands, not convinced, indecision etc.	Training, instruction, persuasion etc.
3.	Physically unsuited (Hearing, sight, age, sex height etc.)	Placement and medical attention etc.
4.	Improper mechanical or physical environment, space and light, heat, arrange, ventilation, materials, tools, equipments, procedures and company's policy etc.	Engineering revision.

To prevent repetition of anyone given unsafe act, may require revision of mechanical procedures, provision of providing PPEs. like goggles/hearing aids, training, instruction, practice, education, placement and ensuring strict discipline.

Application of remedy (corrective action)

Who should apply corrective method ?

1. The management must have a sound corrective policy.
2. The Safety Inspector, Supervisor, Safety officer, Engineer are the designated personnel to enforce these corrective methods.
3. The Safety Inspector, who is very close to the worker or the concerned foreman who is always present with the worker,should have strict watch on activities of the worker and in case of any deviation must immediately get it corrected.
4. In any case, the whole responsibility should not be placed on the safety Deptt/personnel alone. This responsibility, primarily lies with the line managers. As told earlier, statutorily, responsibility for safety begins with the occupier and the factory manager of the organisation, who is held accountable for preventing accidents. They delegate the responsibility along with authority to respective heads of Departments and inturn, to line managers, supervisors, inspectors, foreman. Thus , people at every level of line management in on organisation are responsible for controlling losses due to accidents in their respective areas.

5. The concern of Safety officers with preventing injury and damage must be sustained continually, rather than becoming a temporary involvement (during his round only). Safety and health must be recognised as essential ingredient of commercial success and must be managed effectively at par with other organisational activities such as production, cost-control and quality. This is possible only when safety is fully integrated with work processes. Line manager is in a key position in the organisation to carry out management orders and policies to prevent injuries to workmen.

Safety Psychology

Opportunity exists for professional psychology to be introduced much more completely in the field of practical accident prevention. Accident control measures based on better knowledge of human behaviour as it is affected by the traits, feeling process and attributes collectively of the mind. The reference is made to trained professional psychology.

Psychology and Safety Officer

The safety officer is to perform the duty of the psychologist. He can create a much wider and quicker acceptance of Industrial psychology as a practical aid in safe and efficient production. A table given below briefly indicates known performances leading to immediate personal causes resulting psychological subclasses or causative back ground.

Sl. No.	A	B	C
	Known unsafe performances/Action	Probable immediate personal causes	possible psychological sub-clauses or causative background'
1.	Consciously violates safe practice rules	Uniformed	Inadequate training, education or low IQ.
2.	Consciously violates	Inexperienced	Impulsive, intolerant, adventurous
3.	Destroys, cheats misadjusts	Lacks full appreciation of danger	Feeling insecure or anxiety
4.	Indulge in horse play, teasing, practical joking, distracts attention of others	Slow mental physical reaction	Concerned with health of family
5.	Bungles the task	Lacks co-ordination	Rebellious, insubordinate, resentful
6.	Quarrels with others	Fatigued, physically unsuited	In considerate, destructive, jealous
7.	Hasty action, excessive speed	Improper diet	Monotony of job, home life etc.
8.	Fails to warn others	Inattentive or absent minded	Work not recognised or appreciated
9.	Exertion beyond physical capability	Finds unsafe method convenient	Harried, pushed around, severely criticized.
10.	Undertakes task beyond mental capability	Exaggerated sense of humor	Over anxious to perform.

Broadly, remedies could be enumerated in the following ten categories:

1. Engineering revision
2. Elimination of injurious substances or sources
3. Reduction of the original volume and use of injurious substances.
4. Strict control of injurious substances at the point of origin.
5. Provision for washing clothes, changing and luncheon away from unsafe areas.
6. Removal of injurious substances or sources after use.
7. Isolation and guarding of injurious substances, sources (stores)
8. Provision of personal protective devices.
9. Control of unsafe personal acts.
10. Periodic physical examination, adoption of personal hygiene.

GENERAL ACCIDENT PREVENTION

The safety norms and procedure avidly prescribed that Safety and Fire prevention starts on the drawing board, where, for safety errors in the original design can be corrected. The architect, machine designer, the engineer, who are initially associated with the drawing of the building, infrastructure of all hues, should look to the following aspects.

1. Provision of a water tank with required capacity to feed the First aid hoses in case of fire.
2. Fire escape stairs (Fire exit).
3. Provision of fire damper (Steel curtain type both for vertical and horizontal operation) is there.
4. Provision of fire break walls and other fire barriers.
5. Test holes provided.
6. Provision of fresh air intake and outlet louvers.
7. Provision of flexible connections, air inlets and outlets, diffusers, floor grills and under cuts, access doors, sound attenuators and garbage chute.
8. Provision for firefighting sprinklers, complete with pipe work in all floors of the building.

   So far civil construction is concerned, all precautionary measures are to be meticulously followed.
9. In respect of Electrical Section, 240 volt A.C., supply for use of temporary lighting and power tools should be provided in the drawing.
10. Fire alarm and detection as per BS-5839, BS-5445, BS5446 requirements for accurate identification of source of heat, smoke, fire (soft addressable type) are to be provided.
11. Built in key board and FACP recording the time, date and place of occurrence in FACP memory, which could be displayed on video or printed. There should also be provision for silencing the alarm.
12. There should be provision for sprinkler flow switches in each level and the tamper switches for the fault conditions.
13. There should be provision for location of fire alarm panel and the PAVA Pannel in the fire command centre room. Menu function keys for users to know software based features are there in the drawing.
14. Provision for smoke, censors, heat censors, interface units and NICC cables so far fire alarm installation is concerned.

15. The most important is the provision for public alarm system, fire phone system (Fire phone jack outstation, wired to the central exchange need be there so that the fire havoc could be effectively controlled.
16. Provision for installation of portable fire extinguishers, fire blanket, CCTV and monitor etc.
17. Provision for commissioning sufficient number of elevators, loading, speed, steps and openings, power required, door safety, electronic infrared door, detection system, lighting protection system.
18. Provision for alternative power supply system. (Installation of power generator with transformer to be built away from the main building to be there.

Application of Remedy for Slips and Falls on Floors

1. Non-stop floor maintenance procedure to be adopted.
2. *Static electric and conducting flooring*: National Fire Protection Association (NFPA) code-56 "Safe practice for hospital operating rooms" specifies a minimum floor resistance of 25000 ohms and a maximum of 1-megahm (10 lakhs ohms) as a permissible range of floors, in areas where static electricity can be generated. The purpose is to provide a path of moderate electrical conductivity, between all persons and equipments, the floor and ground. Resistance is measured between two electrodes placed 3 feet apart and also between an electrode placed on the floor and ground connection. The range specified permits static flow without sparking and at the same time prevents flow of injurious current through the body to the floor and ground.

   In nutshell, the following essential factors to be considered for safety of plants at the stage of plan layout.

   (*a*) Minimization of pipe work
   (*b*) Segregation of pipe work
   (*c*) Containment of accidents
   (*d*) Efficient and safe construction.
   (*e*) Minimization of personal injury
   (*f*) Safety control room design
   (*g*) Facilitation of process operation
   (*h*) Efficient and safe maintenance
   (*i*) Emergency control facilities
   (*j*) Access to emergency services
   (*k*) Security.

More details on prevention of accidents would be elaborately discussed while dealing with fire prevention and safety in individual industries.

All the above provisions of safety should be ensured at the drawing stage, while preparing plan of the building.

In respect of designing the machineries, the matter has been discussed earlier. The designer should see that protective and self guarding system is provided.

Personal Protective Equipments (PPEs)

PPEs play significant role in accident prevention. The table briefly indicates use of PPEs in various situations, operations.

	Functions of device	Hazardous agencies and conditions	Typical devices (PPEs)
1.	Protection against injurious mechanical contact with outer surface of the body	Heavy, rough sharp abrasive or moving object	Hard hat, helmet, hard toed shoes, tampered and laminated lens, goggles, leather glove, apron, pads, shin, guards
2.	Protection against injurious chemical, electrical, burning, freezing, contact with outer surface of the body	Corrosive, irritating hot, or cold substance, acid, caustics, alkalies, hot metals and electrically charged equipments	Rubber and asbestos gloves, suits aprons, boots, conductive and non-conductive shoes, cup type goggles, face shields
3.	Protection against injurious absorption or non-chemical penetration through the outer surface of the body	Nuclear and other radiant energy, higher frequency sound, light and heat radiation and radio frequency energy and temperature extremes	Dasimeters, special coverings, welder, goggles and masks, sunglasses, eye shields, ear defenders, protective clothing
4.	Protection against inhalation and ingestion of injurious substance	Toxic, irritant and otherwise harmful gas, dust, vapours, fumes and irradiated particles	Masks and respirators of all kinds.
5.	Protection against injurious forces and circumstances	Air and water on pressure, vacuum body parts or members under strain, hazardous and weakened conditions	Diver's suit, trusses, bandages, elastic stockings, Safety ropes and belts, hair nets etc.
6.	Contaminated atmosphere. Highly toxic fumes, having concentration of more than 20,000ppm	Respiratory	Totally enclosed encapsulating suit with self contained air breathing system
	Dusty, Toxic/poisonous gas of 20.000ppm and a minimum cone of 16% O_2 in atmosphere.	Respiratory	Dust guard respirator. Full vision mask with Jumbo canister
	Nuisance dust, toxic gases, cartridge and vapour upto 1000 ppm cone. provided 16% O_2 is available in atmosphere	Respiratory	Mask with gas and dust respirator
7.	Handling of corrosive chemicals	Non-respiratory	Face shield, pvc apron and gloves Gumboots, safety goggles
	Grinding, Welding, cutting, chipping	Non-respiratory Non-respiratory Non-respiratory	Safety goggles and face shield wielding gloves (leather), wielding goggles, leather leg cover welding helmet, safety goggles

8.	Handling sharp edged object	Non-respiratory	Canvas gloves
9.	Handling hot object	Non-respiratory	Asbestos gloves
10.	Handling live electrical equipments	Non-respiratory	Electrical rubber gloves
11.	Working at heights	Non-respiratory	Safety belt, helmet, chrome leather coat and pant with fibre lifeline.
12.	Sand blasting	Non-respiratory	Glass rubber moulded helmet
13.	Around hot furnace	Non-respiratory	Protective mask of vulcanised fibre
14.	Sound polluted area	Non-respiratory	Ear muffler
15.	Machinery fitters' job shifting of items etc.	Non-respiratory	Safety boots with STC
16.	Fire/excess heat due to fire	Non-respiratory	Fire proximity/Fire entry suit

Besides the above, the following may be added.

1. Laminated bake lite protective helmets used in oil industry.
2. Electrical Line men is provided with non-conductive hard hat, gloves and sleeves as well as rugged safety belt.
3. Life line, window cleaner belt, Manila rope, life jackets, nets, sure grabrope, full body type harness, tripod and lifeline hoist rope, emergency life saving apparatus, resuscitator equipment, life jackets and nets to protect fall from great heights.
4. Hair protector, flexible caps, bump caps, body belt, chest belt, knee shield, retrieval chest belt, face shield, earmuffs, earplugs.
5. Apron, fire resistant covered apron, air supply respirator, air purifying respirator, self contained breathing apparatus, airline respirators, encapsulating suits, gas masks, etc., against fire hazards and rescue operations, ladders operators induded.

Besides the above, use of P.P.Es have been discussed in more details, while dealing with industry wise safety functions in succeeding chapter which may be carefully studied.

SAFETY EDUCATION AND TRAINING

Training the employees no doubt is imperative for organisational growth and performance. But most important is how to train them. Sometimes employees view the training as the mundane chore and thus there is not much learning unless the employer devises an out of the box unique method to train the employees. However the smarter ones have seen a clear advantage of having unconventional form of training and are proactively adopting such training methods. The death of six workers at a Delhi metro rail construction workers for instance, are supposed to be provided with sturdy helmets, gumboots and safety belts. But in reality, most workers wear slippers, and climb heights without safety belts. This is because the contractor knows that he will get away unscathed. He would rather pay the insignificant fine than provide for safety of the worker. The Buildings and other Construction Workers Act 1996 provides a safety net for workers but its provisions are rarely implemented. Similar is the case of Professional Sedentary jobs. They are not safer either. Uncomfortable chairs and long hours of work at the computer can cause life long ailments and even paralysis. Orthopaedic doctors treat patients with repetitive strain injuries (RSD) say that 60% of the 55000 RSD patients treated between 2000–2009, had received no training in ERGONOMICS at office. An other 30% got incorrect ergonomic advice. As per law, RSD is not an illness for which, the worker would get compensation. It is the unorganised sector, which is the worst affected. No safety guide lines or laws govern it. Labour welfare is the last thing on mind of employers,who cut corners to improve profitability. Therefore, with more awareness about safety standards, Indian industries are enforcing stricter safety norms.

Broadly, the safety training shall consists of the following types.

1. Induction training should be given to all personnel, meant to work on the site or in the factory premises, office or in any establishment. The training should include the following:
   (*a*) General safety swareness.
   (*b*) First aid.
   (*c*) Emergency procedure.
   (*d*) Use of personal protective equipment
   (*e*) Specific site hazard and job safety.
   (*f*) Ergonomics in general.
   (*g*) SOP & SMPC (Standard operating and Standard maintenance Practices).
   (*h*) Disaster Management.
   (*i*) Occupational hazards (Diseases).
   (*j*) Fire prevention and Fire-fighting including operation of fire extinguish equipments.
   (*k*) Various statutes related to safety and working environment.
   (*l*) Observance of post-accident legal formalities.
   (*m*) Maintenance of accident datas.
   (*n*) Industrial health.

The training, including refreshers training should be imparted by qualified trainers, specialised in the above mentioned subjects/disciplines. Some of the more commonly used educational procedures are listed below.

1. Placing responsibility on the foremans inspector/supervisor for the investigation of the accident and for correcting unsafe practices and mechanical hazards.
2. By conducting safety meetings where the lapses could be more elaborately discussed, and problems, constraints located.
3. Through class room lectures, exhibitions, and through Audio-visual appliances.
4. Use of simulations with elaborate pops, which range from card-boards to kites and balloons and eggs to create games and activities in order to make the training interesting. People tend to learn more and faster through games, videos, than boring lectures.
5. Developing an online competency tool, where employees would be able to see scores of their competencies using 10 questions and 10 competency grid. For weak areas in competencies, employees can build a development module themselves online. The development module would encompass, reading relevant books, CDs, attending training programmes in-house, website links etc. The best part of this initiative, is the fact that the module is given to the employees without any monitoring and intervention, purely as a self development "Do it yourself" kind of tool.

6. Through publicity, safety bulletins, posters, notices, special letters, pamphets, leaflets, inserts, payroll envelop inserts, slides, films, CDs, TV announcements etc.
7. Safety Plays.
8. Employee Rule Books.
9. Hiring and Training Programmes.
10. Use of loyal employees in "Setting good example."
11. Sending Safety Messages in work orders.
12. Organising Quiz Contests, Seminars, Conferences, workshops, games and sports.
13. Distributing Prizes, Shields, Cups, Caps, T. Shirts, Bags etc., as a token of recognition and encouragement to the individuals for good safety performances.
14. Organising safety competitions among different departments in the factory like power plant, electrical, mechanical, chemical, instrumentation, etc., and award certificates for achieving accident free year.
15. Preparation of Safety Code, booklets in local language and distribute among all the employees.
16. Putting hoardings, boards, pieces, posters, prepared by National Safety Council of India, in various departments and factory premises particularly warning signals against smoking and fire prevention.
17. Encouraging Safety Officers/Engineers to take part in National and International Safety Conferences, Seminars and Workshops to brush up their knowledge on SHE.
18. Organising essay competitions on Safety topics among the employees and the best one be felicitated, praised and prizes awarded.
19. Featuring specific safety practice rules.
20. Training Programmes must highlight the Safety Audit initiatives.
21. Regular safety, Industrial Health and Environment campings/Bill Boards/Banners/ Glowsigns should be devised, with posters displaying. Safety and health related issues, being displayed around the factory/worksites as a part of the effort to raise safety awareness amongst the workforce.

3

SAFETY ORGANISATION

BASIC ELEMENTS OF ORGANISED SAFETY

The safety organisation is a "*clearing house*", prevention of accident is a continuous process.

1. Creating and maintaining interest.
2. Fact finding through periodical inspections of surveys of structures, machine, tools, equipments, process and procedures and by accident investigation and analysis for determination of casual facts.
3. Selection of remedies and creative action with regard to the unsafe practices and mechanical hazards described above.

Personnel chart of safety organisation

(Sample of an Indian industrial establishment/organization)

Chairman

↓

Managing Director Cum Chief Executive, Safety Committee

or

His designated Member

↓

Seniormost General Manager (Member)

↓

Executive Safety Committee

↓

Plant Safety Engineer/Officer

(Convener of Safety Committee)

Members

1. MD or his designated person.
2. Plant Manager (Factory Manager)
3. General Manager

4. Safety officer Ex-officio Secretary/Convenor
5. Supporting clerical staff, who will prepare agenda, fix date of meeting etc. On direction of Safety Officer. If the workload of Safety Officer is more, a secretary can be appointed.

Operation Safety Committee

Members

1. The factory manager (In the rank of General Manager)
2. The Safety Officer (Secretary/Convenor)
3. The G.M. HRD
4. The Chief Security Officer
5. The Superintendent/Supervisor/of all departments
6. The foremen of all departments.
7. The Safety Inspectors
8. Supporting clerical staff.

Responsibility of the Executive Safety Committee

(*a*) Review the report and recommend action of the operational safety committee.

(*b*) Take action as deem necessary.

(*c*) Review periodical consideration of trends and progress

(*d*) Approval of expenditure

(*e*) Approval of major changes in safety organisation and of activities affecting matters of policy of the company.

(*f*) Decide matters on engineering revision, repair, replace and installation of new machineries, removal of unsafe materials, purchase of labour saving machineries (Automation policy), expansion programmes, removal of health and environmental hazards etc.

Responsibilities of the Operational Safety Committee

(*a*) The Chairman of this committee will directly report to the Executive Safety Committee.

(*b*) To review the actions on the report of the Safety Officer, and act as per his recommendation, review accident investigation reports submitted by the supervisors. Periodically check all authorised safety procedures and their proper functioning. Approval of proposed new construction and installation of equipments from the safety point of view.

(*c*) Hear the grievances/constraints faced by the safety personnel and redress them if minor in nature.

(*d*) Co-ordinate with all departmental heads for organising Training Programmes for personnel including induction and refresher courses.

(*e*) Implement decisions of Executive Safety Committee.

(*f*) Review the weekly, monthly accident reports, master sheet, working sheet and accident summery furnished by the Safety Officer and take remedial measures where ever feasible and recommend to the Executive Committee for action.

DUTIES OF THE SAFETY OFFICER

The first and foremost duty of the safety manager/officer/engineer in what ever name and designation he works, must prepare a safety master plan for the plant, site infrastructure of the organisation he works. In framing the plan, he should look in to the following expects. Before preparation of the plan, he is to study, the project plan and drawings etc. to acquaint himself of the location of the project, site, plant, area, administrative building, the residential area (Temporary and permanent), the personnel engaged etc. He should observe the various aspects of Safety provisions made in the plan and drawings.

(*a*) Positioning of Fire water tank and its capacity.

(*b*) Fire escape stairs (Fire exit)

(*c*) Provision of Dampers.

(*d*) Provision of Fire break walls and other fire barriers

(*e*) Provision of Test holes, provision of fresh intake and outlet louvers, flexible connections air inlets and outlets, diffusers, floor grills and undercuts, access doors, sound attenuators and garbage chute.

(*f*) Provision of Fire fighting sprinklers complete with pipe work in all floors.

(*g*) Electrical installations like use of temporary lighting and power tools.

(*h*) Fire alarm and detection system, requirements, for accurate identification of source of heat, smoke, fire.

(*i*) Built in key board and FACP recording, sprinkler flow switches and the tamper switches for the fault conditions, location of Fire Alarm Panel, and PAVA panel in Fire Command Centre room, menu function keys for users to know software based features.

(*j*) Smoke censors, heat censors, public alarm system, fire phone system, fire extinguishers/fire blanket, CCTV system, monitor etc. His next step is to study the past records of accidents both major and minor and fire hazard if any and accordingly acquaint himself with the magnitude of the hazards already happened in the past and plan for the future. Before drawing the plan, he has to conduct surveys as detailed below.

While preparing the master plan, he should include the following aspects:

1. List-out activities (routine and non-routine)
2. Break the activities in sub-activities
3. Identify the activities in each sub-activity.
4. Evaluate the risk
5. Quantify the Risk (low, medium, high) using risk metrics table.
6. Practice and strength existing controls to maintain or even further reduce the low risk.
7. Develop and implement additional controls/management programmes to bring down medium and high risk to low or medium risk respectively.
8. Safety Sampling
   (*a*) Isolation of the job
   (*b*) Analysing the job
   (*c*) Step by step checklist
   (*d*) Observation and work through observation
   (*e*) Tabulation
   (*f*) Computing the percentage of the state act.
9. Safety Audit
10. Safety education and training (in detail)

11. Fire prevention and fire fighting
12. Occupational health hazards
13. First aid and treatment
14. Insurance and various possible compensatory agencies
15. Statutory Compliance
16. Requirement of Safety personnel if any and their detail development chart
17. Specific norms of Accountability
18. Formation of Safety committees.

Once the master plan is prepared, the Safety Officer will submit the same to the management for approval.

His Specific Duties

The Safety officer is to perform the duty of a psychologist. He can create a much wider quicker acceptance of Industrial Psychology as a practical aid in safe and efficient production. His duties include the followings as detailed below.

1. Weekly inspections of all structures, yards, machines, tools, fire fighting equipments, first aid and report to the operational Safety Committee.
2. Report orally to the foremen about the unsafe practices adopted by workers found during his inspection/round.
3. Prepare a weekly report and with his recommendations for correction, send to the Committee.
4. He shall investigate cases of accidents both major and minor, it's causes and remedies, arrange treatment and payment of compensation etc.
5. He shall prepare monthly annual summaries of accidents, showing type, causes, frequency and severity.
6. He shall co-ordinate the entire Safety Programme.
7. Impart knowledge, arrange meetings, conferences training through publicity, safety bulletins, posters, notices, special letters, pamphlets, leaflets, payroll envelop inserts, slides, films, TV etc. Organising safety plays, observation of national safety weeks, rewarding honest and good employees of various departments, sections, organise competitions, contests on safety related matters.
8. Impart training on first aid, resuscitation, artificial respiration etc., fire prevention, fire fighting.
9. He prepares Safety master plan, safety code and employee rule books.
10. He shall make suggestions for improvement.
11. He shall try to eliminate hazardous conditions with regard to occupational diseases.
12. He shall minutely check machinery parts to detect unsafe usage, suggest maintenance, oiling, lubrication of wornout machineries and parts, suggest removal and engineering revision.
13. He is to ensure compulsory and mandatory use of personal protective equipments.
14. He is to remove ingredients of environmental pollution and take steps to control them.
15. He is to remove all the elements leading to health hazard to the employees.
16. Conduct safety audit at intervals.
17. He is responsible for statutory compliances, answerable to the Government authorities on lapses found during the latter's' visits/inspections.

18. He is responsible for hospitalisation and in case of death, accomplish all legal formalities, payment of compensation, insurance amount etc.
19. In respect of habitual erring workers, he is required to suggest for taking disciplinary action, stoppage of wages/salary and discharge from service.
20. He is considered as the supreme authority so far as safety is concerned inside the plant and can direct the head of the organisation for observance of Safety Code.
21. He is responsible for total house keeping including garbage disposal of the plant/organisation
22. He is to train all employees of the plant/organisation on Ergonomics for creating better and improved safety awareness.
23. He is to attend refresher's courses, training and retraining sessions, organised by various agencies in and outside the country at intervals.
24. He is to attend Conferences, Seminars and Workshops organised by Government, agencies on safety, health and environment.
25. He is to ensure not only machine safety but also product safety and in nutshell he is responsible for the whole gamut of loss prevention in the plant/organisation.
26. He is to prepare on side and off side emergency plan as per section 41B (4) of The Factories Act 1948.
27. He is responsible for compilation of record of accident safety statistics and safety related documents.
28. Make programme for tool box talks for the site
29. He is to ensure Safety Protocol.
30. He is surely responsible for issue of work permits to the workers to work in confined spaces like tanks, vessels, boilers, silas etc. He is to follow statutory provisions for such operation in confined spaces.
31. He is expected to organise mock fire drill once in a month, which would include the following
    (*a*) Emergency call for fire fighting
    (*b*) Shifting to casualty for first aid.
    (*c*) Introduction address regarding fire safety by safety officer/engineer
    (*d*) Brief description of fire classification and use of fire extinguishers as per classification by safety officer
    (*e*) Brief address by Chief Security Officer on Security
    (*f*) Views of President/MD present and his address to the employees.
32. Observation of National Safety Day on March 4th each year.

    Agenda:

    (*a*) Flag hoisting by MD/President/Chairman
    (*b*) Safety pledge taking administration by all the employees in one go.
    (*c*) Speech by Chairman/ MD
    (*d*) Prize distribution to the winners of various competitions on safety related issues held earlier.
    (*e*) Vote of thanks by Safety Officer/Engineer.

Duties and Responsibilities of Supervisor

Formula for Supervising: Guides

1. He is to deal with people as human being not as machine
2. He is to lead never to drive, push.
3. He is to get people to like and respect him.
4. Create loyalty, win co-operation, instill confidence and build morale.
5. Give credit when due.
6. Explain changes in advance.
7. Give orders clearly and precisely
8. Ask for opinions and suggestions.
9. He should be patient and impartial, consistent, friendly and courteous.
10. Display personal interest in home life, the hobbies, avocations, recreations and personal problems of workers.
11. He should not argue or become dogmatic when disagrees with his workers.
12. He should get to know his personal characteristics so as to avoid irritating and antagonising others.
13. He should get to know personal characteristics of his workers.
14. He should get to know, the personal characteristics of his boss.
15. He should recognise his responsibilities both for management and workers.
16. He is to run his department/section as business.
17. He is to find out what the workers really want most.
18. He is to test his men on attitude and ability
19. He is to infuse his team with competitive spirit to work
20. He should learn to recognise symptoms of trouble.
21. Correct the misdemeanours only when a man has cooled of.
22. Anticipate difficulties and remove obstacles in advance, plan ahead and organise.
23. Interest and encourage workers on quality of production.
24. He should maintain sound physical health and develop a saving sense of humour.
25. He should continuously lookout for hazardous condition of unsafe practices on the part of his men in his section, in addition to periodical inspection. He should maintain a suitable interval, in periodical inspection, which is much more necessary.
26. He should always point out whether hazards in his section are properly (safe guarded) and that safe processes are being followed.

22. He should make periodical and special inspection of equipments of his deptt/section.

Duties of Safety Inspector/Foreman

1. He is always alert to contact the Safety Supervisor/Safety Officer for their help in need, when serious deviation of Safety norms by workers detected.
2. He is to observe all conditions for compliance with established standards.
3. He is to observe all operations for unsafe acts or violation of safety rules.
4. Prepare a systematic report in the prescribed formats of all hazards detected and corresponding remedial measures taken. The report should include good suggestions of plant,

people and the recommendations may be discussed with the safety officer concerned. The report should be clear and concise. Copies sent to concerned people for compliance and follow up actions should be continued till full compliance is achieved and reported to Safety Officer.

SAFE WORK PRACTICES

It is an evolution from injury prevention to the control of all business losses by the application of management principles. It is a system of reporting and controlling all accidents, however the loss is small or large. All incidents are examined and recommendations are made and acted upon to avoid repetition. Activities relating to safe guarding corporate investment and continuing of operation need, and co-ordination with the Safety efforts is a must. Such related functions as for example,. Personal safety, and fire protection must-be closely integrated, which are meant for loss control.

SAFETY SAMPLING AND INSPECTION

This is a technique of analysing minor and near miss accident and it provides analyst with improved feedback methods. Following are some of the terms used in accident prevention literature.

(*a*) *Safe behaviour performance:* It is the safe performance obtained by the company personnel of their assigned task.

(*b*) *Unsafe-behaviour performance:* It is the unsafe performance obtained by the company personnel of their assigned task.

(*c*) *Safe behaviour outcome:* It refers to the end result obtained by the company behaviour.

(*d*) *Unsafe behaviour outcome:* It refers to the result of the company personnel actions and the end result of this behaviour.

Technique of safety sampling

There are seven steps procedure use to carryout the safety sampling procedure successfully:

(*a*) *Isolate a job*: Where there is an accident frequency or other safety problems,

(*b*) *Analyze*: Using past accident data related to that job carefully and making a detail analysis to determine cause and remedical measure to be taken.

(*c*) *Step by step check list*: A specific step by step checklist is made up on type of behaviour, needed to perform the job safety.

(*d*) *Observation*: It is made by the analyst-observe the operation for a time to set that the checklist items are easily observed.

(*e*) *Work through the observation*: Safety observation, who are thoroughly familiar with the process, will work through it. Then they will observe a particular operation for the department using only these items on the checklist. By repeating the procedure several times a week, a number of observations can be made in a short period.

(*f*) *Tabulation*: The checklist marks are tabulated and total number of unsafe and safe Act is determined.

(*g*) *Compute the percentage of safe Act*: To compute the percentage of Acts, the no. of safe Acts should be divided by the total number of observation and that figure is multiplied by 100.

Percentage of safe action = No. of safe observation ÷ Total No. of observation × 100 .

As has been said earlier, a plant safety inspection programme will not be completed unless observation of the unsafe work practices are also made.

During the visit, count is made separately of the no. of case and both, where safe practices are followed and not work more than operation from the observation recorded the percentage is determined.

Plant Safety Inspection

It is an oldest method of discovering potential accidents risk, and is one of the most adopted methods of accidents prevention technique. Any dangerous unit, there is possibility of accident to any person who are involved in the job and the supervisor also, who visit it. Thus, plant safety inspection is an important tool in accident prevention right from early days. Development of statutory regulations has covered safety and inspection programme. Plant safety inspection is an important function in quality control; of the production manufactured. Plant safety inspection plays vital role in accident prevention.

- Their purpose is to find and remove hazard in jobs before accident occur.
- It may not be possible to spot and remove all the unsafe conditions, unsafe practices in the entire plant in one stroke but ultimate aim is to make all the operations safe and efficient.

Sometimes plant safety inspection is wrongly taken as detection of only physical and mechanical conditions in view of the emphasis on statutory requirement. Also an unsafe action is a causative factor — of all accidents and it may not be possible to make the physical condition; safe in all circumstances. So it is very important that unsafe action or work practices are also carefully noted during plant safety inspection.

The safety officer, plant superintendent or other supervisory personnel should conduct safety inspection the better way, as they can consult the person concerned in the shop and enlist their cooperation.

A system of discovering the hazard in various jobs, taking prompt action and taking corrective measures, to eliminate those hazards, in the shop and motivating them (workers) should be devised. That's why, they give the correct hazard data associated with their job.

Responsibilities of inspection lie with:

1. Safety officer.
2. Line management personnel.
3. Front supervisor.
4. Engineer for maintenance.
5. Senior plant manager.

Types of Inspection

There are four types of inspection.

1. Periodical
2. Intermittent
3. Continuous
4. Special.

- **Periodical inspection:** These inspections are well planned and made at regular intervals, of machinery, equipments, tools and all parts of plant. It is made by management and cover the entire plant for thrust area or any other inconspicuous places. Such are pressure vessel, crane, hoist, lift, elevator, chains, cable, ropes, power trucks etc. These inspections are carried out by experienced and qualified persons.

- **Intermittent inspection:** These are unannounced inspections made at irregular intervals to check new equipment installations procedures, modifications, and accident analysis etc. Some partial inspections are also carried out at uncertain intervals.
- **Continuous inspection:** Many companies have set up a system of continuous inspection so that selected employees spend all their time observing certain equipments and operations. Maintenance men, electricians and others, whose job is to keep equipments in good mechanical condition, assigned to specific groups of machines. They have no other duties than to roam about the department, continuously observing operations and making adjustment and minor repairs.

  These men are responsible also for safe operation of sign machines and help train employees in safe practices.

  This inspection method makes these men familiar with the exact condition of each part of the machines, covered. It enables them to point out weaknesses and defects long before they become serious or a hazardous in nature. Continuous system of inspection of PPFs is especially desirable.
- **Special inspection:** Special inspections are some times necessary because of the installation of the new equipment or new process, the construction of the new building or remodeling of old ones because of new hazards. Also, special inspection may be made during special campaigns, such as fire prevention week or waste elimination campaigns. Accidents investigation requires special inspection by the investigation committee and the safety professional. These should be made with the same thoroughness and determination to control the accident causes, as are periodic inspection.

JOB SAFETY ANALYSIS (JSA)

A job safety analysis or (job hazard analysis) is a procedure to make a job safe by:

(*i*) Identifying the hazard or potential of accidents associated with each step of the job.

(*ii*) Developing a solution for each hazard that will either eliminate or control the exposure. If some job hazards are known, proper solution can be developed. Some solutions may be physical changes that control the hazard, such as placing a guard over exposed moving machine parts. Other hazard may be job procedure which can be eliminated or minimized. Job safety analysis is also helpful in discovering elimination or safe guarding of motions, positions actions, and data hazards. It adds the determination of equipments and tools needed for safety.

A detail analysis of each step in the operation of any job, helps to decide the personal characteristics such as: Age, health, physique, skill, height, weight and reaction time necessary in selecting the operation of the job. It also helps to give proper instructions to the selected employees on the various steps of operation in the job, so that he may do his work better, faster, and more safely and with minimum supervision and follow up.

Techniques of job safety analysis

There are four basic steps in job safety analysis.

(*i*) Select the job.

(*ii*) Break the job down into successive steps

(*iii*) Identity the hazards and potential accident.

(*iv*) Develop ways to eliminate the hazards and prevent the potential accidents. Select the job:

The most hazardous and those with the worst accidents experience, should be taken first. The following are suggested as guiding factors for selection of the job.

(a) *Frequency of accidents:* The greater, the no. of accidents, associated with the job, the greater it's priority claim for job safety analysis.

(b) *Production of disabling injuries:* Every job that has produced disabling injuries should be analyzed.

(c) *Severity potential:* Some jobs may have no history of accidents but may have the potential of severe injury.

(d) *New jobs created by changes in equipment or process:* Job safety analysis should be made on every new job.

(e) Legal requirement

(f) No. of employees at risk.

Identifying the Hazards and Potential Accidents

Each state is analyzed for hazards and potential accident. The purpose is to identify all the hazards by observation and then studied further to determine how it could be possible to happen and so on, until the beginning or sources of the chain of events has been highlighted, for each branch of the fault tree.

Certain standard symbols are used in the construction of a fault tree. In a factory every department will have its own challenges, hazards and situations, It may consider to be unfair to compare their safety status (*e.g.*, frequency rate) with others. So at the middle management level, we concentrate on results measure.

Near-miss accident

The term "Near-miss accident" is defined as an occurrence such as a major emission of fire or explosion resulting from uncontrollable development in the course of an industrial activity, leading to a serious danger to a man or may not involve personal injuries or disablement, immediate or delayed inside or outside establishment and to environment and involving one or more dangerous substance.

Examples: Bursting of vessel, falling of materials or equipments, collapse of crane, explosion, collapse of any roof/floor/chimney/bridge/tunnel etc.

Fault tree analysis

Fault tree analysis is a technique that may be utilized to trace back through the chronological progression to posses and effects that have contributed to a particular event, whether it be an accident (industrial safety) or failure (system safety).

The fault tree analysis is a logic diagram based on the principle of multi casualty that traces all the branches of event. That would contribute to an accident or failure. In constructing a fault tree to assist in cost analysis, firstly the event, the accident or failure must be identified. Secondly all the proximate causes (contributory factors) must be investigated and identified. Thirdly, each proximate cause (each branch of a contributory factor) must be traced back to identify and establish all the conceivable ways in which each might have occurred. Each contributing factor or cause, thus identified is available at some point to return to work for his employer, and given the difficulties involve in obtaining employment for workers still under medical care, compounded by the probability that the new employment will be temporary.

Plant safety observation

It is a process to analyse and survey the potential hazards by means of accident prevention. To access the overall standards of safety performance, a concept of plant safety observation is required. It aims to prevent accident at every stage and control hazards during maintenance, operation and to minimize the loses.

Collect details of past accidents and near-miss accidents, find out the route causes of accidents. Publicity, propaganda, education and training of safety, environmental control, establishment of standards, code, regulation and the views of plant safety observation etc., are to be worked out to eliminate risk and hazardous accidents in work area.

Work permit system

The work permit system is one of the most satisfactory methods of ensuring positive controls over hazardous operation performed in unfavourable conditions. Moreover, the system helps to affect greater control on education of jobs, maintenance and repair. It is approved written information on the prevention or hazards connected with the job performance. It is an essential document that categorically spells out the task, equipment involved, its location, personal involvement, time limitation, precautionary measures to be taken together with likely hazard to be encountered if any. It also indicates the various PPEs to be used at different stages of work. It is also served as checklist for various works and what type of safety cautions to be taken.

Depending upon the hazardous operation carried out, there are mainly six types of permits and to work system. Permits are issued by that personnel having proprietary responsibility of that areas and equipments.

Severity rate

Only when the company compare its injury experience with that or similar company with its own previous experience, it can obtain meaningful evaluation of its safety achievement. One of the most useful basis of making measurement is the Severity rate (S.R.).

It is a measure of how severe the accidents are. The severity rate gives the true position regarding the severity of the injuries highlighting the need of concentrating efforts in specific areas. Because of the heavy time charges, a number of serious disable cases can distort result of severity rate even when exposure is large.

It is the ratio of total man-days lost due to accident to the total man-hour worked.

$$\text{Safety rate} = \frac{\text{Man-days lost due to reportable loss time injury}}{\text{Man-hours worked}} \times 10^6$$

SAFETY SURVEY

Once the key problem area is identified by safety audit, the concerned officials must conduct safety survey specifically to that key problem area. Safety survey is nothing but the detail examination of thorough inspection and analysis of various management techniques. A team of experts involving the safety survey will prepare formal reports and submit it to the competent authority in the organisation. An action plan indicating specific responsibility, schedule of time for completion of each activity etc., will be charted out for attending the problem area. The reports generated after safety surveying will also be helpful in determining the training need of workers, supervisors and other line management personnel. Specific norms of accountability can also be devised for them, so that they may deliver their duty in a result oriented and committed manner.

Product safety

It is the safety of product before starting of production. It is started and maintained from design stage of the production. It is a legal responsibility of every manufacturer, seller, supplier of each product to render it safe other wise if any harm is caused to a consumer, buyer, user, by that product, legal damages are payable over and above any statutory compensation for accidental injuries. The manufacturer is obliged to:

- Exercise care in planning, designing and producing his product, reasonably safe for all employees.
- Provide adequate warning with precaution with the product.

Product safety is the thought of safety to achieve safe and quality products. Following points are important for product safety:

- Safety design
- Safety process
- Safety material
- Safety production
- Safety environment
- Safety work force.

Reportable accident

Where in any factory, an accident occurs which causes death or any bodily injury by reason of which a person injured is prevented from the working for a period of 48 hours or more, immediately following accidents, the manager of the factory shall send a notice thereof, to such authority in such form and within such time. This is called reportable accident. During preparation of an accident report, it should be devised to include many facts such as name, age and address of the injured person, place and time of accident, cause of accident, nature of injury recorded and also nature and type of work at the time of accident.

Event tree analysis

An event tree graphically shows the possible outcome of an accident that results from an initiative event.

An event tree analysis considers the responses of the safety system and an operator to the initiating events when determining the accident potential outcomes. The results of the event tree analysis are accident sequences. An event tree analysis is well suited for analyzing complex process that has several layers of safety system to respond specific initiating event. Event trees are used to identify various accidents that can occur in a complex process. After this individual accident sequences are identified, the specific combination of failures that can lead to the accident, can then be determined using event tree analysis. Accident sequences depicted in an event tree represent logical combination of events, Thus these sequences can be put in to the form of event tree model for further qualitative analysis. Analysts used these results to identify design and procedural easiness and normally provide recommendations for reducing the likely hood consequences of the potential accident.

Total loss control

It is an evolution from injury prevention to the control of all business loses by the application of management principle. It is a system of reporting and controlling all accidents, however the loss is small or large. All incidents are examined, potential loses estimated and knowledge of the hazard whether produced by environment or connected with job procedure.

Breaking the job down in to successive steps: The job should be brokendown in to basic steps. The steps must not be too general or too limited. Each step should be described what is being done. Step should be taken under the ambit of occurrence.

Developing solutions to eliminate the hazards and potential accident. The principal solution are:

(*a*) Find an entirely different way to do the job.

(*b*) Change the physical condition that creates the hazards.

(*c*) Revise the job procedure.

(*d*) Reduce the frequency of the jobs by correcting the conditions that make it necessary.

ON-SITE AND OFF-SITE EMERGENCY PLAN

On-site Emergency Plan

The hazard and risk assessment of plant brings out improvement to eliminate or reduce hazard. The management should prepare an on-site-emergency plan based on residual risk. This plan must be specific to the needs of the plants. There are certain elements essential to every plan and these are:

(*a*) Assessment of size and nature of events and the probability of occurrence.

(*b*) Assessment of resources available on site and outside agencies.

(*c*) Procedure for raising alarm and communication within and outside works.

(*d*) Appointment of key personnel and defining their duties.

(*e*) Provision of emergency control centers.

(*f*) Action required on-site.

(*g*) Action required off-site.

Purpose: The purpose of this plan are:

- To protect persons and property of the factory in case of all kinds of accidents, dangerous occurrences, emergencies and disasters, happening in or affecting plant at any time.

(*a*) To inform people and surroundings about above happening if it is likely to adversely affect them.

(*b*) To inform authorities including helping agencies (doctors, hospital, police, transport etc.) in advance and also at the time of actual happening.

(*c*) To identify, assess, foresee and work out various kinds of possible hazards, their places, potential and damaging capacity and area in case of above happenings. Review, revise, redesign, replace or reconstruct the process plant and control measures so associated.

(*d*) To work out a plan with detailed instructions to cope with above happenings based on your personnel requirements and resources. Necessary requirements shall be added if not already available. Hazards substances shall be minimized to the extent possible.

Elements of an on-site-emergency plan

- Alarm and communication system declaring of major emergency.
- A reliable and prompt communication system is extremely important in controlling a major emergency.
- Alarm system will depend on the size and specific needs of the plant. There should be number of points from where alarm can be actuated.
- A communication system using predetermined short message to indicate type and scale of emerency should be available to inform the following personnel.
- All personnel inside the works.
- Outside emergency services and neighboring areas included.
- All the key personnel, outside duty hours.
- Nearby hospitals.

Off-site-Emergency Plan

New sec-41 B(4) of the factories Act requires on-site-emergency plan and detailed disaster control measures, for the safety of the factory workers and the general public living in the vicinity of the factory.

If the accidents take place in an industry/unit and its effects are felt outside its premises, the situation thus created is called as an off-site emergency. Off-site emergency plan is an integral part of any major hazard control system.

Purpose: To save lives and injuries. To present or reduce property loss to provide for quick resumption of normal situation or operation.

The off-site-emergency plan is to be prepared by the government— A central control committee shall be formed under the chairmanship of the area head. Example District Collector, DDC = District Central committee. Other officers are police, fire, factory, medical, engineering; social welfare, publicity, railway transport and requisite departments shall be incorporated as members.

Emergency Preparedness Plans

Emergency Situations: Every firm shall formulate an Emergency Preparedness Plan for each of his sites. These plans will address foreseeable emergencies that may arise during the construction activities. Examples of activities for which plans should be prepared include amongst other things:

(*a*) An accident which results in death or major injury. (Major injury as defined in Section 9.17. of The Factory Act).

(*b*) A serious fire that threatens life.

(*c*) A flood that threatens life.

(*d*) Leakage of any dangerous materials or chemicals.

(*e*) Leakage/short circuit of any electrical supply.

(*f*) Major engineering failures such as:

- collapse of tunnel, or structures
- major utility collapse
- unintended explosions
- subsidence causing damage to structures or services.

An emergency preparedness plan should include details of the following:

(*a*) The name, location and phone number of the emergency co-ordinator;

(*b*) Designated personnel with locations and phone numbers;

(*c*) Details of the emergency response team with locations and phone numbers;

(*d*) Functions of the emergency response team;

(*e*) The means of escape;

(*f*) Communication with the emergency services;

- Police
- Fire service
- Ambulance and hospital services

(*g*) First-aid facilities;

(*h*) Site plans;

(*i*) Suppliers of emergency equipments such as sump pumps, lighting, cranage, etc.

Copies of the emergency procedures and the Contractor's rescue organisation (reviewed without objection by the Employer's Representative) should be displayed at each place of work and notice boards. This information should be regularly reviewed and updated,

REPORTING OF ACCIDENTS AND DANGEROUS OCCURRENCES

Contractor's Responsibility

All accidents and dangerous occurrences shall be recorded, regardless of whether or not personal injury occurs.

The Employer and the Employer's Representative shall be notified by the quickest possible means, for example by telephone of the following classifications of accidents and incidents and by subsequent written notification within twenty four hours by the contractors in accident and incident reporting form, containing the following:

(*a*) Fatal accident

(*b*) Major injury accident (see definition in 9.4.1). (The factory Act 1948)

(*c*) Dangerous occurrence (see definition in 9.4.). (The factory Act 1948)

(*d*) Any incident involving a member of the public.

The site safety officer shall conduct 'in depth' investigations into all fatal accidents, major injury accidents, incidents involving a member of the public, dangerous occurrences, and selected over three-day lost time injury accidents. Copies of these investigations shall be forwarded to the Employer's Representative within seven days.

The Contractor shall rseport immediately, orally and in writing, all fatal accidents, and other occurrences requiring reporting, to the police, at the police station in whose jurisdiction the accident occurred.

Reportable Accidents

An accident shall also become reportable to the Employer's Representative if it causes incapacity for more than three days excluding the day of the accident. The Contractor must submit a report in form SAF 001 to the Employer's Representative within seven days.

The following information is required in reporting an accident to the Employer's Representative:

(*a*) particulars of the Contractor or Sub-contractor employing the injured person;

(*b*) particulars of the deceased or injured person: name, address, occupation, sex, and age; the date.

(*c*) cause of circumstances of the accident; and

(*d*) the nature of the injury, stating whether death or incapacity was caused by the injury.

Dangerous Occurrences

The Employer's Representative requires that all dangerous occurrences on site must be reported in writing to him within 24 hours, irrespective of whether there are casualties or not. The following information has to be provided

(*a*) The time of the occurrence,

(*b*) Damage to any building, machinery or plant; end

(*c*) The circumstances in which the accident occurred.

A copy of the standard 'Dangerous Occurrence Report form, may be used.

If no one is injured, the above notification is sufficient. In the case of death or serious injury, the accident reporting procecure outlined in Section 9.1.2 must also be followed.

Definitions of major injury and dangerous occurrence

For the purposes of reporting, a "Major Injury" means:

(*a*) any fracture, other than to the fingers or toes;

(*b*) any loss of a limb or part of a limb;

(*c*) dislocation of the shoulder, hip, knee or spine;

(*d*) loss of sight (Whether temporary or permanent);

(*e*) penetrating injury to the eye; or

(*f*) any other injury that:

- leads to unconsciousnes;
- requires resuscitation;
- requires admittance to hospital for more than 24 hours.
- or which causes more than 30 days certified absence from work.

A 'dangerous occurrence' is defined as follows:

(*a*) Bursting of a revolving vessel, wheel, grindstone or grinding wheel moved by mechanical power.

(*b*) Collapse or failure of crane, derrick, winch, hoist or other appliances used in rasing or lowering persons or goods or any part there of (except the breakage of chain or rope silings), or the overturning of a crane;

(*c*) Explosion or fire causing damage to the structure of any room or place in which persons are employed, or to any machine or plant, resulting in the complete suspension of ordinary work.

(*d*) Electrical short circuit or failure of electrical machinery, plant or apparatus, attended by explosion or fire; causing structural damage involving its stoppage or disuses.

(*e*) Explosion of a receiver or container used for the storage at a pressure greater than atmospheric pressure of any gas or gases (including air) or any liquid or soild resulting from the compression of gas;

(*f*) Collapse in whole or part from any cause whatsoever of any roof, wall, floor, structure or foundation forming part of the construction site in which persons are employed.

(*g*) Total or partial collapse of any overburden, face, tip or embankment on the construction site;

(*h*) The overturning of or collision with any object by any bulldozer, dumper, excavator.

In Brief Safe Work Practices

All working personnel should make a habit of following under mentioned general safety rules.

- make safety habits a part of life.
- remember safety is *i.e.*, everybody's concern, be always safety conscious.
- do not take any chance, without knowing its consequences.
- Follow a good house keeping.
- Use proper tools.
- Use suitable personal protective equipment even for a short period although it may be un-comfortable
- Take immediate action, even though injuries may be minor in nature.
- Remember, over confidence always suppress the safety sense.
- Follow methods while working with hazardous chemical.
- Remember, accident does not happen it is always caused.

Safety in handing and storage of chlorine

- Handle all containers as though they are full.
- It is necessary to provide adequate unloading and storage facilities.
- All storage, handling and use areas should be such designed so that personnel can quickly escape in emergencies.
- Storage location must be cool, well ventilated from heat.
- Chlorine must always be stored above ground because it is heavier than air.
- Ton container should be stored laying on trundles or wedged with rails, so they can not roll freely.
- Full and empty cylinders should be marked and be stored separately to avoid confusion which lead to an accident.
- Protective covers for valves should always be securely in place even when ton cylinders are empty.
- Do not put water over leaking chlorine gas as it's solubility is very low. The leakage will increase due to high rate of corrosion.
- Never heat cylinder beyond 70 centigrade otherwise volumetric expansion of gas will cause the bursting of cylinder.
- Do not weld any piping or equipment until it is thoroughly purged with air.
- Make sure that piping is dry before admitting chlorine.
- Check leakage with ammonia solution , it gives white dense fumes.
- Workers should always carry escape type respirators.

Safety in material handling

- If the store is very heavy and delicate in nature, ensure availability of proper handling of equipments.

- Ensure load test has been carried out shortly before handling any store.
- The sling rope chains d-shackles etc., shall be in good condition and should have been load tested periodically.
- Do not leave the load in hanging condition.
- Care should be taken to see that no person comes under the load, while handling.
- In case forklift no counter weight shall be used by driver cabin to balance load.
- While handling load with EOT crane, working personnel should wear helmets and shoes.
- Ensure effectiveness of brake systems before using EOT crane.

Safety in Workshop

Welding and gas cutting

- Ensure connection given to welding transformer is as per standard procedure. Terminals to be properly marked.
- Use always welding/gas cutting goggles worn together with gloves helmet and leather shoes while welding/gas cutting.
- Ensure earthing of welding transformer.
- Ensure there is no loose connection at any stage while working with welding transformer, checkup tightness of terminal after switching off the power supply.
- Use only approved oxygen and acetylene cylinder during welding.
- Acetylene cylinder shall be kept vertical while welding.
- Do not use LPG cylinder for gas cutting.
- Switch off welding transformer at the end of work.
- Use proper helmet during overhead welding.
- Use cotton dress while welding.
- Remember not to do welding/gas cutting near by inflammable materials.

Safe working on lathe drill etc

- Use safety goggles to protect your eyes.
- Never clean the machine while it is in motion.
- Switch off main switch and allow the rotation to stop before commencement of cleaning.
- Do not clean chips by hand, use brush only.
- Wear tight and half sleeve dress.
- Avoid using wrist watch ring etc. since these may cause injury.
- Ensure availability of safety guard for motors gear boxes etc.
- Never grind lathe tools without wearing safety goggles.

Safety in handing and storage of hazardous chemicals

- To be stored in a dry and well ventilated room.
- While preparing solution caustic soda should be added slowly and in small amounts to water to avoid violent boiling and splatter, hot water.
- Local exhaust ventilation controls should be provided for process such as transfer operations where caustic soda dust can contaminate work environment.

- Small spill and leaks can be washed away with large quantity of water. Drainage should be directed away from work area. Liquid spillages can be neutralized with weak acid.
- Steed in removing the chemical is of primary importance. In case of 15 minutes, the physician be consulted.
- If ingested, the chemicals may be diluted by large quantities of drinking water, contact with skin, safety shower should be used immediately, clothing can be removed later under the shower.
- Contaminated skin areas should be washed with very large quantities of water for as long as 1-2hrs.
- If it enters eyes, they should be copiously irrigated with water for at least 15 minutes.
- Contact lenses should not be used where caustic soda exposures exist, wearing of chemical goggles, full face shield, mechanical filter respirator, rubber gloves, aprons and boots is recommended.
- The mode of entry (caustic soda) into human system can be through inhalation, ingestion of skin contact, its corrosive action on tissues causes burns and deep ulceration. Prolong contact with dilute solutions has a destructive effect on tissue.
- Hydrogen peroxide should be stored in a cool dry well ventilated area away from heat and flame,
- Containers must be well covered.
- Route of entry is inhalation, ingestion , eye and skin.
- Contact with skin should be avoided but immediate flushing with water will prevent any reaction incase of accident, contact eyes; flush with plenty of water for 15 mins.
- Remove the contaminated clothes and shoes. Flush the affected area with plenty of water.
- Inhalation = Remove the victim to fresh air area Ingestion: Have the victim drink water.

Safety in handling and storage of diesel

- Diesel to be stored in an isolated place under cover.
- Shall be stored in separate rooms with proper ventilation.
- "No Smoking" boards should be displayed inside as well as outside the storage buildings.
- No naked flame is allowed near storage.
- Smoking should not be permitted within 500m of storage area.

Safety in storage and handling of gas cylinders

Storage

- To be stored safely in a dry well ventilated room.
- Oxygen cylinder to be stored separately.
- Cylinder to be stored always in a vertical condition.
- Combustible cylinder should not be stored near flammables.
- Always check the last pressure test data of the cylinder as every three years the cylinder should be pressure tested.
- No naked fire should be allowed near the storage site.

Handling

Any careless handling of filled gas cylinder may result in serious accident. Hence, utmost care should be taken during handling of cylinders. Some of the safety precautions to be adhered to are enumerated below:

- Ensure that there is no gas leakage from cylinders before they are transported.
- Don't carry the cylinder from one place to other by lorry.
- Don't unload the cylinder by dropping from trolley or truck.

Use of cylinders

- Use pressure regulator, with upstream and downstream gauge, with high pressure cylinder.
- Don't use leaky cylinders, though leak may be minor in nature.
- Don't use leaky cylinders with electric current, sparks.
- Open the gas cylinder with proper key.
- Don't use oil or grease on valves.
- Use acetylene cylinders in upright position only.
- Oxygen should not be used for running pneumatic tools.
- Don't attempt transfer of gas from one cylinder to other in your working site.

Safety in storage of handling of acid

- Acid should be stored in container which is compatible to that particular type of acid.
- The acids are of highly corrosive in nature
- Protective clothing viz pre hand gloves, shoes, face masks should be used while handling the acid.
- In case of spillage, wash the area with a large quantity of water.
- A first-aid box containing medicines should be always available at work-site.
- The storage handling area should have suitable drainage system to clear the neutralized acid.

Electrical safety

Electricity and Fire are friends of none, thereby avoid electrical accidents with proper knowledge of electrical safety. Electrical accidents mainly happen due to many of the following reasons:

- Unsafe equipment/installation.
- Unsafe work-place.
- Unsafe work-practices.
- Protect people from electrical shock caused by:
- Improper insulation.
- Improper guarding of electrical installation.
- Improper grounding.
- Improper safety devices viz fuze, auto shut off device, protective relays etc.
- Display warning for any installation dealing with more than 220 volts.
- Don't replace fuze in line circuit and without knowing the cause of blowing.
- Don't attempt to remove a person who is in contact with the wire without insulating yourself or switching off the power supply.
- While working in high voltage line, use high voltage safety shoes and gloves.
- Don't throw water on line, electrical item, in case the dress is wet.
- Don't operate electrical circuit/switches when hand/dress is wet.

- In case of fire, fight fire after switching off the ckt.
- Never remove earthing from an electrical installations, it is an essential part.
- Ensure non-availability of power supply in socket while connecting to it.
- Don't allow unauthorized/untrained personnel to work on electrical installations.

Personal Protective Equipments

To protect working personnel from direct contact of harmful and toxic substances, personnel protective equipments are used. Hence the responsibility of wearing by an individual during working:

	Contaminated atmosphere	Type of respirator
1.	Dusty	Dust guard respirator ,mask.
2.	Nuisance dust, toxic gas	Mask
3.	Handling of corrosive chemicals	Face shield pvc apron, and Gloves, Gum boots, safety, goggles.
4.	Handling sharp edge object	Canvas gloves
5.	Welding/cutting	Welding gloves (leather), welding leather leg cover.
6.	Handling of hot object	Asbestos gloves
7.	Handling the electrical equipments	Electrical rubber hand gloves.
8.	Machinery/ First aid shifting of items etc.	Safety boots with STC

Safety audit

Safety audit is an useful technique to undertake a systematic, critical appreciation of the effectiveness of the companies'safety management system. An audit provides guidelines for verifying improvement in the system and take corrective actions as per guide lines,

Objectives: The main objectives of safety audit are as follows:

(*a*) To study existing systems, procedures, plans and programmes of safety and health.

(*b*) To review implementation status of the existing safety systems, Procedures, Plans and programmes.

(*c*) To recommend the measures for improving effectiveness of implementation of the safety system, procedures of setting up of new procedures required.

Safety audit is nothing but systematic independent critical examination of each area of the activity in the industry mainly with a view to prevent injury, ill health, damage to property, plant equipment and the environment.

The most commonly used modern and important tools of accident prevention is the safety audit, which is now a statutory requirement dealing with hazards of material and process. Safety operation is achieved reasonably and practically by:

(*a*) Recognition of the hazard

(*b*) Elimination of the hazard

(*c*) Containment of the hazard, where elimination is not feasible.

Procedure of safety audit

Process of safety audit has some specific procedure enumerated below as follows:

(*i*) *Preparation:* It is the first step of audit in which the documents containing audit questionnaires and other requirements in a specific format is prepared. Herein, it is defined that what is to be audited and how. This is prepared by auditor team.

(*ii*) *Preliminary discussion:* It is done to access the details of the questionnaires and the contents of the audit. Preliminary discussion is mainly performed with the higher management and the key person of the organisation.

(*iii*) *Field work:* It is done inside the organisation at the site or deptt. or unit or any part of the process where the audit is to be done. Its objective is to examine the documents and data. For this, auditors interview the key persons and evaluate them.

(*iv*) *Close conference:* lt is conducted with the management to highlight the findings of audit report and the recommendation from management and auditiees.

(*v*) *Audit report preparation:* The audit report should reflect all the facts of the audit. It should be detailed and signed by the leading auditor and should contain the following:

(*a*) An executive summery of the report, presenting the introduction, objectives and methodology.

(*b*) Over view of site plant description, management and occupational health system.

(*c*) Scope and objective of the audit.

(*d*) Details of audit team.

(*e*) Identification of reference documents against which audit was conducted,

(*f*) Observation of non-conformities and good practices.

(*g*) Audit team's satisfaction of the extent of complaints.

(*h*) The recommendation for improvement.

Presenting the report to management

The audit team's report, covering all the informations is presented to the chief executive of the company.

Implementation of recommendations: The audit organisation would prepare action plan on the recommendation of the audit report. The action plan should include the time to cover the action plan with specific responsibility.

Review of effect of audit recommendations: The implementation of auditor's report in the line suggested are prepared in a modified manner.

4

FIRE HAZARD AND FIRE FIGHTING

FIRE

The relation between man and fire is very old. From the early stages of life on the earth, fire has been of great assistance to the mankind. Life would not have remained life, had the element of fire been taken away. But at the same time, if the fire become uncontrolled, everything is brought to naught. Thus, in one way, the fire is the best friend of mankind and on the other hand, it is also the worst enemy. To stop that fire does not pose an enemy, it is very essential to control and fight the fire most effectively and timely. What type of extinguishing a particular type of fire is a moot question for all of us. Fire engineers have selected the extinguishing media as per the classification of fire. To combat different types of hazards associated with different types of Fire, water, foam, CO_2, DCP, Halons and TEC powder have been found suitable extinguishing agents and their detail description has been given in the subsequent paragraphs.

A material may be toxic, reactive or explosive under certain conditions but while we measure the hazard potential of any material in our context, we mean the fire hazard. Each of these materials carry known and unknown risks. Today, fire science is concerned with the properties of plastics, cryogenic materials, new and more lethal insecticides, radioactive materials, new fuel and rocket propellants. These substances are part of our modern day today life and the members of fire safety services must learn to handle them during an emergency with due caution.

Definitions

1. **Fire hazard:** The relative danger of the start and spread of fire, the danger of smoke or gases being generated, the danger of explosion or other occurrence potentially endangering the lives and safety of the occupants of a building or structure.

Note: (*a*) Fire Hazards in respect of buildings is the fire risk in connection with the external or exposure hazard and internal material hazard.

(*b*) Fire Hazard may be expressed qualitatively as the product of potential for harm and the degree of exposure.

2. **Toxic:** Poisonous, destructive to body tissues and organs or interfering with body functions.
3. **Reactivity:** The readiness with which a substance takes part in a chemical reaction.

4. **Explosive:** A substance capable of sudden high velocity reaction with the generation of high pressures. Explosives are classified as deflagrating (low explosive) and detonating (high explosive). Depending upon their sensitivity, explosives are set off by fire (heat), by friction, by concussion, by percussion or by detonation.
5. **Fire science:** The systematic body of knowledge and principles drawn from physics, chemistry, mathematics, engineering, administration, management and related branches of arts and sciences required in the fraction of fire protection and prevention.
6. **Cryogenics:** The study of extremely low temperature phenomena.
7. **Radioactivity:** Emission of energy in the form of alpha, beta or gamma radiation from the nucleus of an atom.
8. **Fire prevention:** Precautionary fire protection activities carried out before fires occur, aimed at preventing outbreak of fire, early detection and minimizing life and property loss if fire should occur. Activities undertaken include public education, inspection, law enforcement and reduction of hazards.
9. **Fire protection:** The theory and practice of reducing life and property loss by fire, fire extinguishment, fire control and fire prevention. Narrowly, the field concerned with the detection or extinguishment of fire.
10. **Combustion:** Is a chemical reaction or series of reactions, in which heat and light are involved. When the rate of reaction is very slow only heat is evolved and a slow oxidation occurs, such as rusting. Combustion represents a rapid rate of reaction in which light is emitted as well as heat.
11. Flame is a reaction having the ability to propagate (or spread) through an atmosphere with emission of heat and light.
12. Fire is a rapid chemical combustion reaction producing smoke, heat flame and gases.
13. **Fuel:** A material that yields heat through combustion.
14. **Starvation:** Oxygen deprivation. Fire extinguishment by limiting the fuel or oxidizer.
15. **Smothering:** To extinguish a fire by blocking the oxygen supply or limiting it to a point below that required for combustion.
16. **Cooling:** A process by which fire can be controlled by reducing the temperature of the fuel and its immediate environment.
17. Flash Point is the lowest temperature at which there is sufficient vapourisation of the substance to produce a vapour which will flash momentarily when a flame is applied.
18. Fire Point is the lowest temperature at which the heat from the combustion of a burning vapour is capable of producing sufficient vapour to enable combustion to continue.
19. **Spontaneous ignition temperature:** The lowest temperature at which the substance will ignite spontaneously, that is the substance will burn without the introduction of a flame or other ignition source.
20. Portable Fire Extinguisher is an item of fire fighting equipment containing a fire extinguishing medium which is expelled by internal pressure, either stored or chemically induced, or by the action of a self-contained hand pump.
21. **Fire fighting:** The activity and procedures necessary to diminish or halt combustion and flame from fuels. Fires are extinguished by covering the burning material, removing or interrupting the fuel supply, blocking the air supply, halting the combustion process by chemical quenching materials, or by a combination of other measures.
22. **Activation energy:** An energy barrier to a process, ego to chemical reactions, to the promotion of an electron to the conduction band.

FIRE FIGHTING TECHNIQUES

The success or failure of fire fighting will largely depend on the Fire Protection arrangements catered for in the particular establishment. Fire protection is that facet of fire engineering, primarily concerned with the provision of fire fighting equipment and early warning devices and the conduct of training to optimise the application of fire fighting equipment. Fire fighting equipment may come in the form of small extinguishers, beaters etc., which may be portable or wheel carried to operate in a fire situation. These are generally termed as minor fire fighting equipments. There are other fixed fire fighting installations like Sprinkler, Drenchers, Mulsyfire. Remote Controlled fixed Foam Monitor systems, Halons, Dry Chemical Powder or Inert gases. Such systems are major fixed fire fighting equipments which may be integrated to automatic fire/smoke detection devices or can be manumally operated or both.

Portable Fire Extinguishers

It is estimated that out of total fire load in a building or industry, 90 percent is covered by First Aid Fire Fighting Equipments and only 10 percent by fixed fire systems and Fire Services. This is because all fires are small to begin with and portable extinguishers are the only effective tools to halt its spread and its extinguishment at the initial stages itself. A portable fire extinguisher is generally hand portable or wheeled device that can be used to extinguish small or incipient fires. Typically, it is a container filled with any of a variety of fire extinguishing agents and means for expelling them. Some fire extinguishers are effective only on one class of fire; others are suitable on two or three classes. None are suitable for all the classes of fire.

Fire Disasters Why?

Despite the advancement in technology, it has been seen that disastrous fire are on the increase. It has been established that in most cases extinguishers are used by untrained operators, often resulting in more harm than good, for *e.g.*, use of an upright type extinguisher in an inverted position further fans a fire. In some instances portable fire extinguishers did not work because of poor quality of refilling, care and maintenance.

Another important facet is the standardisation, correct method of distribution and finally the selection of portable fire extinguisher for a particular fire risk in respect of type of fire and size. It may grow in the early seconds. Today, even in a small fire point one can see 5 to 6 types of extinguishers, viz. water, sand buckets, soda acid, water CO_2' foam, CO_2, DCP etc. It particularly leads to confusion in an emergency. Certainly an endeavour should be made to reduce these numbers and still cater for all kinds of risk.

Essentials of Fire

For proper selection of fire extinguishers, it is essential to know "Fire" and "Extinguishing Chemicals" contained in an extinguisher.

Fire is a chain reaction and the activated free radical from a fuel acts as a catalyst of the chain. Practically in all fires (except metals), the hydrogen and hydroxyl group are responsible for flaming. A fire will continue unless radicals are diverted into other reactions by anti-catalytic agents.

A fire thrives on the following three essentials, as represented in the triangle of fire, and successful extinguishment depends upon the removal of anyone of them.

(*a*) **Fuel:** Any material that yields heat through combustion. It may be noted that fuels can be gaseous, liquid or solid material capable of igniting and burning.

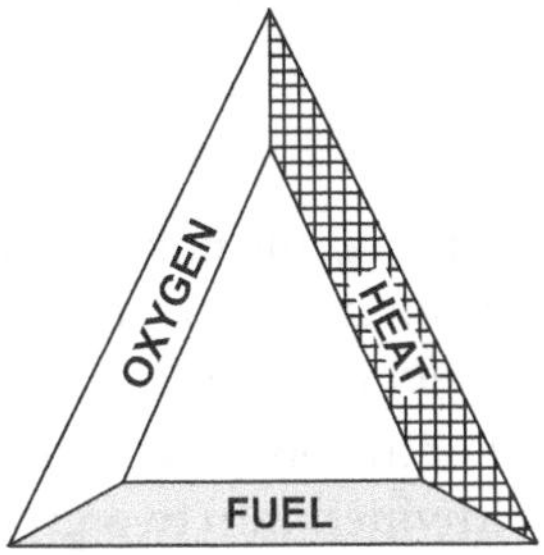

The triangle of combustion

STARVATION

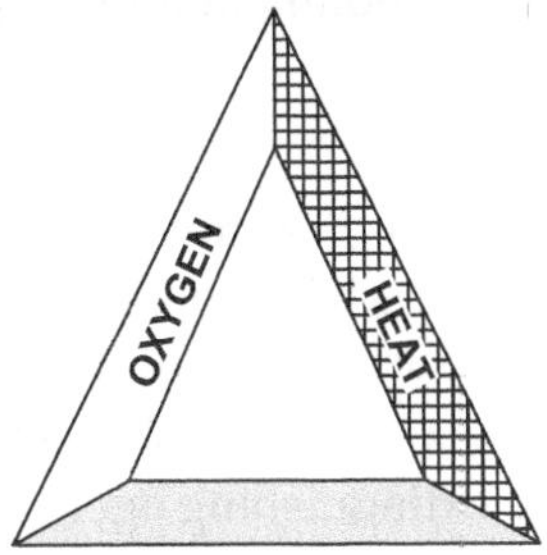

Fuel removed

SMOTHERING

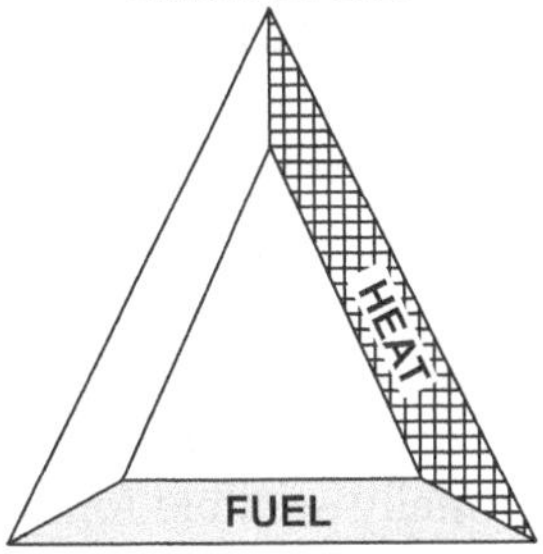

Oxygen removed

COOLING

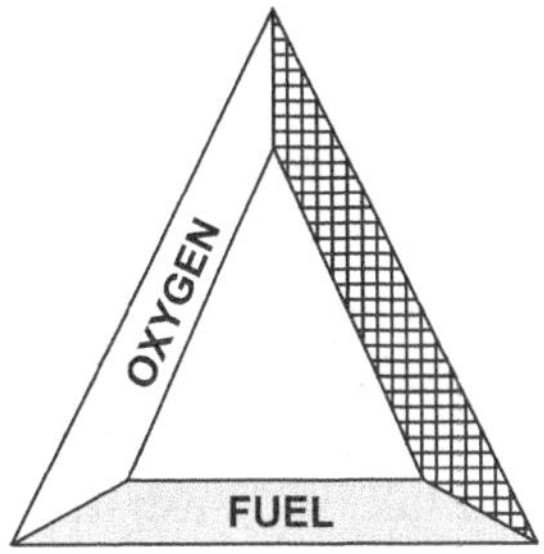

Heat removed

(*b*) **Oxygen:** Air and oxidising compounds in contact with fuel are sources of oxygen for a fire. Nitrates, chlorates, oxides, peroxides etc., whenever (mono-fuels) present within the fuel, pose potential fire hazards. Therefore, special extinguishing agents are employed under these circumstances.

(*c*) **Heat or ignition energy:** A fuel requires minimum of heat energy for ignition. This is specific in nature. Ignition usually takes place when the temperature of a fuel is raised by some mechanism, such as a flame, electric spark, arcing, resistance and static electricity. Spontaneous ignition occurs when chemical reaction between fuel vapours and oxygen causes a temperature increase that leads to self ignition of a material.

Extinction of Fire

Fire extinction, in principle consists in the limitation of one or more of the three essentials mentioned above. Therefore, the methods of extinguishing fire may be classified under the following headings:

(*a*) Starvation is essentially the limitation or removal of fuel. Starvation is applied in three ways:

- (*i*) By removing combustible material from the neighbourhood of the fire.
- (*ii*) By removing the fire from the neighbourhood of the combustible material.
- (*iii*) By sub-dividing the burning material when the smaller fires produced may be left to burn out or to be extinguished more early by other means.

(*b*) Smothering is essentially the limitation of oxygen. If the oxygen content of the atmosphere in the immediate neighbourhood of burning material can be sufficiently reduced,combustion will cease. However, this principle is ineffective where as in the case of celluloid, the burning material contains within itself, in a chemically combined form, the oxygen it requires for combustion.

(*c*) Cooling is the limitation of temperature. If the rate at which heat is generated by combustion is less than the rate of which it is dissipated through various agencies, the combustion cannot persist. The application of a jet or spray of water to a fire is invariably based on this simple but fundamental principle.

Classification of Fires

Class A: These are fires involving solid materials normally of an organic nature such as wood, paper; textiles, rubber etc., Class A fires are the most common and the most effective extinguishing agent is generally water in the form of a Jet or Spray.

Class B: These are fires involving liquids or liquefiable solids. For the purpose of choosing effective extinguishing agents, flammable liquids may be divided into two groups:

(*a*) Those that are miscible with water

(*b*) Those that are immiscible with water.

Class C: These are fires involving gases or liquefied gases in the form of a liquid spillage, or a gas leak and these include methane, propane, butane etc. Foam or dry chemical powder can be used to control fire involving shallow liquid spills (water in the form of spray is generally used to cool the containers).

Class D: These are fires involving metals. Extinguishing agents containing water are ineffective, and even dangerous; Carbon dioxide and the bicarbonate classes of dry chemical powders may also be hazardous if applied to most metal fires. Powdered graphite, powdered talc, soda ash, limestone and dry sand are normally suitable for class 'D' fires. Special fusing powders have been developed for fires involving some metals, specially the radioactive ones.

Class E: Electrical fire: CO_2 (carbon dioxide extinguisher) (DCP) is used to extinguish Electrical fire.

Portable Fire Extinguishers

Portable fire extinguishers can be divided into four categories according to the extinguishing medium they contain, viz:

(*a*) Water type

(*b*) Foam type

(*c*) Powder type

(*d*) Gas type containing Carbon dioxide and Halon extinguishers.

Extinguishers are normally operated by the use of gas pressure in the upper part of the container which forces the extinguishing medium out through a nozzle. The required pressure is produced by one of the following methods:

(*a*) *Chemical reaction:* Two (or more) chemicals are allowed to react to produce an expellant gas when the operating mechanism is activated.

(*b*) *Gas cartridge:* The pressure is produced by means of compressed or (more commonly) liquefied gas released from a gas cartridge fitted into the extinguisher.

(*c*) *Stored pressure:* The expellant gas is stored with the extinguishing medium in the body of the extinguisher, which is thus permanently pressurised. In the case of carbon dioxide extinguishers, the expellant gas is itself the extinguishing medium.

CATEGORIES OF PORTABLE FIRE EXTINGUISHERS

Water type : Soda acid, water CO_2 stored pressure (upright; inverting) come under this class. They expel water on to the fire and cool. It to a point where combustible vapours no longer evolve. The various types of water type extinguishers are discussed below:

(*i*) **Soda acid extinguisher is-934-1976**

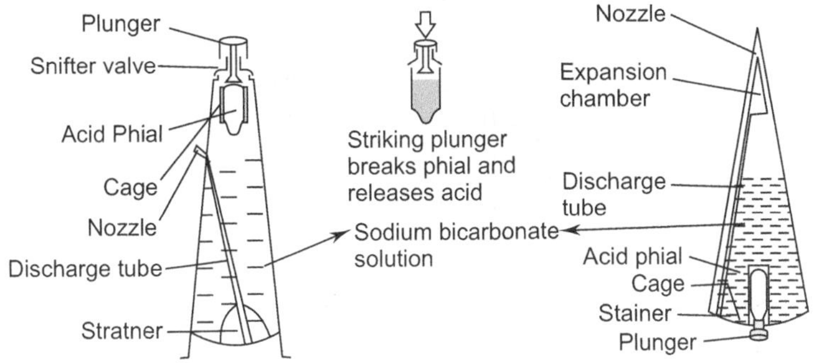

Sections through conical type soda-acid extinguishers

Left: with plunger at apex. **Right:** with plunger at base.

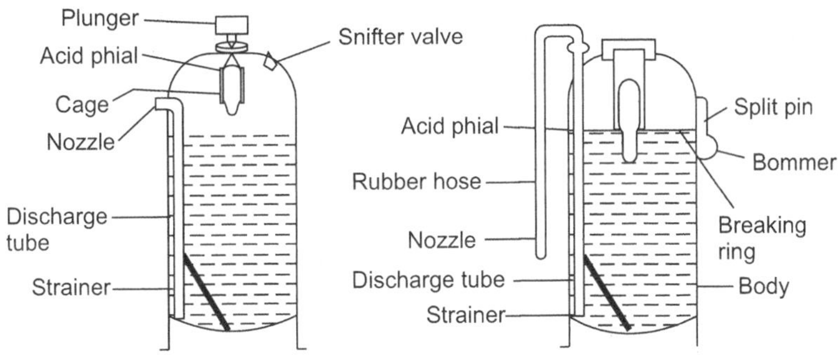

Sections through cylindrical type soda-acid extinguishers

Left: with plunger at top. **Right:** with hammer at side.

Cylindrical type

(*ii*) **Water (gas cartridge) type extinguisher**

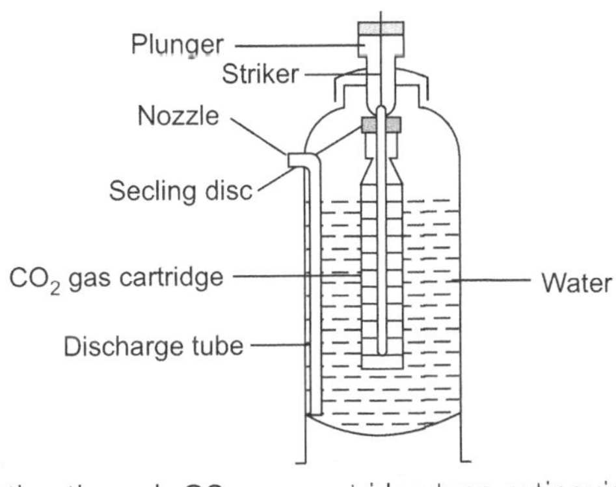

Section through CO_2 gas cartridge type extinguisher

(*iii*) **Water stored pressure extinguisher**

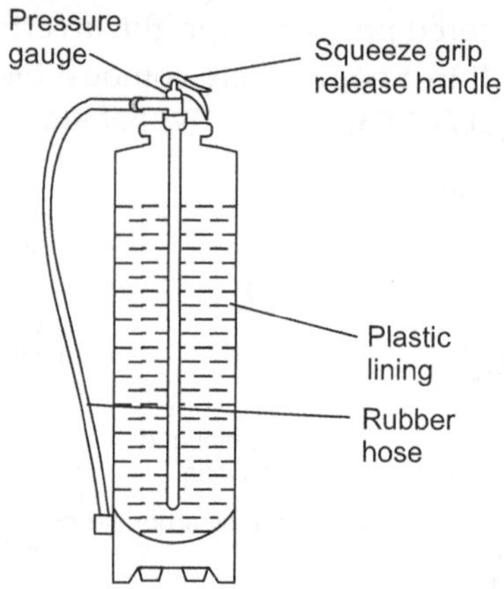

Section through water (stored-pressure) type extinguisher

Foam type: Foam is the most efficient extinguishant for flammable liquid fires. As against water, foams do not sink in the lightest of fuels, and floats on the surface of burning liquids. Fire extinguishment is achieved primarily by blocking of the oxygen supply followed by vapour sealing and cooling of the fuel and surroundings with the release of water. The mechanism of foam formation is : Water (94 parts) + Foam Compound (6 parts) + Air = Foam or Froth. There are basically two ways of producing foam from portable fire extinguishers.

(*a*) By Chemical reaction

(*b*) By Self-aspiration of a foam solution through a foam branch pipe.

There is also the method of applying AFFF as a non-aspirated spray.

A recent development is a fluoroprotien film forming foam (FFFP) extinguisher, which can handle Class A and B fires including those Involving "polar" products (alcohol etc.) on which most foam extinguishers are ineffective. This is usually applied as an aspirated foam.

(*a*) **Chemical foam extinguisher** is the oldest in this class and is operated by inverting the extinguisher and foam is produced by intermixing of two separately stored chemical solutions. These have now largely been replaced by more modern types.

Polyethylene valve

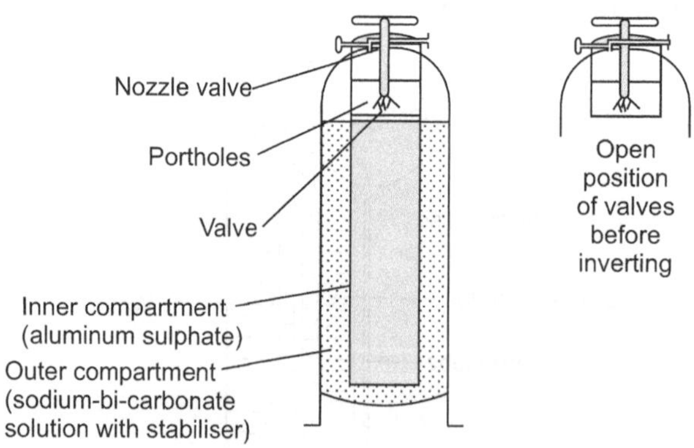

Sections through chemical foam extinguishers type

Is: **933-1976** (Manufacturing Contents)

1. Mild Steel sheet - 1.25 mm
2. Neck Ring clear opening - 75 ± 1 mm
3. Inner container - polyethylene -1.6 mm
4. Effective length of neck - 16 mm
5. Vent Holes - 3 Nos. 2.5 mm
6. Effective length of jet - 6 m
7. Discharge time 30–90 sees
8. Proof pressure Hydraulic Test for 5 minutes - 25 Kg/cm^2
9. Burst lest- 35 Kg/cm^2
10. Static Pressure (Expansion space) - 15 Kg/cm^2

(*b*) **Mechanical foam extinguishers** is of recent origin. They work in upright position. Either stored pressure or gas cartridge is used for foam making and its expulsion. Various types of foaming concentrates in the ratio of 6 parts V/V with water are used in them. They are Synthetic Alcohol Resistant Protien Foam, Fluoro Protien or Fluoro-Carbon Surfactant. The most common and efficient mechanical foam uses Fluoro-Carbon commonly known as the AFFP extinguisher.

These extinguisher are also called in self-aspirating foam extinguishers. Here a foam solution (foam concentrate in water) is stored in a container; ususlly of 6 or 9 litre capacity, and either discharged by a gas cartridge of CO_2 fitted in the body of the extinguisher or by compressed air or nitrogen which has been pumped into the container. In both cases the solution is forced out through the delivery hose and aspirated by the specially designed branch, which is of the low expansion type. The operating pressure is usually about 10 bar, and all types can be fitted with some kind of low freeze depressant. Operation is either by striker or lever to pinch the gas-cartridge pressure disk or, in the stored pressure type, a lever to open a valve. Extinguishers operated by a striker do not have a containable discharge. .

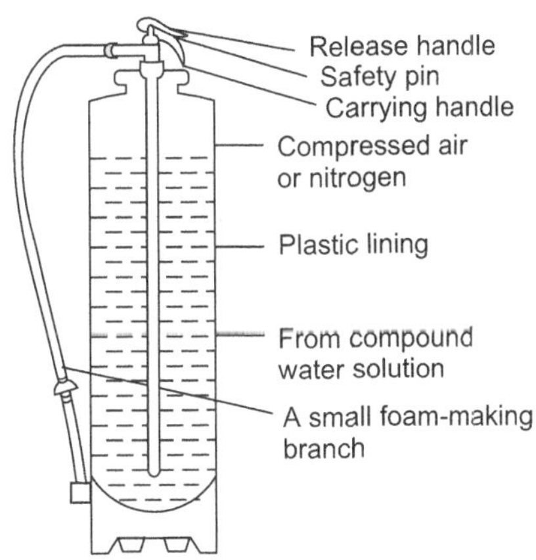

Mechanical foam extinguisher (stored-pressure) type.

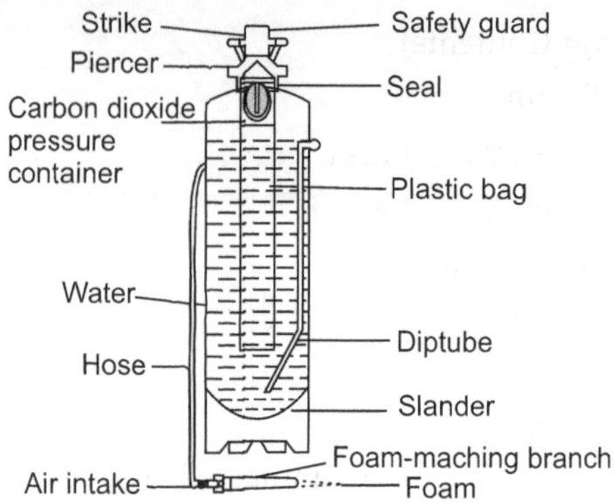

Mechanical foam extinguisher (gas-cartridge) type

Is : **10204-1982** (Manufacturing contents)

1. Mild Steel sheet - 1.25 mm
2. Syphon Tube -1 mm thick
3. Hose - bursting pressure 50 Kg/cm^2
4. Syphon tube dia - 16 mm
5. Outer cylinder dia - 175: ± 5 mm
6. Neck Ring - 75 mm
7. Neck Ring length - 16 mm
8. Cartridge Holder - Female left handed threads corresponding those of cartridge, with port holes.
9. Impact resistance from 1 meter height - 5 Kg weight fall
10. Effective length of jet - 6 m
11. Discharge time - maximum 90 sees.
12. Gas cartridge -As per **IS : 4947 - 1977**
13. Proof pressure hydraulic test for 5 minutes - 25 Kg/cm^2
14. Burst test-35 Kg/cm^2
15. Control of fire - 1.4 cm^2 in 40 sees.
16. Anti-corrosive thickness = .012mm.

(*c*) **Non aspirating AFFF extinguisher:** This is usually a gas - cartridge - operated type with a modified nozzle which sprays the AFFF onto the fire.

Dry chemical powder extinguishers: Portable powder extinguishers are available in sizes ranging from 1 to 12 Kg of powder content. The method of expelling the powder from the containers are similar to those of water and foam extinguishers *i.e.*, stored pressure or gas cartridge type. They are upright operation type and they extinguish a fire by anti-catalytic action.Powders are of various compositions such as Sodium and Potassium Carbonates and bicarbonates, chlorides, ammonium phosphate and borates etc. The latest and a particularly effective powder developed is a complex combination of urea and Potassium Carbonate, also trade named as Monnex, has non: caking characteristics and greater effectiveness than either Sodium or Potassium Carbonates. Though these powders are effective, if not more than Halons, they suffer from the following drawbacks:

(*a*) Powders leave a mess on equipment and materials

(*b*) Poor visibility during fire fighting.

(*c*) Flash back of fire is common.

(*d*) They neither do inerting action nor cooling

(*e*) Though powders are of minimum toxicity, some, especially those employing chlorides *e.g.*, TEC can be dangerous if inhaled for any length of time.

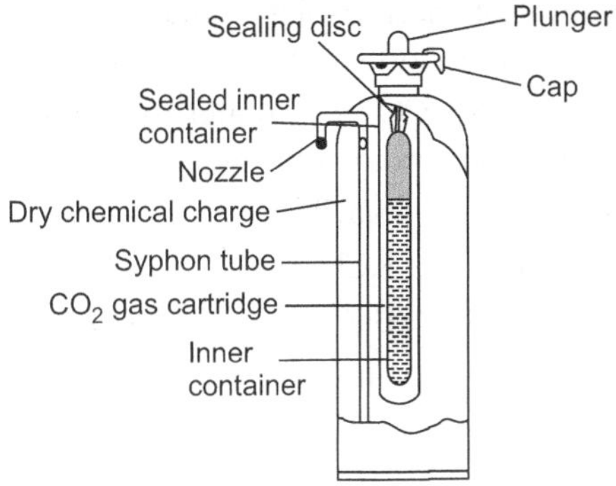

Dry powder (gascatridge) type extinguisher

Dry Chemical Powder Actions

(*a*) **Dry chemical powder (action mode on flame matrix)**

(*i*) $2NaHCO_2 + Heat = Na_2CO_3 + H_2O + CO_2$; $Na_2O + H_2O = 2NaOH$; $NaOH + H = Na + H_2O$; $NaOH + OH = NaO + H_2O$.

(*ii*) Free radical quenching mechanism: $2KHCO_3 + Heat = H_2O + 2CO_2 + K_2O$; $K_2O + H_2O$ (from the combustion) = 2KOH also $KOH + H = K + H_2O$ and a regeneration of KOH occurs as $2KO - K_2O$; $K_2O + H_2O$ (from the reaction) - KOH; $KOH + OH = KO + H_2O$.

(*iii*) Sodium Bicarbonate (basic salt) 97% (Zinc) Magnesium Stearate: 1.5% (to prevent caking) promote flow properties. To make it foam (protien) compatible siliconized.

Magnesium Carbonate: 1 %

Tri.Calcium phosphate: .5%

(*iv*) NH_4H_2PO (Acidic Salt) (for Class A and B). Monoammonium phoschate; Ammonium dihydrogen phosphate break down to yield glassy coating of HPO_3 (meta phospheric acid). Glassy residue on the solid combustible halts further combustion.

(*v*) Super 'K' - Potassium Chloride based - less efficient than $KHCO_3$

(*vi*) Monnex, an addition product, twice as efficient as $KHCO_3$ based powder for Class 'B' fire. $KHCO_3$ (urea) + $NH_2 - CO - HN_2$ (inclusion complexe).

The particle size breaks down (only under extinguishing processing condition from 25 micron to .01 micron.

(*b*) **Combustible metal extinguishing sgents (Class D' fire)** Extinction (mode) principle

(*i*) withdral of oxygen (smothering or interfacing) or

(*ii*) cooling in some way, be it through radiation or heat removal with cooling agents. In order to lower the temperature of the mass below its ignition temperature (point).

1. G-1 Powder Proprietary mixture composed of graphite and foundry taken carefully, sized to facilitate easy handling and packing. An Organic Phosphate is added. The graphite ignites with difficulty, a good heat conductor for absorbing heat and radiating heat from the burning area. The organic Phosphate decomposes to yield

a non-combustible smoke that penetrates the spaces between the graphite particles, thus excluding the oxygen. The mixture is reducing in action, moderately effective in all metal fires.

2. Met-L-X Powder, a proprietary mixture, containing NaCl + $Ca_3(PO)_{4-2}$ ± Zn stearate. A small amount of thermosplastic material A (PVC) to fecilitate fusing and holding NaCl for cooling and smother, suitable for all metaltics.
3. Na-X Powder, chiefly Na_2CO_3 for greater resistance to higher temperature. They may be designed for sodium metal fires.

4. T.E.C suitable for	NaCl	-20%	
Mg and all Radioactive	KCl	-29%	Fusing at 580 °C
metals	$BaCl_2$	-51%	
5. Bitumen	-45%	Suitable for light.	
State Powders	-45%	alloys	
Ai Stearate	-10%	(size 20 to 30 Microns)	

Gas Type Extinguishers

(*a*) **Halon extinguisher:** Halons (a contraction of 'halogenated hydrocarbons') are a group of agents which are stored under pressure in liquid form and released in such a way as to vapourise rapidly in a fire zone. They extinguish fire mainly by interfering with the chemical reactions involved in the propagation of flame and they posses little cooling effect. Halons are hydrocarbon compounds of a group of 5 non-metallic elements known as halogens, *i.e.*, fluorine, chlorine, bromine, iodine and astomine. For various reasons, Iodine and astomine are not suitable as extinguishants. BCF is the most common in portable extinguishers. BTM & DBE/DTE in fixed installations for total flooding systems. BCF Extinguishers are available in various sizes and have superiority over other gas extinguishers for following reasons:

(*i*) *Long throw:* Since BCF is a mixture of gas and liquid at equilibrium, it has a longer projection. Even from a 1 Kg extinguisher fire can be approached from a distance of 2.5 m. as compared to CO_2 only 0.5 m. Fire fighting is therefore safer and creates confidence for operation.

(*ii*) *Light weight:* From the weight point of view, Halon extinguishers are lightest in weight of Agent to container ratio. 1 Kg BCF weigh only 2.5 Kgs as against 6 Kg for CO_2 extinguishers. Light weight makes it easy to carry and operate.

(*iii*) *Penetrates obstructed fires:* Because BCF vapourises on contact with heat/fire or heat to the fire, it can find its way to various obstructed places such as engine and machinery spaces.

(*iv*) *Does not leave residue:* It vapourises 100 percent, hence does not leave any residue in electrical and electronic equipments.

(*v*) *Effectiveness:* In comparison, very effective to its counterpart and a concentration of about 3 to 5 percent is sufficient for any fire as against 25 to 35 percent of CO_2.

(*b*) **Extinction mechanism:** Though exact mechanism of Halons is complex chemical inhibition, their anti catalytic action is responsible for high efficiency. They interfere in the fire chain reaction as under:

(*i*) For BCF

CF_2ClBr (BCF) + Heat of Fire = $CF_2 + Cl^x + Br^x$

$H^x + Br^x = HBr$

from fire

$H^x + Cl^X = HCl$

(x = Activated radicals)

(*ii*) For BTM

$BrCF_3 + Heat - Br.CF_3$

$CF_3 + Heat - F +$ (other CF Compounds)

With CH_4 as fuel.

$2CH_4 + O_2 = 4H + OH + CO +$ (other compounds)

The active Brand F from equations 1 and 2 could also attack the native fuel molecule such as:

$CH_4 + Br = HBr +$ other CH compounds

$CH_4 + F = HF +$ other CH compounds

Active atomic H from equation 3 react with active Br.

$H + Br = HBr$

Regeneration of Br and F

$OH + HBr = Br + H_2O$ (inert)

$OH + HF = F + H_2O$ (inert)

(*c*) **Characteristics of Halons**

	Agent	BTM	BCF	DTE/DBE	CBM	CT C	MIB
(*i*)	Halon No.	1301	1211	2402	1011	104	1001
(*ii*)	Chemical Formula	$CBrF_3$	$CBrClF_2$	$C_2Br_2F_4$	CH_2BrCl	CCl_4	CH_2Br
(*iii*)	Molecular Weight	149	164.4	259.8			
(*iv*)	Boiling Point	-57.6° C	-4°C	47.5° C	67.8°C	76.7° C	45°C
(*v*)	Freezing Point	-168°C	160.5° C	-110.5° C	-86° C	22.8°C	-93.7°C
(*vi*)	Vapour Density	5.2	5.7	9			
(*vii*)	Specific Gravmy	1.8	1.8	2:1			
(*viii*)	Dangerous Concentration - By volume	50	24	8.5			
	- By weight (Relative CO_2)	12.5	9.2	5.1			
(*ix*)	Extinguishing Concentration - By volume	4.9	5.2	3.9			
	- By weight (Relative to CO_2)	0.61	0.72	0.76			
(*x*)	R = EC/DC	0.098	0.22	0.41			
(*xi*)	lethal Concentration						
	-(ppm) Vapour -Rre cleanup vapour	832.000 14.000	-324.000 -7.650				

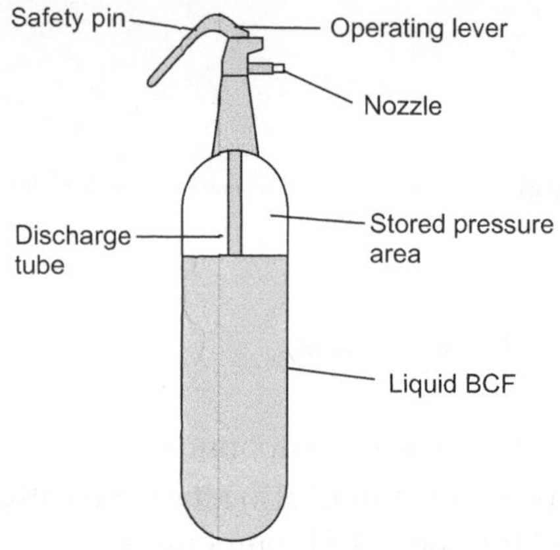

Halon Extinguishers

(*d*) **Carbon dioxide extinguishers:** CO_2 has a number of properties making it useful for extinuishing fire. It does not react most substances, it provides its own pressure for discharge from an extinguisher and since it emerges as a gas, it can penetrate to all parts of a fire area. It is a non-conductor of electricity and will not damage electronic equipment. It's uses are generally similar to those of halons, except that it is not recommended for class 'A' fire. Some of the properties of CO_2 are:

(*i*) Molecular weight	44
(*ii*) Soecific gravity	0.8
(*iii*) Freezing Point	– 56.6° C
(*iv*) Sublimation Point	– 78.5° C
(*v*) Density	1.5 times as heavy
(*vi*) Cooling Effect	279 Kilojoules per Kg of gas
(*vii*) Compressed to liquefy at	20°C
(*viii*) Critical pressure	73.1 bars
(*ix*) Critical temperature	31.1° C
(*x*) Filling ratio	0.667
(*xi*) Discharge time for 907'gms	16 secs & 6.8 Kg 30 secs
(*xii*) Expansion ratio	450 : 1

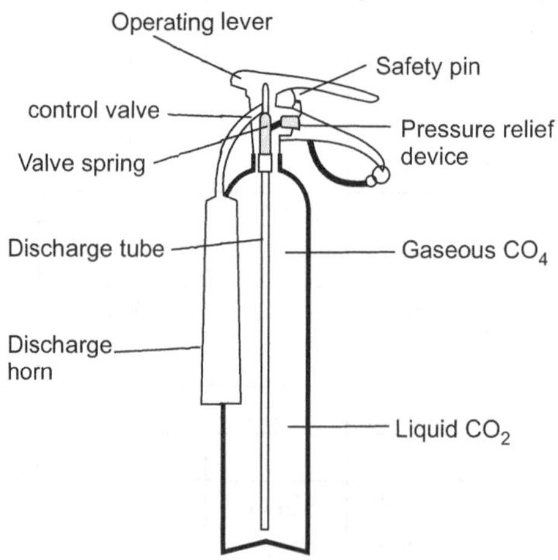

Carbon dioxide extinguisher (small size)

SELECTING A PORTABLE FIRE EXTINGUISHER

Hundreds of combustibles are used for manufacturing of various types of products, thereby implying that fire has to be treated on its merits. A number of factors such as nature and reactivity, flash point, explosive limits, miscibility, toxicity of fire products as well as extinguishing agents and proximity / approach to fire etc. are important in selecting a most efficient and safe fire extinguisher. Today fire types are classified alphabetically, and each deserves an extinguisher with an appropriate agent, be it water, foam, powder or gas or a combination of these.

Faced with such a bewildering classification of fires, numbers and types of extinguishers, how does an inexperienced person avoid using the wrong type of agent leading to a possible disastrous effect? The **Defence Institute** of **Fire Research** had taken up a study to reduce the number of extinguishers to a minimum possible and standardise their mode of operation. Their study aimed at selection of a portable extinguisher to cover atleast 3 to 4 classes of fires. It also aimed to standardise method of operation be it upright position, stored air pressure or gas cartridge operated to be the squeeze grip type on their operation head. Based on this, table below is prepared for selection of a suitable type of fire extinguisher for a particular fire risk.

Type of extinguisher	Class	Cause of fire	Hazard areas	Extinguishing methods	Warning
Water	A	Due to burning of organic solids *e.g.*, wood, paper, cloth etc.	Offices, shops, theaters, banks restaurants etc.	Penetrates and cools even deep seated a class fires	Do not use on live electrical equipment
From	B	Due to burning of inflammable liquids	Petrol storage depots, manufacturing units of paints inflammable liquids	Forms a thick foam blanket over the burning liquid cutting off oxygen supply	Do not use on live electrical equipment
Dry Chemical Power	B,C	Due to burning of inflammable liquids and gases	Storage areas of LPG, acelylene petrol, industrial solvents and electrical equipment	Chemically interferes with the combustion chain	Do not use in very confined spaces as cloud of powder reduces visibility
ABC Powder	A,B,C	Due to burning of organic soilds inflammable liquids, gases	Mixed hazard areas like open storages, generator rooms, heat treatment shops	Chemically interferes with the combustion chain	Do not use in very confined spaces as cloud of powder reduces visibility

Carbon Dioxide	B,C	Due to burning of inflammable liquids, gases and electrical hazards	Motor pump rooms, laboratories museums and process control area	Aim at the base of fire close range. CO_2 being heavier, settles below air and cut off oxygen supply	CO_2 can cause asphyxiation in a confined space Ventilate area after extinguishing fire
BCF	B,C	Due to burning of inflammable liquids, gases and sensitive electrical equipment	Laboratories, museums, computer room, archives, electrical control panels	Chemically interferes with the combustion chain	Ventilate area after extinguishing fire

Rating

The rating of extinguishers shall be shown by a numeral designation in which the numeral indicates the approximate relative extinguishing potential of the extinguisher under prescribed test conditions.

On selecting an extinguisher from table, next problem that a Fire Officer will face is its size.

In lines of Under Writers Laboratories Inc. USA, Defence Institute of Fire Research has set up the facilities for studying comparative efficiency of various extinguishers.

Bigger fire extinguishers

1. Trailer Fire Pump : (water)

   Capacity : 1800 litre water per minute throwing capacity to be attached to a water source nearby. Distance 50 ft. (location of trailer from the burning site)
2. Water tender (mark-2) Maximum throwing capacity 300 liters per minute. Distance -100ft. Contains Foam and Water,

   Throwing capacity : 2250 litre per minute
3. Foam Tender (CO_2) for Electric Fire: Contents : 800 liters
4. Extinguishers (BCF types) Oil and electrical fire.

   BCF : Broncho chloro diafloro mithane
5. Fire hydrant should be around 60 meter distance from the site of occurrence.
6. Crash fire Tender: (For Airport fire)

   Engine power: 320 HP make, MAGIRUS (Germany)

   Controlling personnel: Five Fire man

   Pump capacity: 3200 litre per minute.

   Monitoring throw: 55 meter

   Contents capacity: 6000 litters water in the Tank
7. SCBA (Self Contained Breathing Apparatus) Cylinder: Carried by Fireman, wearing entry suit/approach suit/ chemical suit to enter inside.
8. Minimax, ceasefire extinguisher Capacity : 1 kg to 50 kg powder type (BCF)

9. Truck fire fighting extinguisher (large) (Dry powder)
10. Mobile water canon used for riot control Capacity : 1200 litre to 16000 litre
11. Mulsifire: (oil and gas extinguishers)

Merit of water as an extinguishant

1. It has got solvent ability
2. It can be easily pumped and conducted in hose and pipes from 1°C. to as high as 99°C.
3. It has high latent heat of vaporation of water that is 2260 kj/kg., which means water can absorb a tremendous amount of heat when transformed from boiling state to steam. Thus it is most efficient heat removing and cooling liquid.
4. It passes at a property of flashing in to large volumes-an expansion of about 1700 times the liquid volume.
5. The high surface tension of water at ordinary temperature enables it to issue in a consolidated stream or in discrete water droplets from the so called "fog" nozzles or spraying devices.
6. It has got high density quality
7. It has high molecular stability, which enables it to avoid break down or dislocation.
8. It is easily available.

Demerits of water

1. Due to its freezing quality, difficult to use at places of sub-zero temperature
2. It has high surface tension quality and for which it becomes impenetrable in wetting tightly packed combustibles.
3. It is of low viscosity
4. It has problem of high friction loss.
5. It is of higher density. It sinks in to the fuel.
6. It is good conductor for electricity. For which it's use is restricted in electrical fire
7. It is also metal reactive. It's use is therefore restricted in metal fire.

Fire Fighting Foams

Types of Foam

(*a*) **Chemical/foam:**

Composition :	Sodium bicarbonate	Sodium Sulphate
	+	+
	Aluminium sulphate	Aluminium hydroaide
	+	+
	Saponion	Carbon dioxide

(*b*) **Mechanical foam (Air Foam):** Foam is generated by a turbulent mechanical action by mixing air in to the foaming water solution. These are following five categories.

(*a*) Protein type foam
(*b*) Fluro protein type foam
(*c*) Aqueous film forming foams (AFFF)

(*d*) Synthetic medium and high expansion type foam

(*e*) Alcohol resistant foam.

Carbon dioxide as an extinguishtants (in detail)

It does not react with combustible materials. It is classified as inert gas. Helium, Argon, Krypton, Nitrogen and Carbon dioxide are used to produce inert atmospheric ambience. But among all, carbon dioxide and nitrogen are principal inert gases in modern use.

Main characteristics

(*a*) It has very high expansion ratio

(*b*) It is easily liquefied and maintained in that state for a longtime

(*c*) It produces cooling effect

(*d*) It is mildly toxic, no lasting toxic effect at 5% concentration.

(*e*) It is heavier than air and dilutes the concentration of oxygen there by kills the fire by smothering

(*f*) It is bad conductor of electricity, hence can be applied on live wires.

Dry Chemical Powder

This is a powerful three static surface of fuel and prevent evolution of flammable vapour various uses:

(*a*) *Sodium Bicarbonate BC Powder:* It is very useful in 'B' & 'C' type fires.

(*b*) *Potassium Bicarbonate B C Powder:* It is twice more powerful than Sodium bicarbonate powder. It has the power to distribute the powder relatively timely, spreading over the area in short time.

(*c*) *Potassium Sulphate foam Compatible BC Powder:* Here foam and powder are used together.

(*d*) *Potassium Chloride B C Powder:* This is rather more powerful than Potassium bicarbonate.

(*e*) *CI Monnex BC Powder:* This is most powerful of the three. It is in addition to urea and potassium bicarbonate. Very useful in chemical fire.

(*f*) *General purpose ABC Powder:* All the fire fighting powders deal with flame in the air borne condition when used in class 'A materials like wood, paper, fabric, plastic etc. Smouldering of the material often continues after extinguishment of the flame and the material then burst in to action (flame) after powder is settled. ABC powder overcomes this drawback by forming fire proof coating on the burring material and this has the effect of arresting of smouldering combustion. This is used in 'A', 'B' and 'C risks.

Metal Extinguishing Agents

Metals burn in air. Here oxidation become major when big quantity is involved. To halt the oxidation, readiness of burning metals, reliance is placed upon withdraw L of oxygen and cooling. The following extinguishing agents are there.

(*a*) G-l Powder: Composition : Mixture of graphite and foundry coke. Organic phosphate is also added. Suitable for all types of metal fire.

(*b*) Met-L-X Powder: Composition : Sodium chloride mixed with tricalcium phosphate powder, zinc sterate, small quantity of Thermoplastic material is added for cooling and smothering the burning area.

(*c*) Na-X Powder

(*d*) Inert Gas

(*e*) TEC Powder (Ternary Eutectic Chloride) very effective in metal fire. Sodium chloride, Potassium chloride and Borium chloride are its ingredients.

Causes of Fire

Fire hazards occur due to the following costs.

(*a*) Smoking

(*b*) Electrical defect/fault mainly short circuit

(*c*) Open flame

(*d*) Electrical and Mechanical spark

(*e*) Spontaneous ignition

(*f*) Hot surface, sparking

(*g*) Welding/Gas cutting

(*h*) Petroleum and Petroleum Product spill

(*i*) Pirofuric chemical.

In a fire, there are two different phases, in which combustion proceeds

(*a*) Source Combustion model ,

(*b*) Flaming mode.

The source combustion mode in corporate the following essential elements (fuel, heat, oxygen). The flaming mode incorporates four basic requirements (fuel, heat, oxygen and chain reaction)

Tetrahedron of Fire

These two modes may occur either single or in combination. Flammable liquid and gels burn in flamming mode only.

Example: Both the modes exist in coal, wood etc. With these materials, the early stage of combustion starts and gradually transits towards the surface combustion made during which both the modes are simultaneously act. Ultimately, the flamming mode is terminated with residual surface combustion mode existing alone.

Example: Fire involving pure carbon and other readily oxidisable non metal like Sulphur, Phosphate etc. and certain other metal like magnesium and Aluminium etc.

Flash point: Flash point of an flammable substance may be defined as the lowest temperature at which it gives up sufficient vapour so as to form a momentary flash with air on application of a small flame.

Ignition Temperature

Ignition temperature of a substance is the temperature at which spontaneous ignition can take place. The self ignition point some times called as auto ignition point or combustion point. Also refers to the temperature at which, a substance must reach before it ignite in the absence of a flame.

Example : Auto ignition point for

Coal	: 125–130
Jute	: 103
Cotton	: 228
News Paper	: 184

Fire point: The lowest temperature at which the heat from the combustion of a burning vapour is capable of producing sufficient vapour to enable the combustion to continue. It is generally above the flash point.

Effects of Fire to a Human Skin

The effect of fire on people take the form of skin burn due to exposer to thermal radiation. The severity of the burn depends on the intensity of heat and exposure time. Heat Radiation is inversely proportional to the square the distance from the source. In general terms, the skin withstands a heat energy of $10kw/mt^2$ for 0.4 sec before pain is felt.

Fire: Fire can take several different form:

(*i*) Jet fire

(*ii*) Pool fire

(*iii*) Flash fire

(*iv*) Bleve fire (boiling liquid expanding vapour expansion) Fire is nothing but a chemical reaction. Oxidisational fuel + heat = Fire.

Explosion: Explosions are characterised by a shock wave which can be heard as a bang, which can cause damage to buildings, breaking windows, ejecting missiles over distance of several hundred meters. The injury and damage are in the first place cause by shock wave of the exposure, The effect of shock wave vary, depending on the characteristic of the material, the quantity involved and the degree of confinement of the vapour cloud. The peak pressure in an explosion therefore, vary between a slide over pressure and few hundred kilo pascals (Pressure unit-KPA). Direct injury to people occurs at the pressure 5-1- KPA. The loss of life occurring at a greater area pressures where as dwellings are demolished and windows and doors of shock wave decreases rapidly with increase in distance from the source of explosion.

Example: The explosion of tank containing 50 tones of propene, result in a pressure of 14 KPA at 250 mts and the pressure of 5 KPA at 500 mts. from the tank.

Deflagration and Detonation

Explosion can occur either in the form of a deflagration or detonation, depending on the burning velocity during the explosion.

Deflagration occur when the burning velocity/flame speed is relatively slow of the order of 1mt/sec. In a detonation, the flame travel as a Shockwave of velocity 2000-3000mt/sec. Detonation generates greater pressure and is far more destructive than a deflagration. The peak pressure caused by deflagration in a close atmospheric pressure reaches about 70-80 KPA, whereas, a detonation can easily reach a pressure of 200 KPA.

Boiling Liquid Expanding Vapour Explosion (BLEVE)

It is sometime referred to as fire ball. A bleve is a combination of fire and explosion with an intense radiant heat emission, with in a relatively short time interval. As implied by the term the phenomena can occur within a vessel or tank in which a liquefied gas is kept above its atmospheric boiling point. If a pressure vessels fails as result of weakening of its structure, the contents are instantaneously released from the vessel as a turbulent and dispersing in air as a cloud. When this cloud is ignited, a fire ball occur causing an enormous heat radius. This heat intensity is sufficient to cause severe skin burns and death at several hundred mts. from the vessel. A Bleve can therefore, be caused by a physical impact on a vessel or tank which is already over stressed or damaged.

Example: From a traffic accident with a tank car or derailment of tank wagon or it can be caused by fire and thus weakening its structure. A Bleve involving 50 tons of propen tank can cause 3rd degree injury at a distance of approximately (200mts. and blister at approximately 400 mts.

Unconfined Vapour Cloud Explosion (UCVCE)

In chemical industry, flammable liquids may be processed or stored at temperature above the acb boiling point. A sudden uncontrol release of such liquids in atmosphere can readily vapourised creating an explosive mixture defusing in air. Fine mist from such a released remain suspended in air which readily ignites in droplet sizes from 0.5 to 10 micron. In variably, such vapours are denser than air and drift over a wide area covering plant, equipment and buildings where the release have been massive (huge quantity) ignition occur within 15–20 sec from the pheriphery of the cloud. Rapid flame propagation through vapour air mixture in the form of a pre-mist flame travelling very rapidly through the cloud attains deflagration velocity with severe shock wave.

	Lower Explosive Limit		Upper Explosive Limit
H_2	------4%	⟷	75%
LPG	------2%	⟷	10%
Accitelene	------1%	⟷	09%

Fire Detector

1.

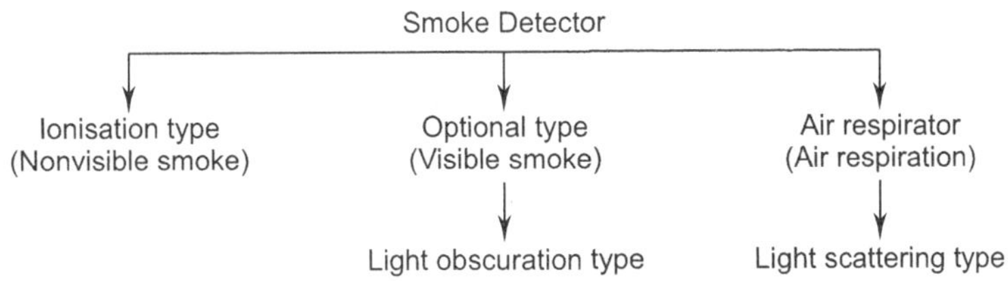

2.

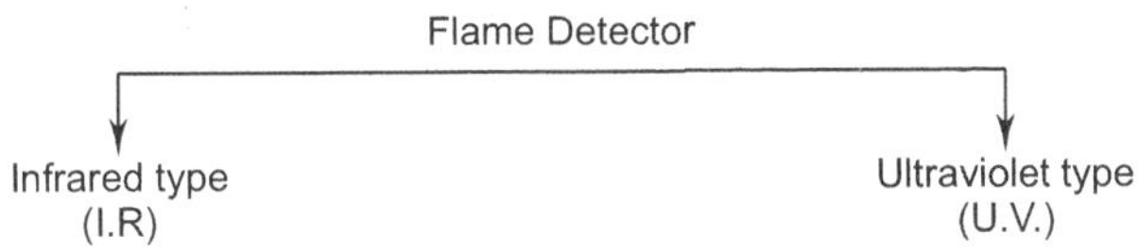

3.

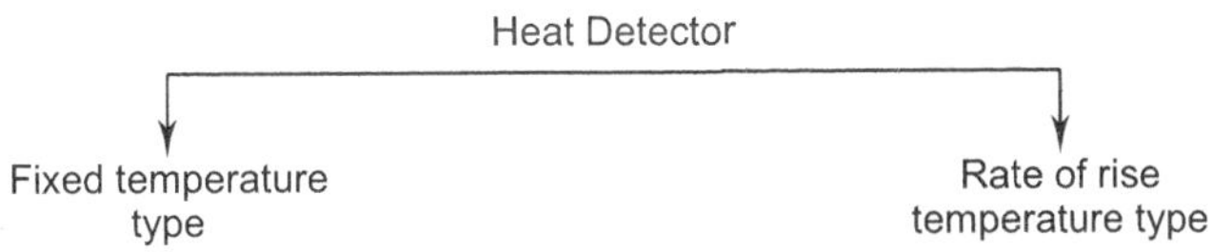

Start of fire goes through following four steps:

(i) Insipid/Childhood state
(ii) Smoke state
(iii) Flame State
(iv) Heat State

Smoke Sensing Fire Detector

Smoke detector may be further divided as following type.

(i) Lonisation smoke detector
(ii) Photoelectric light scattering detector
(iii) Photoelectric light obscuration detector
(iv) Air assorted sampling smoke detector
(v) Linear beam smoke detector
(vi) Duct detector.

Ionisation Smoke Detectors

It contains small amount of radio active material that ionise the air between a positive and negative electrode. The conductance between the electrode is measured for the ambient cleaned air state. The introduction of smoke to the detectors sampling chamber/reduces the conductance between the electrode. Because ionise smoke particles move slowly and their attraction to the electrode is less than would be leakage for cleaned ionised particle when the conductance falls between below a signal to the fire alarm control pannel.

AUTOMATIC FIRE DETECTION SYSTEMS AND FIRE FIGHTING SYSTEMS

Digital India Group established since 1981, has been rendering most reliable and efficient services to our valued clients in the field of Security Electronic Devices. Now, they have introduced 'Automatic Fire Detection System' and 'Fire Fighting System' towards saving precious lives and property from the most damaging hazards of 'FIRE'.

Automatic Fire Detection System' has been developed indigenously by R and D Engineers conforming to ISI-2189 standards and manufactured to withstand Indian conditions for longevity of life-span for uninterrupted service. Besides being most-economical, it has the following features for an automatic functioning:

- Automatic detection of smoke or fire
- Automatic dialing of telephone Nos. of Fire Brigade/Civil Hospital and Residence of Owner/ Official incharge for calling immediate assistance to extinguish fire, round the clock.
- Automatic sounding of local alarm inviting attention of nearby available persons to extinguish fire by using fire extinguishers/equipment kept in the premises.

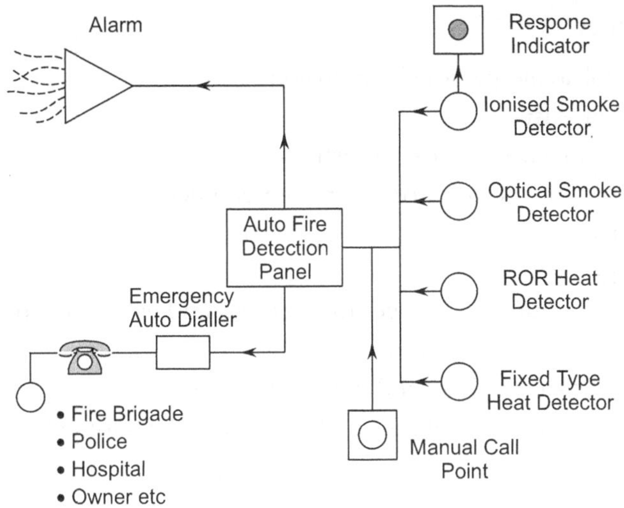

Fire prevention rules

Prevention is better than cure.

1. Maintain good housekeeping.
2. No smoking in bed.
3. Never throw burning cigarette or match stick in to dustbin.
4. Never smoke in no smoking area, use designated smoke room only.
5. Keep all ashtrays filled with -1/4 of water to extinguish burning match stick/cigarette.
6. Never carry cigarette lighter on board tanker/liquid gas carrier.
7. Do not use faulty electrical equipment. Repair, replace immediately.
8. Never overload electrical connections by using multiplugs.
9. Take proper precautions at the time of welding/cutting.
10. Clear oil spill immediately
11. Never spray aerosol container near naked flame or electrical fire.
12. Check fire detection and fire protection system regularly.
13. Do not cover an electrical bulb with combustible material shade, leave ample space for air circulation.
14. Never lay or hang clothes, where it can fall or touch electrical fittings.
15. Read safety notices/guide lines before switching on/off any machinery.
16. Never store flammable gas cylinder with other cargoes.
17. Carry out testing of electrical wirings insulation and earth faults.
18. Never check leakage, leak of gas (LPG) such as propane or butane with naked flame. Always use soap solution.
19. Burning of waste paper and other combustile material should be carried out in an approved place.
20. All vehicles must be fitted with spark arrester at the exhaust.

In case of fire emergency

1. Keep your head cool, avoid panic and confusion
2. know the location of exist
3. Know the location of nearby fire extinguishers
4. Know how to report afire occurrence of other emergency
5. Follow exit instructions
6. Walk to your assignment exit.
7. In case of factory fire, shut off all power to machines and electric connections.
8. Turn off gas under heat treating ovens.
9. Close windows and clear the long and narrow passages.
10. Form a double line in the centre of aisle and follow monitors for exit.

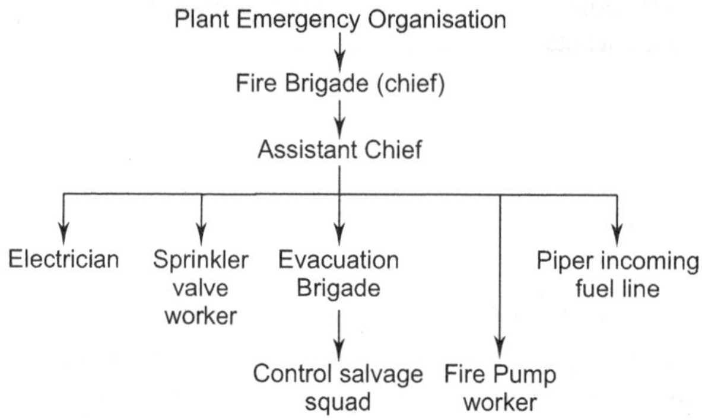

This is a model of the emergency organisation and present, to man the various responsibilities on receipt of fire havoc call. The chief would immediately deploy the staff for operation. The electrician would disconnect the supply line, the worker would activate the sprinkler system, the evacuation squad would start evacuation work, the control salvage squad on direction from the evacuation squad chief, start salvage work and the pump worker would remain busy in activating fire pump network and the worker incharge of fuel line would control flow of fuel.

Fire first aid equipments

The following are some of the fire first aid equipments:

(*a*) Door breaker (8 kg weight)
(*b*) Sledge hammer
(*c*) Excess pick
(*d*) Shovel
(*e*) Axe
(*f*) Crow bar
(*g*) Fire beater
(*h*) Fire hook
(*i*) Emerger
(*j*) Fire bucket (sand filled)
(*k*) Water bucket
SCBA (Self Contained Breathing Apparatus),
(Alarm system attached to it)
(Sound system would be there for supply. Gazometer fitted)

Emergency life saving apparatus

1. Gas-Contents (normal) 2 kg quantity:
   Duration -10 minutes to last
   Gazometer fitted
   No alarm system
2. En capsulating suit:
   (*a*) Fire entry suit + SCBA
   Composition : Silver containing wool, heat resistant (make).
   Duration : 20 seconds to last inside the burning fire.
3. Fire man helmet:
   Standard make of Apparatus
   1. SCBA : EN 137
   2. Arms and leg covering : En 345
   3. Hard Helmet : EN 443
   4. Gloves : EN 659

Class A Fire: Soda extinguisher, 9 litre capacity

(*a*) Composition : Sodium bicarbonate + Sulphuric Acid

(*b*) CO_2 only water : 9 litre and 5 litre capacity

(No regulator. (Bottle to be upside down) It is exhausted once operated)

Class B Fire: Foam

(*a*) Composition : Sodium Bicarbonate, aminophosphate, foaming agent and water

(*b*) Capacity : 9 litre

No upside down. Straight operation of the cylinder. Hose is attached.

(*c*) Capacity : One litre also
Film formed compound (FFFC bottle) Extinguisher: upside down operation. Solution is to be prepared, inserted and date of refill ing must be noted.

Class C Fire: (*a*) BCE : Bottle from one litre to 50 litre capacity

Composition : Bromin + Chlorine + Diflorine

Operation : Upside down (one litre)

(*b*) Dry chemical powder

Composition : Sodium bicarbonate potassium, urea base sterioate hose fitted. No upside down operation. Straight operation. All the bottles have safety clips

Class D Fire: DCP Type

Class E Fire: : (*a*) Extinguisher: 1 kg to 50 kg

Composition : Carbon tetra chloride

(*b*) BCF + CO_2 gas more appropriate

(*c*) Supply line (power) to be disconnected.

Fire protection materials

The following are some of the fire protection materials:

1. Sprinkler installation
2. Smoke detection
3. Fire alarm system installation
4. Public address system
5. Fire pumps
6. Fire hydrants
7. Fire extinguishers
8. As per montreal protocol, Hallon use is discouraged.

Colour code in accordance with the B S5423–1985

(*a*) CO_2 fire extinguisher: Black, 9 litre capacity
(*b*) Water-fire extinguisher: Blue, 9 litre capacity
(*c*) Powder-fire extinguisher: Red, 4.5 kg. capacity

The fire extinguishers should be tested to the following:

(*a*) Working pressure : 10 Bar
(*b*) Test pressure : 24 Bar
(*c*) Temperature range : 20°C to 60°C.

The extinguishers shall be operated by pulling the safety pin and squeezing the lever. This action shall open a valve in the head allowing the elements in the extinguisher to be expelled under its own pressure to the horn through siphon and nozzle. All the extinguishers should be clearly marked with the operating instruction.

Fire blanket: The contractor shall supply, install and connect to the surface of the wall 1.5 x 1.2 m light weight, high efficient fire fighting agent, fibre glass blanket, complete with a suitable container clearly marked with the operating instructions.

MEANS OF ESCAPE IN FIRE EMERGENCY

It is the structural means forming an integral part of the building, whereby persons can escape from fire by their own unaided efforts to a place of safety. A careful study of each phrase in the above will show that it can form a yardstick against which a solution to a means of escape problem can be judged.

(*a*) Structural and Integral (Good means of escape must be part of the structure)
(*b*) escape (means of escape)
(*c*) Fire
(*d*) Unaided
(*e*) Place of Safety

Definition

It is the structural means forming an integral part of the building where-by persons can escape from fire by their own unaided efforts to a place of safety.

It is the structural means forming an integral part of the building whereby a person can walk away from the products of combustion by their own unaided efforts to a place of ultimate safety at ground level where dispersal can take place.

A careful study of each phrase in the above definitions will show that it can form a yard-stick against which a solution to a means of escape problem can be judged.

(*a*) *Structural and integral:* Good means of escape must be part of the structure of the building and made immovable. The effect of this is to generally rule out the use of most portable self rescue devices, which cannot be relied upon to be in positive when required.

(*b*) *Escape:* This word con be visualised as including jumping from windows, crawling across roof etc. None of these methods is acceptable and it is suggested that the word "proceed" should be substituted. It should be possible to turn one's back on the fire and walk away to a safe place. This should be aimed at from all parts of a building which are occupied.

(*c*) *Fire:* Whilst it is obvious that person should be able to escape from fire, the question of safe escape from heated and toxic smoke and gases must also be considered. Protection against the products of fire is probably the most important part of means of escape.

(*d*) *Unaided:* The use of this word indicates that except in special cases such as non ambulant patients, no mechanical or other devices should ever have to be operated or relied upon. This does not mean that door fastenings should not be used, but would certainly rule out the use of lift or any form of escape which relies upon other persons for its operation *e.g.*, canvas chutes.

(*e*) *Place of safety:* Ideally in the open air where dispersal can take place. Escape routes should never discharge finally into enclosed yards or areas unless the dispersal area is large enough to permit all the occupants to proceed to a safe distance away. Dispersal in the open air therefore constitutes place of ultimate safety.

It is often necessary to device a "half way" place of safety, for example, when evacuating high buildings. This may be defined as a place of comparative safety and includes any place which puts an effective barrier to fire between the person escaping and the fire. Fire resistant enclosed staircases are one example of a place of comparative safety.

FACTORS AFFECTING MEANS OF ESCAPE

There are five factors to be considered when assessing any means of escape.

1. Construction
2. Time of evacuation
3. Occupancy
4. Exits
5. Travel Distance.

Construction

The construction of a building will obviously dictate to some degree how it is likely to behave under fire conditions. This must affect your considerations while evaluating this factor, but building seldom fall into clear cut categories such as non combustible or combustible. When assessing this factor,it is therefore necessary to consider the influence of:

(*a*) Primary construction and (*b*) Secondary construction.

(*a*) *Primary construction:* Primary construction refers to the main fabric of the building *i.e.,* wall, floors, roof and internal dividing wall. The following classifications are awarded according to this primary construction.

Class A: Complete non combustible.

Class B: Traditional constructions *i.e.,* non-combustible walls with combustible floors and/ or roof.

Class C: Combustible construction.

(*b*) *Secondary Construction:* Secondary construction refers to internal partitioning wall and ceiling linings etc. The anticipated effect that the secondary construction would have upon a fire may well cause you to lower your original classification based on the primary construction.

Time of Evacuation

This is the time it takes a person to go from any occupied part of the building to a place of safety, be it ultimate or comparative safety. It is obviously going to vary quite considerably according to a persons speed of travel *e.g.* old persons cannot move as quickly as younger persons.

The time permitted for evacuation to a place of safety should be as short as possible and exists should be arranged to achieve a time between two and three minutes. These are arbitrary figures derived from the standards set for occupancies of high density such as places of public entertainment, where some figure had to be set to determine the number of exits required. The principle is that evacuation should be as immediate as the occupancy permits and.the escape routes should retain their integrity for the evacuation time envisaged.

Your initial assessment of the time should be based upon your construction classification thus:

Class A: Three minutes

Class B: Two and half minute

Class C: Two minutes.

It cannot be over stressed that these times are a guide only and may be varied according to the circumstances. In extreme cases, evacuation time may be limited to seconds if the occupancy hazard is great *e.g.,* a cellulose spray space.

Occupancy

This factor divides into two parts:

(*a*) Population — basically physical condition of people living in the building.

(*b*) Use of Building — basically the processes being carried out in the building.

(*a*) ***Population***

(*i*) Number of occupants
(*ii*) Density Factor
(*iii*) Distribution about the building
(*iv*) Physical condition
(*v*) Expected reaction
(*vi*) Discipline
(*vii*) Asleep or alert.

(*i*) *Number of occupants:* Before you can begin to assess the adequacy or otherwise of any existing means of escape, or decide what additional means are required, some reliable method must be devised to establish how many persons may need to use the means of escape. The obvious thing to do would be to ask some person connected with the building but if this is done, be sure that only a person in a responsible position is asked. It is sometimes possible to count numbers, but this should only be attempted in small premises. Finally, it may be necessary in unoccupied premises to calculate the maximum numbers permissible (or desirable) in any given occupancy.

(*ii*) *Density factor:* This may be defined as the number of persons be per unit of floor space, usually in square meters. Once the correct density factor has been selected, it is only necessary to divide this into the floor area.

(*iii*) *Distribution about the building:* There is obviously a difference between the means of escape required from 100 persons on the ground floor and that required for 100 persons on an upper floor. Basements require special consideration as do 'dead end' condition. A dead end condition is any place from which escape is possible in one direction only.

(*iv*) *Physical condition:* It is important to ascertain the degree of agility that can be expected, therefore it must be established whether or not any aged, infirm, handicapped or very young person may be present. It is worth remembering that an average person will probably not react mentally under fire conditions as he would in normal circumstances.

(*v*) *Expected reactions:* What can you reasonably expect ordinary people to do when confronted with a fire emergency? Experience shows that there are cases when it is really a difficult task to persuade 'people to leave the building which was on fire. On the other hand, there are many cases on record of multiple deaths caused by panic when the danger from fire was negligible.

(*vi*) *Discipline:* The degree of discipline one can expect in any given occupancy will to a large extent, be dependent upon the frequency and thoroughness of fire drills, the degree of familiarity with the premises and of course with the escape routes being kept smoke free.

(*vii*) *Asleep or alert:* This is of vital importance as any sleeping occupancy must have time to react to an alarm of fire and to then respond actively. It should also be remembered that in most such occupancies, there will be a delay in discovering the fire. It is interesting to note that tests in an old person's home have shown that at night, 15 minutes can be required for occupants to travel to a place of comparative safety while 30 minutes would be necessary to ensure total evacuation.

(*b*) ***Use of buildings***

(*i*) Nature of contents
(*ii*) Furnishings
(*iii*) Goods stored or used.
(*iv*) Processes carried on.

It must be established what kind of fire is anticipated, what effect on the spread of fire the contents may have and whether or not smoke in large volumes is expected. Processes of a hazardous nature *e.g.*, low flash point liquid, obviously require special consideration. The location of such processes adjacent to escape routes or the access, there to will also have to be carefully considered upon a consideration of the foregoing. It will be seen that occupancies fall generally into one of three groups. Each group has its own characteristics and common hazard as shown in the following table:

Group	Examples	Characteristics	Hazards
Assembly buildings	Places of public entertainment large canteens, public halls, restaurants Deptt. stores.	High density poor discipline, all age groups, strangers.	Panic
Trade, commercial, Industrial	Offices, Shops Factories, Ware houses	Lower density, disciplined constant number`	Rapid spread of fire and smoke
Residential, institutional	Schools, Hospitals, homes, hotels.	Infirm or handi-, capped,very young to very old, supervised sleeping.	Delay in discovering fire, delay in evacuation.

Exits

Entrances, exits and circulation areas are provided in all buildings for normal everyday use and a means of escape should be utilized where possible on existing arrangements. The first approach to this factor should therefore be a question of disposition, number and width of the existing exits. Only if they are insufficient in some respects, should further steps be taken.

When the question of adequacy in number or width arises, a more detailed study of the movement of persons become necessary. This is usually so when dealing with places of public resort or those having a high density factor.

In most other cases the width of exits is not a critical factor as the number provided for normal use is generally adequate to cope with the number of persons involved.

In premises with high density factors, formulae for determining the number and width of exits required become essential and these are based on two constants which have been decided by experience. The first is a "Unit of Exit Width" . This is the width of opening required for persons to travel through in single file and is 525 mm. After the first two files, this space is reduced to 450 mm due to "Over-lapping"

Units	Width
1. Units not less than 525 mm	525 mm
2. Units not less than 525 mm + 525 mm	1050 mm
3. Units not less than 525 mm + 525 mm + 450 mm	1500 mm
4. Units not less than 525 mm + 525 mm + 450 mm + 450 mm	1950 mm

The second constant is the 'rate of flow'. Many assessments of this figure have been made after experiment, but the generally accepted one is 40 persons per minute per unit available. With these factors it is possible to calculate the number of units of exit width and subsequently the number and width of exits required for a given number of persons'

1. Number of person

$$= \frac{\text{Floor Area}}{\text{Density Factor}}$$

2. Number of units of exit width

$$E = \frac{N}{40 \times T}$$

N = Number of person
T = Time Factor in minute
U = Number of units required.

Note: Where a fraction of 0.3 or over results, the next whole number be taken.

3. Number of exits:

$$E = \frac{U}{4} + 1^{*}$$

or

$$E = \frac{U}{4} + 2^{**}$$

E = Number of exits or stairs required.
* = Absolute minimum requirement
** = When Accidental Blockages of extra exit cannot be ruled off.

Note: Where there is fraction of 0.75 or over results, next whole number be taken. Whilst the use of these formulae will give minimum requirements concerning the width and number of exits, It must be noted that the question of siting cannot be decided by any formula. This is a matter for personal judgement taking into account travel distance and siting of occupancy hazards.

Travel Distance

"The distance to be traversed in order to reach a place of Safety". Place of safety in this context can be either ultimate or comparative. If evacuation times are to be maintained, then some limit has to be placed on the travel distances acceptable. It is not possible to set out maximum distances which cover a variety of occupancies. The distances vary according to assessment of the risk present by fire prevention officers. The increase/decrease in distances are affected by. such as, very high ceilings, automatic fire detectors etc.

A further point to bear in mind is that if a corridor for instance is say 24 meter long, it is not possible to reduce travel distance by the addition of self closing fire check doors across the corridor. This action will of course reduce the area of smoke logging but the overall distance remains the same.

As a 'rule of thumb' method of assessing acceptable travel distances, it is pointed out that normally ambulant persons can be expected to travel at a rate of 12 meter per minute in congested conditions (a fairly normal means of escape expectation). Bearing in mind that you will have already determined the evacuation time to be achieved, it is only necessary to multiply 12 by time in minute to arrive at a reasonable travel distance for those particular circumstances.

Maximum TD = 12 x Time of evacuation of particular construction.

$$\text{Maximum Dead End} = \frac{TD \times 2}{5} \text{ meters.}$$

TD = Time distance.

Exit signs notices

All notices indicating exits from any part of the premises to which the people are admitted be illuminated by safety lighting.

The "EXIT" sign should as far as possible be guided by directional ARROWS preferably by green tinted/illuminated light followed with words "TO EXIT"

These should be done with proper meaning of the following words:

(*a*) DISTINCTIVELY
(*b*) CONSPICUOUSLY
(*c*) ADEQUATELY
(*d*) CLEARLY
(*e*) INDICATED
(*f*) ILLUMINATED.

This chapter has been dedicated solely to the General Fire Prevention and Fire Fighting. Fire hazards relating to particular Industrial Sectors have been dealt more precisely in the subsequent chapter while dealing with individual industrial establishments, which may be carefully gone through for acquisition of comprehensive knowledge on the subject.

5

SAFETY IN CONSTRUCTION INDUSTRY

ROAD CONSTRUCTION

Injury is a disease defined as damage caused to the human body by the exchange of energy. Interestingly, even "Shushrutra Samhita", the old Indian text on medicine, considers injury caused by accidents to be a disease. Approximately, 8 lakhs of people die of injury in India every year of which 85% are below the age of 50. The number people killed per year due to road accidents alone is approximately 70.000 and those seriously injured ranges between 15 to 25 lakhs. If you compare this figures with the number of people, who die in every from diseases like cancer, pneumonia, AIDS and heart disease, you realise the latter get importance for beyond the actual loss to society. Surveys have revealed that nearly 70% of the people are willing to accept the low rating to road safety. They are averse to disobedience of traffic rules and this is evident in their retaliatory action in burning buses, trucks and severely beating the drivers, whenever they cause accidents. When we talk of "Road safety", we must bear in mind that lack of improved design or model of vehicles as per exhotic standard is not enough to curb road accidents. It covers death rate to the extent of one percent only, since car occupants comprise meagre less than 5% of all road accidents in India. Here, we have to take into account issues ranging from helmet design and safety of three wheelers, trucks, buses to pedestrians and cyclist safety. Efforts of Govt, in initiating traffic safety programmes like observation of safety weeks, declaring "year for road safety for the child" etc., no tangible improvement has been achieved. Such programmes only create awareness about the problem but do not provide any solutions. Behavioural change comes about by change in the design of the environment and change in regulations, which are enforceable, if we want to turn the crossings in to round about. The traffic automatically slows down at the round about, that kind of design change, promotes behavioural change and in turn, reduces accident rates in that area.

An easily enforceable law like making it mandatory for two wheelers to keep their head lights on even during the day. This simple action facilitates their fellow drivers to notice them in their peripheral vision. This will bring about drop of accidents to 10% less. This has been experimented in Malaysia and Signapore and proved a success. Similarly, enforcing by cyclist to use reflectors and painting them yellow for greater visibility on the roads. The two agencies, National Institute of occupational safety and health in Ahmedabad and factory advisory service in Mumbai are working on the subject of occupational accidents. The centre for road research institute is engaged in evolving ways and means to reduce the accidents figure. A separate department for traffic safety in the ministry of surface transport needs to be set up for any meaningful action in view of the gigantic magnitude of the road accidents in the country.

Prevention and control of road accident starts on the design and drawing boards. The road alignment should be as per approved standard. Next comes execution, which comprise the following.

Blasting, hill cutting and removal of slides and maintenance: While planning, the visibility factor (hills, bends, both horizontal and vertical curves) should be designed for the likely speed of the vehicles using the road. Visibility be improved by batter benching. It is generally seen that sight distance is reduced by buildings, hedges, trees, bushes, walls and other structures that grow after road is constructed. So it should be properly planned and authorities should clear such dangerous obstructions.

Blasting:

(*a*) Training of staff to handle explosives and methods of using explosives is a must. These should be at safe distance, care be taken in storage and carriage.

(*b*) Smoking or use of naked light be prohibited.

(*c*) Sentries be posted with red flags around the site to prevent access of men and animal to the area whistles or other suitable signals should be used for warning. It should be encircled by warning tapes.

(*d*) In case of electric firing method being adopted, operation should be stopped as soon as an electric storm approaches the area of work.

Slide clearance

(*a*) In the event of land slide on the road, barricades should be placed sufficiently far away from the slide in both sides, indicating "Danger, Road block under clearance" board. The traffic checkpost be informed.

(*b*) Watch and ward be posted on the site of clearance to give warning to the working party by prearranged signals.

(*c*) Before starting clearance after the blast, both the flanks of the land slide should be examined thoroughly for further potential land slide as the tension cracks, developing along peripheral area of the slide, which may not be apparent and a further slide may cause casualty.

(*d*) All the workers should wear PPEs and 1st aid kits during operation.

(*e*) Proper lighting arrangements be made in the night or if the atmosphere is foggy.

(*f*) Crowding of working party be avoided.

(*g*) In case of habitated area nearby, blasting operation be controlled by adjusting the spacing, depth and direction of bore hole, type of explosives used and delay fuse. To prevent flying materials, wire mesh can be placed over the area of explosion, suitably anchored or heavy stones or sand bags placed over it.

Road cuttings

(*a*) Vehicular traffic be regulated.

(*b*) Trenches barricaded and red flag placed in both the sides of Trenches and Excavations. In the night illuminated red Lamps be placed all through the length of the cuttings and trenches. Diversion boards caution notices be placed.

(c) *Excavations:* Should be planned against the following.

(i) Collapse of the sides

(ii) Materials falling onto workers and vehicles.

SAFETY SIGNAGE

Safety Signs

All safety signage, that is displayed in and around the sites shall be both in English and local language. Examples of signs that shall be required, include amongst others are the following.

(a) Wear Safety Helmets

(b) Permit to Work areas

(c) Wear Safety Footwear

(d) Wear Hearing Protection

(e) Wear Eye Protection

(f) Danger Electricity

(g) Danger Crane Overhead

(h) Stop Look and Listen

(i) No Smoking

(j) First Aid

(k) No Entry

(l) Fire precautions

All safety signs shall comply with the internationally recognised Safety Colours as indicated below.

- Blue = Mandatory
- Yellow = Danger
- Red = Prohibition
- Green = Safe Condition

Besides Traffic norms are to be strictly enforced, good maintenance, hoarding display for diversion, speed limit of vehicles, warning tap fixation, road repair and maintenance of vehicles, road worthiness and load factor enforcements are some of the very essential elements to be enforced for a perfect Road Safety. Importance of road signs-mandatory, cautionary and informatory, emergency information pannel for the vehicles, transporting dangerous goods, correct methods of overtaking and reversing, hazards combining drinking and driving should be kept in mind by the drivers. Importance of taking rest, pauses and breaks during long drives and maintaining correct posture of the driver, while driving are to be observed. It should be born in mind that one accident is enough to ruin the life of a driver. A strong AIDS message is also to be conveyed to the drivers, who in their long drive fall victim to this dreaded disease, which they must avoid.

References

Indian standards

IS 9457 Standard for colours of Safety Signs

IS 12349 : 1988 Fire Protection Safety signs

SAFETY IN BUILDINGS INDUSTRY

As discussed earlier safety and fire prevention of a building starts on the drawing board where for safety errors in the original design can be corrected. The safety officer may meticulously scrutinise the building plan and satisfy himself if the following provisions are made.

(*a*) Positioning of the water tank at least having a capacity of about 200 cubic meter to feed the 1 st aid hoses in case of fire.

(*b*) Fire escape stairs (Fire exit)

(*c*) Provision of Dampers. (Steel curtain type both for vertical and horizontal operation.)

(*d*) Provision of fire break walls and other fire barriers.

(*e*) Provision of Test holes.

(*f*) Provision of fresh air in take louvers

(*g*) Provision of flexible connections, air inlets and outlets, diffusers, floor grills and under cuts, access doors, sound attenuators and garbage chhute.

(*h*) Provision of Fire fighting sprinklers complete with pipe work in all floors of the building.

(*i*) Temporary lighting and power tools provided.

(*j*) Fire alarm, fire detection, fire censors, accurate identification of source of heat smoke, (soft addressable type,) provided.

(*k*) Built in key board and FACP recording the time, date and place of occurrence in FACP memory and provision for silencing the alarm.

(*l*) Provision for sprinkler flow switches in each level and the tamper switches for the fault conditions.

(*m*) Provision of public alarm system, fire phone system, portable fire extinguishing system, fire blanket, CCTV monitors.

(*n*) Sufficient number of elevators, loadings, speed, stops, openings, closing, door safety, lighting protection system, alternative power supply systems are there.

(*o*) Design errors like providing canteen with the workshop be corrected.

(*p*) Provision of sufficient ventilation system, building utility system, fire cut off, interior finish and other fire safety features are there.

The fire code subdivides the various types of buildings into seven major classes.

1. Fire resistant Type 'A'
2. Fire resistant Type 'B'
3. Protected non combustible
4. Unprotected non combustible
5. Use of heavy Timber
6. Ordinary type of buildings
7. Wood framed buildings

Examples

1. Type A: Covers buildings made of cast iron, brick tiles etc.
2. Type B: Buildings with covering of Asbestos sheets.
3. Buildings protected by non-combustible elements, chemicals, coatings, covering and use of less non-combustible elements/materials.
4. Buildings made of unprotected non -combustibles,
5. Heavy Timber used in construction but fire protection system heavily provided.
6. Ordinary buildings with provision of fire protection.
7. Only in wooden houses, fire protection provision are heavily provided.

Fire Hazard in Buildings Under Construction

Site preparation stage

(*a*) Non clearing of rubbish/garbage from the spot creates fire.

(*b*) Accumulation of debris that blocks access to the part, that is affected by fire or to the hydrant for use in fire fighting.

(*c*) Temporary structures like shops, eating houses, stores located close to the main proposed building under construction causes fire hazard. Example: Fire may percolate from the eating house kitchen and spread to the proposed building.

(*d*) By defective heating apparatus.

(*e*) Poorly installed temporary wiring.

(*f*) Inadequate ventilation. Example: Paint and Engine shops.

(*g*) Lack of providing fire protection system at the initial stage of construction.

(*h*) Careless handling of flammables like paint chemicals, acid, gasoline, liquids and lubricating agents used in temporary garages for vehicles, wooden frames etc.

Fire protection should go hand in hand as the construction progresses as per below:

1. Limiting the amount of combustibles stored and use of minimum practicable.
2. Use of non-combustible scaffoldings in place of wood or bamboo.
3. Stand pipe and sprinkle system should be in service not more than two floors below the highest construction.
4. The openings in the new construction should be carefully plugged.

Hazardous Operations

(*a*) Cutting and Welding stage: The fire may come from sparks. This section may be located away, sealed with protective barriers.

(*b*) Flammable liquid be safely stored.

(*c*) Trash disposal should be prompt.

(*d*) Tar kettles and roofing mops: kettles need not be placed on wooden roofings.

(*e*) Temporary wirings should be fool proof.

(*f*) Gasoline powered equipments be kept far away.

(*g*) Exhaust systems of machines be stretched far away into open spaces.

Fire Cut Off

The fire cut off system be adopted along with the progress of construction. This is of two types:

(*a*) vertical

(*b*) horizontal.

Horizontal system

1. Placing fireproof shutters from outside. There should be provision to operate these shutters from outside too, to cutoff fire from inside.
2. Rolling steel fire doors, which are excellent cut off lines be provided.

3. Exit doors: No steel doors be fitted in these exit doors. In case of emergency, those could be broken.
4. Sliding-fire doors.
5. Fire curtains be provided.
6. Pipe chases: The workers remove the fire stops for work but forget to place them later, which sometimes cause fire hazard.
7. Conveyor opening.
8. Provision of self supporting fire wall at the building expansion joints. All recommendations of the fire officer/safety officer must be cost effective. It should not cost more than the losses, they design to prevent.
9. Concealed spaces be properly plugged as they are potential fire carriers. These should be covered with fire resistant Verma culite and plaster, wire mesh to sheet metal to flow density fire board.
10. Fire is caused by combustible ceilings. Flammable items like oil, chemical, vaporous items, paintings, hangings and ferrying methods adopted for room decoration, be avoided. Fire is caused by supports like wiring, steam and hot water pipping, unless proper distance is maintained between piping and wood framing. The wood gradually carbonize and ignite.
11. Air conditioning, heating and ventilating ducts should be provided with fire dampers. There should be enough ventilation provision.
12. Interior finish like paint, wall paper, plastic films, floor coverings are basically fire motivators. Fire resistent coverings be used. Fire resistant chemical coated plastic sheets may be used but these sometimes cause toxic fumes. In room separators, plywood is used but thick plywood with adhesive become fire resistant.
13. Smoke towers in the building be there.
14. Electrical firing: Short circuit fusion transformer should be fool proof.

Safety in Scaffolding

1. No wood scaffolding be used on the following reasons.
   (*a*) Timber forms sometimes give way. It strays and used many times earlier cause danger.
   (*b*) Ballies used are of sub standard quality.
   (*c*) The earth support sometimes collapses.
   (Therefore always steel frame work scaffolding be used.)
2. Known characteristics of each component enabling complete calculation of loading must be adhered to.
3. In steel frames, safety requirements with regard to working platforms, clear and unobstructed walk ways, guard or nets for working at heights be always there.
4. Sills properly placed and of adequate size.
5. Have screw jacks used to level and plum scaffold, instead of unstable objects such as bricks, stone pieces, concrete boulders.
6. To see if they are in firm condition.
7. All scaffold legs braced with braces properly attached.
8. Guard railings in place, in all open sides and ends 10 feet above, should be fitted.
9. Proper access to platform for work has to be provided.

10. Overhead protection or wire screening to be provided.
11. Scaffold to be tied to the structure atleast every 30′ in length and 26′ in height.
12. Have free standing towers been tied in every 26′ in height
13. Brackets and accessories nave been properly placed.
14. Is it free from makeshift devices or ladders used to increase height.
15. Are working level platforms fully planked between guard rails.
16. Toe boards installed properly
17. Anti hazardous conditions provided for power lines.
18. Wind loading calculation has to be made while fixing scaffolding system.
19. Washout of footings be watched and remedied.
20. All operators are instructed to use personal protective equipments (P.P.E.S.) like hard helmet, chest belt, life lines, gloves, manila ropes and below safe nettings.
21. Working at height is the largest single cause of serious accidents in the construction Industry and therefore the Contractor shall take all appropriate measures to ensure that whenever workers are working at a height, from which they can fall more than two meters, suitable safety precautions have been taken.
22. All scaffolds should be erected and dismantled by workmen who are thoroughly experienced in the erection and dismantling of scaffolding.
23. All scaffolds should be inspected on a regular basis and the results of inspections, recorded and the records shall be kept available for checking by the Employer's Representative.
24. All scaffolds shall be constructed of sound materials free from patent defect.
25. The following measures shall be taken;
    - (*a*) the scaffold shall be constructed for the correct use (Light or Heavy Duty)
    - (*b*) securely fixed to existing structures or adequately buttressed;
    - (*c*) the use of barrels, boxes, loose tiles or other unsuitable material shall not be used as supports for working platforms;
    - (*d*) all working platforms shall be fully boarded;
    - (*e*) all working platforms Shall have guard rails at one meter height and shall also have an intermediate rail at half height;
    - (*f*) all working platforms shall be provided with toe boards,
    - (*g*) a working platforms shall be kept free of unnecessary obstruction or rubbish
    - (*h*) secure ladder access shall be provided;

Use of ladders

1. All ladders shall be of sound construction and shall be free from patent defect.
2. Ladders should be checked weekly and defective ladders shall be promptly and properly repaired or replaced.
3. Ladders shall not be used as working platforms but may be used for work of short duration of up to thirty minutes.
4. Metal ladders shall not be used near or adjacent to overhead power lines.
5. Ladders shall:
    - (*a*) Be 3 planks of 9" width placed side by side properly secured.
    - (*b*) Be secured at the top or footed at the bottom to prevent slippage.

(*c*) Not be used if any rung is missing;
(*d*) Not be used for any other purpose other than to provide access;
(*e*) Be set at an angle of seventy five degrees unless designed for vertical access.

Safety Harnesses/Fall Arresters

Where it is not possible to provide a safe working platform, then the use of safety harnesses (lifeline) may be considered. If safety harnesses are used they should be of the full body type and secure anchorage points shall be provided and used. For some of the more difficult situations, where work at height is necessary and the provision of working platforms is not practicable, Contractors should give consideration to the use of fall arrest devices to provide a safe system of work.

Indian standards

IS 3696 (Part 1): 1987 Safety Code for Scaffolds and Ladders, Part 1, Scaffolds

IS 3696 (Part 2): 1991 Safety Code for Scaffolds and Ladders, Part 2, Ladders

IS 13416 (Part 1): 1992.Recommendations for preventive measures against hazards in the workplace, Part 1, Falling material hazards protection

IS 13416 (Part 2): 1992 Recommendations for preventive measures against hazards in the workplace, Part 2, Fall protection.

British standards

BS EN 354: 1993: Personal protective equipment against falls from a height,

BS EN 355: 1993: Personal protective equipment against falls from a height. Energy absorbers

BS EN 358: 1993: Personal equipment for work positioning and prevention of falls from a height work positioning Systems

BS EN 360: 1993: Personal protective equipment against falls from a height. Retractable type fall arresters

BS EN 361: 1993: Personal protective equipment against falls from a height. Full body harnesses.

BS EN 362 1993: Personal protective equipment against falls from a height. Connectors

BS EN 363: 1993: Personal protective equipment against falls from a height. Fall arrest-systems

BS EN 364: 1993: Personal protective equipment against falls from a height. Test methods

BS EN 365: 1993: Personal protective equipment against falls from a height. General requirements for instructions for use and for marking

BS EN 795: 1997: Protection against falls from a height. Anchor devices. Requirement and testing.

Excavations

General

(*a*) Excavation is one of the important phases of any construction activity. Due to insufficient attention to the safety aspects, it frequently becomes the cause of many accidents.Contractors are therefore required to plan and execute all excavations in a safe manner.

(*b*) The contractor shall ensure that all excavations are supervised by workers with thorough knowledge and experience of excavation work.

(*c*) The integrity of the excavation and the support system shall be inspected prior to the commencement of any work on a daily basis with the results of the inspections being formally recorded. All such records shall be kept available for inspection by the Employer's Representative.

(*d*) Where there is the possibility of any ingress of water, then pumping sumps shall be established with pumps being readily available for use and additional ladders placed for use in the event of an emergency evacuation.

Planning

The correct planning of excavations is essential for safety and before digging any excavations. Contractors should plan against the following:

(*a*) collapse of the sides;

(*b*) materials falling onto people working in the excavation,

(*c*) people and vehicles falling into the excavation;

(*d*) people being struck by plant;

(*e*) undermining nearby structures;

(*f*) contact with underground services;

(*g*) access to the excavation;

(*h*) fumes and toxic gases emission.

(*i*) Make sure the necessary equipment needed such as trench sheets, props, baulks etc., are available on site before work starts.

General Precautions

The following precautions should be observed:

(*a*) Prevent the sides and the ends from collapsing by battering them to a safe angle or supporting them with timber, sheeting or proprietary support systems.

(*b*) Do not go into unsupported excavations.

(*c*) Never work ahead of the support.

(*d*) Remember that even work in shallow trenches can be dangerous. You may need to provide support if the work involves bending or kneeling in the trench.

(*e*) Prevention of materials falling into excavations

(*f*) Do not store spoil or other materials close to the sides of excavations. The spoil may fall into the excavation and the extra loading will make the sides more prone to collapse.

(*g*) Make sure the edges of the excavation are protected against falling materials. Provide toe board where necessary.

(*h*) Wear a hard hat when working in excavations.

(*i*) Take steps to prevent people falling into excavations. If the excavation is 2 m or more deep, provide substantial barriers, *e.g.*, guard rails and toe boards.

(*j*) Keep vehicles away from excavations wherever possible. Use brightly painted baulks or barriers where, necessary.

(*k*) Where vehicles have to tip materials into excavations, use stop blocks to prevent them from over-running. Remember that the sides of the excavation may need extra support.

Undermining nearby structures

The following precautions should be taken to prevent the undermining of nearby structures;

(*a*) Make sure excavations do not affect the footings of scaffolds or the foundations of nearby structures. Walls may have very shallow foundations, which can be undermined by even small trenches.

(*b*) Decide if the structure needs temporary support before digging starts. Surveys of the foundations and the advice of a structural engineer may be needed.

Avoiding underground services

The following precautions should be taken to avoid underground services;

(*a*) Look around for obvious signs of underground services, *e.g.*, valve covers or patching of the road surface.

(*b*) Use locators to trace any services. Make the ground accordingly.

(*c*) Make sure that the person supervising excavation work has service plans and knows how to use them. Everyone carrying out the work should know about safe digging practices and emergency procedures.

References

Indian standards

IS 3764: 1992 Excavation Work - Code of Safety

IS 13430: 1992 Safety during additional construction to existing buildings - Code of

Practice,

IS 2314: 1986 Steel Sheet Piling sections,

IS 5121: 1969 Safety Code for Piling and Other Deep foundations.

Lifting Operations (Cranes)

1. **Lifting appliances**

(*a*) The Contractor shall ensure that all lifting appliances, including synchronised mobile Jacks, pit jacks, mobile cranes, tower cranes, gantry cranes, launching beams and lorry mounted cranes, prior to being allowed to work on site, shall be available for inspection by the Employer's Representative.

(*b*) All lifting appliances with a lifting capacity of more than one ton shall, where ever practicable, be fitted with Automatic safe Load Indicators and audible Warning Devices, which shall be kept in an operable condition at all times. Checks should be made to ensure that the Automatic Safe Load Indicator is properly calibrated and is functioning properly. Overloading of materials beyond capacity be avoided.

(*c*) All lifting appliances shall be maintained in accordance with the manufacturers instructions and shall be subject to a regular preventative maintenance programme.

(*d*) The operators of lifting appliances shall conduct daily inspections of their respective lifting appliance with the results of the inspections being recorded and kept available for inspection by the Employer's Representative.

(*e*) The Contractor shall ensure that only thoroughly trained and experienced personsaged twenty-one years and over, are allowed to operate lifting appliances.

2. **Lifting gear**
   - (*a*) Lifting gear includes chain slings, rope slings, or similar gear and a ring, link, hook, plate clamp, shackle, swivel or eye bolt.
   - (*b*) The Contractor shall ensure that all lifting gear shall be in good condition and shall be tested and certified every six months, with the Safe working Load being stamped or clearly displayed upon it. Records of test shall be kept available for inspection by the Employer's Representative.
   - (*c*) All lifting gear shall be visually inspected before any use and if any defects are found then it shall be removed from site or dismantled / disabled in order to ensure that it is not used in a defective state,
   - (*d*) All lifting gear shall be properly stored and not left lying on the ground where it could be damaged or used in an unsafe manner.
3. **Lifting operations**

   The Contractor shall ensure that during the course of any lifting operations, the following minimum requirements shall be followed:
   - (*a*) Only thoroughly trained and experienced persons are allowed to sling loads and give directions to crane operators.
   - (*b*) A standard code of hand signals shall be adopted for controlling the movements of the crane and both the driver and tile signaler shall be thoroughly familiar with the signals.
   - (*c*) The driver of the crane shall respond to signals from only the appointed signaler but shall obey the stop signal at any time no matter who gives it.
   - (*d*) Before commencing and lifting operations, the ground conditions on which the crane is to stand shall be investigated in order to ensure that the load bearing capabilities are adequate.
   - (*e*) No loads are to be stewed over public areas without stopping pedestrians and vehicles first.
   - (*f*) No unauthorised persons are allowed into the lifting zone.
   - (*g*) No person is allowed to ride the hook of the crane or the loads being lifted.
   - (*h*) Any areas where a minimum clearance of six hundred millimeters from the rear of the slewing kentledge of the crane cannot be achieved and where persons could be trapped against obstacles, then a fence shall be erected to prevent access.
   - (*i*) All crane hooks shall be fitted with an operable safety catch.
   - (*j*) Wherever practicable all loads shall have tag lines attached in order to ensure that the load can be controlled at times.
   - (*k*) Provision shall be made to ensure that the lifting slings or chains can be safely removed from the loads once they have been landed.
   - (*l*) All lifted loads and stacked materials shall be left in a secure and stable condition at all times.
   - (*m*) Whenever working close to isolated overhead power-lines, the lifting appliances shall be grounded to earth as a secondary precaution against accidental energisation.
   - (*n*) No close working to any live overhead power-lines is permitted without the operation of a strict Permit to Work system being in place.
   - (*o*) Daily check of couplings, shackeles, crane wire ropes and belts. In case of damages of these vital parts, replacement is a must.

Reference

Indian standards

IS 807: 1976 Code of Practice for the design, manufacture and testing of cranes

IS 7293: 1974 Safety Code for working with Construction Machinery;

IS 13583: 1993 Code of Practice for training of Crane Drivers Part 1 General

British standards

BS 7121: Code of practice for safe use of cranes

Part 1: 1989: General

Part 2: 1991: Inspecting, testing and examination

Part 4: 1997: Lorry loaders

Part 5: 1997: Tower cranes

BS 7262: 1990: Specification for automatic safe load indicators

Work in Confined Spaces

1. **General**
   - (*a*) The term 'confined space' has two defining features. Firstly, it is a place, which is substantially (though not always entirely) enclosed and, secondly, there will be a reasonably foreseeable risk of serious injury from hazardous substances or conditions within the space or nearby.
   - (*b*) Some confined spaces are fairly easy to identify, for example, closed tanks and sewers. Others are less obvious but may be equally dangerous, for example closed and unventilated or Inadequately ventilated rooms and silos, ducts, culverts, tunnels, boreholes, bored piles, manholes, shafts, excavations, sumps, inspection pits, cofferdams, and building voids.
2. **The hazards**

   The most likely hazards are as follows:
   - (*a*) Flammable Substances and Oxygen Enrichment;
   - (*b*) Toxic Gas, Fume or Vapour;
   - (*c*) Oxygen deficiency;
   - (*d*) The ingress or presence of Liquids;
   - (*e*) Presence of Excessive Heat,
   - (*f*) Excessive Humidity.
3. **Entry procedures**

   Contractors will ensure that no work will be undertaken in Confined Spaces unless a Permit to Work, see Section 12.3 has been prepared and issued. Only persons who have been thoroughly trained, experienced and are physically fit shall be allowed to work in Confined Spaces. Persons with any of the following medical conditions shall not be allowed to work in confined spaces:
   - (*a*) A history of fits, blackouts or fainting attacks,

(*b*) A history of heart disease or heart disorder,
(*c*) High blood pressure,
(*d*) Asthma, bronchitis, or shortness of breath on exertion,
(*e*) Deafness
(*f*) Meniers disease or disease involving giddiness or loss of balance,
(*g*) Claustrophobia or nervous or mental disorder,
(*h*) Back pain or Joint trouble that would limit mobility in confined spaces,
(*i*) Deformity or disease of the lower limbs limiting movement.
(*j*) Chronic skin disease.
(*k*) Serious defects in eyesight or lack of sense of smell

No smoking shall be allowed in or within 2 metres of the opening to any confined space and suitable warning signs shall be positioned.

Before any confined space work commences, the following equipment shall be available for use:

(*a*) Multi Gas monitor; or other suitable gas monitoring equipment.
(*b*) Sufficient set of Self Contained Breathing Apparatus to enable rescue operation to be carried out;
(*c*) Full Body Type Harness for each worker; (Lifeline)
(*d*) Tripod and Lifeline Hoist Rope; for work in situations where a vertical exit from the confined space is required.
(*e*) Emergency Life Saving Apparatus. (One set per person.);
(*f*) Flame-proof lighting. (Hand lamps not more than 24 volts.);
(*g*) Resuscitation Equipment;
(*h*) Ventilation Equipment.

The persons involved in the confined space working operations shall need to be thoroughly trained and certified as being competent in the use of the above detailed item of equipment.

References

Indian standards

IS 11972 ; 1987 Code of Practice for safety precautions to be taken when entering a sewerage system.

BUILDING SITE ELECTRICITY

General

The Contractor shall nominate a representative whose name and qualifications shall be submitted in writing to the Employer's Representative for review not later than 4 weeks before the appointment and who shall be solely responsible for ensuring the safety of all temporary electrical equipment on Site. The Contractor shall not install or operate any temporary Site electrical systems until this representative is appointed and has commenced duties.

The name and contact telephone number of the representative having been reviewed without objection by the Employer's Representative shall be displayed at the main distribution board for the temporary electrical supply so that he can be contacted in case of an emergency.

The Contractor shall submit schematic diagrams and the details of the equipment for all temporary electrical installations, and these diagrams together with the temporary electrical equipment shall be submitted to the Employer's Representative for review.

All electrical installation work on Site shall be carried out in accordance with the requirements laid down in the Specification. All work shall be supervised or executed by qualified and suitably categorised electricians.

Temporary electrical Site installations and distribution systems shall be in accordance with:

(*a*) Indian Electrical Regulations;

(*b*) The Power Companies' Supply Rules;

(*c*) BS 7671 Requirements for electrical installation the IEE Wiring Regulations (16th Edition);

(*d*) BS 7375 Distribution of Electricity on Construction and Building Sites;

(*e*) BS 4363 Distribution Assemblies for Electricity Supplies for Construction and Building Sites; and

(*f*) BS 6164 Safety in Tunneling in the Construction Industry.

Design considerations

Distribution equipment utilised within the temporary electrical distribution system shall incorporate the following features:

(*a*) flexibility in application for repeated use;

(*b*) suitability for transport and storage;

(*c*) robust construction to resist moisture and damage; and

(*d*) safety In use

All cabling shall be run at high level whenever possible and firmly secured to ensure it does not present a hazard or obstruction to people and equipment.

The installation on Site shall allow convenient access to authorised and competent operatives

Safety of Public Assembly Building

Air conditioning: It's closure leads to stoppage of escape routes of smoker and people die inhaling poisonous gas. False ceiling wall lining, decorative synthetic materials, rubber cushion seats, light electrical cables, telephone cables, water pipes, sanitary lines to various sections which ducts provide easy path for spread of smoke. Many areas like electrical sub-station, oil transformers, switch pannels and motor rooms are not protected by fixed fire protection and warning system. Even whereever is there, they donot function properly due to lack of lead upkeep and maintenance.

Fire Safety Requirements

(*a*) Openings are made in walls and ceilings for laying and repairing electrical cables, telephone, sewage and water supply line, plumbing lines, air conditioning ducts. But these are not covered with fire resistant materials, vermicule with cement plaster.

(*b*) Escape routes and stair cases obstructed with combustible materials and exit door closed. Statutory requirements.

1. Means of Access
2. Underground or overhead water static tank provision should be there.
3. First aid host reels and fire extinguishers

4. Automatic fire detection and alarming system
5. Illuminated exit way exhibiting signs
6. Compartmentalisation
7. Public address system
8. Alternate power supply system
9. Fire lift with fireman switch
10. Wet riser and down corner system
11. Rubber mat, first aid box and electric shock chart be walled.
12. Exit signs illuminated by independent power supply line.
13. More fire fighting and evacuation drill be held regularly
14. Safety audit may be made compulsory. In case hazard in public Assembly buildingings, the modusperandi to be adopted in "Means of escape" has been thoroughly discussed in the previous chapter "Fire hazand and firefighting", which may be studied.

In nutshell, safety in construction industry comes under three board categories:

1. Normal construction up to 40′ height
2. High rise construction above 40′ height (more than 4 storeyed flats)
3. Marine and underground construction.

There are 18 sub-divisions:

1. Demolition and wrecking
2. Excavation and shoring, barricading and pipe lining (Fencing)
3. Scaffolding, ladders & stairways.
4. R.C.C massionary and steel erection
5. High way and Road Constrction
6. Tunnels and caissons
7. Marine equipments
8. Compressed air work
9. Flammable gases and liquid handling, storage and upkeep of equipments
10. Welding and cutting
11. Explosives
12. Hoists, cranes and derricks
13. Heavy equipment-motor trucks, road rollers, mixers etc.
14. House keeping, sanitation and first aid
15. Pipe driving and coffer dam work
16. Accident prevent organisation
17. Inspection
18. Fire prevention

Details of dangers and accidents:

1. Fall (General)
2. Fall from scaffold
3. Insecure walls

4. Unguarded material chutes
5. Work on many different levels and surfaces like wet and slippery.
6. Ladders, platforms, scaffolders even if properly guarded.
7. Falling objects from great height while working in lower levels.
8. Injury from swinging loads
9. Collapse of a section of structure
10. Working in close proximity to heavy moving equipments
11. Unknown uncertainty
12. Explosion
13. Electrical hazards of all types
14. Strain from hard/heavy materials
15. Nail punctures and other hand injury
16. Failure to wear P.P.E.S.
17. Tendency to take chances with inadequate experience and judgment
18. Projected nails
19. Poor lighting
20. Lack of training and apprenticeship

The construction work starts with clearing and leveling of site, excavation, demolition if any, cutting, excavating, trenches etc. Due to absence of proper shoring, the trenches cave in and causes death of workers engaged in digging operation. Excavation and filling process should be done as per proper design with reference to topography of the site. Scaffolding system should be fool proof as described earlier. The planks' strength used for platforms, should be strong enough to withstand a load of 250 to 300 kg. The holes to be made on Asbestos sheets are to be drilled.

Machineries used in building construction

1. Power shovels,
2. Belt and basket converyors,
3. Crane and hoists,
4. Concrete mixers,
5. Forklift trucks,
6. Excavators,
7. Loader trucks,
8. Dumpers,
9. Riggers,
10. Ladders,
11. Welding machines,
12. Lathes,
13. Iron cutting and steel bending machines,
14. Batching plant machines,
15. Machine and hand operated tools,
16. Grinders,
17. Levelling machines,
18. Smothers,
19. Floor smothers and glazers,
20. Vibrators,
21. Dough sprayers,
22. Paint sprayers,
23. Power operated concrete hammers,
24. Air compressor,
25. Demolition hammer,
26. Bulldozer,
27. Wood cutter (electrical),
28. Water pump and Tippers,
29. Blower,
30. Pick up van and different trucks.
31. Shovel loader,
32. Jack hammer,

33. Chemical sprayer,
34. Halty machine,
35. Total station,
36. Shearing machine,
37. Threading machine,
38. Lock forming machine,
39. Sheet folding machine,
40. Wheel boroughs,
41. Floor finishing machine,
41. Fire extinguishers,
42. Vehicle repair machineries like wheel balancer,
43. Dentings welding machines,
44. Jigs,
45. Burners,
46. Radiator Tester,
47. Jointer,
48. Dyes and puncturing machine,
49. Picks,
50. Sharpening machine,
51. Road roller,
52. Bitumen sprayer,
53. Stone crusher,
54. Stone blaster,
55. Sand blasting equipments,
56. Trawler,
57. Generator,
58. Data well used for excavation and road cutting,
59. Tractor,
60. Earth mover,
61. Dynapack,
62. Compactor.

The safety officer to check the following on the worksite:

1. Working and repair of the above mentioned machineries if found un workable. The operators may face accidents.
2. General Safety and health of all workers on the site.
3. Environment and pollution
4. Compulsory use of PPEs
5. Fire prevention and fire fighting
6. Signs, signals and barricades
7. Material handling, storage, use and disposal
8. Tools both mannual and power operated
9. Welding and cutting
10. Whole gamut of electrical operations
11. Ladders and Scaffoldings
12. Floors, wall openings and stairways
13. Cranes, derricks, hoists, elevators, conveyors
14. Motor vehicle mechanised equipments
15. Excavations, trenching and shoring
16. Steel erection
17. Tunnels, shafts and cuissens
18. Cofferdams and air compressors.
19. Demolition
20. Use of explosives and blasting
21. Power 'transmission' and distribution
22. Roll over protection, structures, overhead projection.

References

The ILO, convention held in the year 1938 raised for ratification of convention No 62 of 1937 and the Inspetor (Buildings) recommendation 1937.The Factory Act 1948 does not provide protection steps for construction workers. In U.K. the construction workers are protected by "The building (safety, health and welfare) Regulation 1948, which replaced earlier regislation in 1926. In India "Dangerous Machine Act", "The Building and other Construction Workers Act 1996, provide safety net for construction workers.

MECHANICAL SAFETY

Part or parts of body caught by revolving/rotating machine leading to serious injuries is one of the most common accidents in industry. According to the analysis of Fatal accidents in Plants/Units, during the period of 1990 to 1996,17 (seventeen) fatal accidents occurred due to caught by revolving/rotating machines.

Root cause of occurring such accidents are presumably due to following reasons:

1. Removal of guardings during maintenance and not placing the guards in position after completion of the maintenance job.
2. Guardings were broken/damaged-subsequent action was not taken up by shop floor managers for placing the guards in position after rectification.
3. Shop floor manager (Operation) did not insist maintenance manager for maintaining guards in healthy condition.

Definition of Mechanical Guardings

Machine guarding is a term given for correcting a barrier or a system in between the operator and the machine with a view to protect him from specific danger.

Statutory

Under the provision of Section 21 of the Factories Act 1948, wherein it is stated:

Every moving part or a prime mover flywheel, connected with the prime mover shall be securely fenced by safeguards of substatial strength and constantly maintained. No guard, safety device shall be removed from machinery or made ineffective.

Functional Requirement of Guard

(*a*) The guard should offer positive protection to the operator under all normal working conditions.
(*b*) It should offer protection also to others nearby.
(*c*) The guard should not cause danger itself.
(*d*) The guard shaker, or raffles, one which annoys the operator is not acceptable.
(*e*) The guard should not interfere with the normal movement of the operator or operation.
(*f*) Construction of guard should be strong, rigid and should be rocognised as an integral part of the machine.

COST FACTOR

Protection of the operator can be ensured within reasonable cost if the right guard is selected and installed correctly. Choosing/selecting of guard and function of the machine should be well understood by the danger.

Classification of Guards

Fixed Guards: This is a fixed barrier between the man and the dangerous part of the machine.

Advantage

It is simple in design, construction and use.

Less maintenance is required

- Seldom needs adjustment
- Low cost

Examples

- Grill in front of a fan or main cooler of high r.p.m.
- Covers for gears
- Enclosures for fly wheels
- Belt drive/chain drive
- Couplings/shaft

Selecting fixed guards is the ideal choice when the parts of tile rotating machines are not required to be approached frequently.

Disadvantage

Fixed guard, however, are not suitable where the moving parts are to be approached every now and then.

Interlock Guard

It is a safety device or system which links the function of the operator and machine in such a way that:

(*i*) The machine parts are not accessible when danger is present.

(*ii*) When parts of the machine arc approachable, the danger is not present.

Interlocking can be achieved by Mechanical, Electrical and hydraulic system. Operation of the passenger lift is electrically so controlled that lift does not move when door is open. If door is opened, lift will stop. Mechanical interlock, prevent motion of dangerous parts when guards is opened. Interlock system are effective, only if they are correctly designed, installed and adjusted.

Falling of persons on shop floor due to slip, often cause of coming in contact with the parts of revolving machine resulting serious injuries, if guarding is not correctly maintained. It is a common phenomena.

Clothing

Wearing loose clothing near any rotating/revolving machines may lead to serious consequences due to entanglement with the moving parts of machine.

Taking Out Mortar/Clay, from Running "Pan Mill"

It is often noticed that PAN MILL Khalasi is taking out gun clay/ mortar from Pan Mill with the help of shovel without stopping the machine, which is dangerous and unsafe practice. Therefore, it can be concluded that missing of machine guarding at shop floor means, prevailing of unsafe condition, cause of occurrence of untoward incidents.

Following are the monitoring techniques widely used.

1. Basic monitoring: like corrosion, wear, friction or thermal shock etc.
2. Lubricant monitoring
3. Thermal monitoring
4. Leak detection
5. Strain measurement
6. Corrosion monitoring
7. Vibration monitoring.

Computerised Predictive Maintenance (CPM)

CPM can reduce employee's exposure to hazards, decrease equipment downtime and minimise the effectiveness of maintenance expenditures. By reducing costs due to equipment repair and equipment downtime, an effective CPM programme reduces the total controllable maintenance costs to a minimum. These savings come from the following: Less exposure by employees to malfunctioning equipment, thus fewer accidents;

- Less lost production time;
- Fewer emergency failures of equipment;
- Efficient scheduling of equipment repairs and down time;
- Fewer repairs and lower costs. Improved and safer use of labour;
- Longer life of equipment

Maintenance Culture

A common experience is that over a period of time, in addition to the normal wear & tear, a machine or equipment is hardly maintained in its original form.

What happens is, when the machine is new, it's run loaded to its maximum capacity. Every thing postponing or truncating its scheduled maintenance and when it starts going under breakdown, then wake up for its repair. When repairs are done, all its attachments are not restored back or replaced back either due to hurried repair or due to lack of spares, since it is still possible to "somehow" run the machine to give production, with bare minimum components working. With the limit switches, control gauges, etc. gone situation is ideal ("unsafe condition")' for any mishap to happen. The more complex the machine or the installation, the more chances of hazards created by this "maintenance culture" for accidents to happen.

To prevent this, it is very essential to evolve a culture of carrying out the maintenance of equipment and installation as per schedule and as per its technology, not compromising with quality at any stage.

Training

Maintenance employees should have more thorough training in accident prevention than other workers. Safety fof them involves not a set of pattern of activity, but a complex and constantly changing set of problems. Further more, they must know how to use not only ordinary tools but also ladders, protective equipment, chains, slings, ropes and many other kinds of tools and equipments.

Maintenance crews must be aware of job hazards and receptive to proper training. In industries where irritating, toxic or corrosive dusts, gases, vapours, or fluids are present, give maintenance employees special training to familiarise them with the methods of controlling their hazards.

To prevent escalation of the consequences of an accident and to minimise the loss of life and property, the workers and others need training and education. The Factories Act, has laid great stress on safety training.

Specific areas where training has been made obligatory are listed below:

1. Training of competent person: Under section 2 (ca) of Factory Act
2. Training of workers to get trained: under section 111-A of Factory Act.
3. Training in saftey and health to all workers: under sec 7A (2) (*c*) of Factory Act
4. Training of persons competent to supervise: under sec 41C (*b*) of Factory Act

The concept 'of hazard prediction is becoming highly essential to prevent industrial accident caused due to bad maintenance of plant and equipment. Efficient maintenance organisation is a complex and skilled activity which requires systematic record keeping, item numbering, spare parts stock control and stock keeping, inspection procedures, job analysis, work programming and planning. Whilst safety specialists cannot be expected to be experts in all these activities, they should study the subject in sufficient depth to appreciate them, and evaluate how well they organise and carry out in their own works area.

Standard for mechanical guarding of mechanical power transmission equipments:

1. Guards should consist of filler or expanded metal, perforated or solid sheet metal or wire mesh made of No 14 gauze wire, 3/4inch mess or heavier. The filler should be securely fastened to the guard frame by one of the following methods.
   (*a*) With rivets or bolts spaced not more than 5″ centre to centre.
   (*b*) By welding to the frame in every 4",
   (*c*) By weaving through channel or angle frame,
   (*d*) Where openings in pipe railings are to be filled in with expanded metal, wiremesh or sheet metal, the filler material should be made into pannels with rolled edges or bound with 'V or 'U' edging No 24 gauze or heavier sheet metal fastened to the pannels with bolts, rivets, spaced not more than 5″ centre to centre. The bound pannels should be fastened to the railing with sheet metal clips not more than 5" centre to centre. All joints of frame work should be made equivalent in strength to the material of the frame.

Standard Material and Dimension

Material	Clearance from moving part at all points	Largest mesh or opening allowed	Minimum gauze US standard (thickness)	Minimum height of guard from floor or panel level
Woven wire	under 2" 2"4" under 4" 4"-15"	3/8" 1/2" 1/2" 2"	No16 No16 No16 No12	7 feet all
Expanded metal	Under 4" 4" -15"	1/2" 2"	No 18 No 13	7′ 7′
Pretreated metal	Under 4" 4"-15"	1/2" 2"	No 20 No 14	7′ 7′
Sheet metal	Under 4" 4" 15"	-- --	No 22 No 22	7′ 7′
Wood or metal strip crossed	under 4" 4"-15"	3/8" 2"	Wood 3/4" metal No 16	7′ 7′
Standard	min-15"	2"	wood 3/4"	
Rail	max-20"		Metal No 16	

The framework of all guards should be in angle iron 1″ by 1″ by 1/8″, metal pipe of 3/4″ inside diameter. Guards should be rigidly based every 3 feet or factional part of their height to some fixed part of machinery or building structures. Additional strength be necessary where a guard is exposed to contact with moving equipment.

Guards fastened to the floor

1/2″ by 1/2″ by 1/8″ angle iron, metal pipe $1\frac{1}{2}$″ inside diameter guards 30″ or less in height and total surface area not excess of 10 sq feet may have framework of 3/8″ solid rod 3/4″ by 1/8″ angle wood guards be used, where iron is not suitable.

Prime Movers

Guarding fly wheels

1. Enclosure with sheet metal 7′ or less height from the floor.
2. Guard rails placed between 15″ to 20″ from rim. A standard toe board be provided if the wheel extends to the pit or within 12″of the floor.
3. Less than 5 feet or less, a disc may be fitted to the wheel covering spokes etc.

Conveyor belts

6 feet or less height from the floor, enclosure should be provided extending to the top of the govern ing balls when at their most expanded position.

Shafts

Each continuous line of shafting—horizontal, vertical or inclined, should be secured in position to protect excessive end wise movement. Shafts should run without excessive whipping or vibration. All exposed part of horizontal shafting, seven feet or less from the floor except shafting in run ways, used only for oiling or running adjustment, should be protected by stationary casing, enclosing the shafting completely.

Pulleys

Pulleys having any part 7 feet or less from the floor should be guarded. Pulleys serving as balance wheels, on which the point of contact between belt and pulley is more than 6.5 feet from the floor, may be guarded with a disc covering spokes. A guide be fixed to prevent the belt from leaving the pulley on the side, where insufficient clearance exists.

Belt, rope and chain drives

When both run horizontally, 7 feet or less from the floor, the guards should extend to 15″ above. The belt less than 42″ the belt should be fully enclosed. Overhead belts, 7′ or less from the floor, should be guarded for their entire length on the sides and bottom.

Gears, Sprockets and Chains

Gears

1. Complete enclosure
2. A standard guard at least 7′ high extending 6″ above the mesh point of the gear.
3. A band guard covering the face of the gear and having flanzes need be extended inward beyond the root of the teeth or the exposed sides. Sprocket wheels and chains should be enclosed unless they are more than 7 feet above the floor. The protruding of knob, nuts, screws should be properly cased and protected by sleeves.

Starting and stopping devices

No transmission unit from which machines are driven should extend beyond a single room. Clusters and other emergency control devices should be immediately accessible. Stopping of supply of power to each unit of transmission equipment be subject to remote control in addition to control at the stopping device.

The starting of power by remote control is not recommended. Stoppage of power to prevent accidents, there should be means of controlling and especially of stopping the power from engine, turbine, motor as from the transmission equipment and to the machines driven thereby.

Effective signalling equipment be installed to give ample warnings, that machinery is to be started or stopped. A man's access easily be there.

Machine Safety

- Practical Isolation, (lock and tag procedure)
- Starting and stopping - importance of emergency stop
- Lifting Equipment safety -brief overview of how to properly connect and secure lifting gear
  - eye bolts, shackles, cranes etc. Introduction to "The Machine" and all the ancillaries. General overview of paper processing equipment (brief overview) Machine Ancillaries

Process control
– how they work and general problems

- Material Loader
- Dryer
- Water Heater
- Hot Runner Controller
- Compressor
- Chiller/Air Blast cooler
- Machine Operation.
- Process Control of the Machine.
- Understanding electrical drawings
- Electric (240V to 400V) Brief overview of other machine controls and how they interact with the electrical system:
- Hydraulics, (Valves, Cylinders, Pumps)
- Pneumatics, (Valves, Cylinders), importance of clean dry air.

Safety in Workshop

Welding and gas cutting

1. Ensure connection given to welding transformer is as per standard procedure. Terminals to be properly marked
2. Use always welding/cutting goggles worn together with gloves and helmet.
3. Wear Leather shoes while welding! (gas).
4. Ensure there is no loose connection at any stage, while working with welding.
5. Transformer, checkup tightness of terminal and after switching off the power supply.
6. Use only approved oxygen and acetylene cylinder during welding.
7. Acetylene cylinder shall be kept vertical while welding,
8. Do not use LPG cylinder for gas cutting.
9. Switch off welding transformer at the end of work.
10. Use proper helmet during overhead welding.
11. Use cotton dress while welding.
12. Remember not to do welding/gas cutting near by inflammable materials.

Safe working on lathe drill

1. Use safety goggles to protect your eyes
2. Never clean the machine when it is in motion
3. *Switch* off main switch and allow the rotation to stop before commencement of cleaning
4. Do not clean chips by hand, use brush only,
5. Wear tight and half sleeve dress.
6. Avoid using wrist watch, ring etc. since these may cause Injury.
7. Ensure availability of safety guard for motors, gear boxes etc.
8. Never grind lathe tools without wearing safety goggles.

ELECTRICAL SAFETY

Basic Electricity

- Voltage, Current, Resistance, Power.
- Types of Circuits,
- Fuses, MCB's, RCD's.
- Star Delta,
- Transformers
- How to make safe
- Lock Out Tag Out
- AC and DC.
- Switches, Sockets,
- Single and Three Phase
- Soft Start,
- Safety in Electricity
- Isolation of Supply

Among all the energy resources available, electricity is the most inevitable and commonly used form of energy. It has almost become an impossible thing, to live without electricity. Life can not be imagined without it, in modern society. It serves in industries, offices, homes and in all walks of life. The world would be at its primitive stage if the power had not been invented.

Electricity is a very good servant and serve your need at the Instant, just with a push of a button, but at the same time it is a very bad master too. It can cause instantaneous death, life long disability due to severe burns or devastating fires, turning crores of rupees worth assets to ashes.

3 PHASE, 4 WIRE SYSTEM

Using electricity is very simple. But it is also essential to know how it really works. Considering the extensively used 3 phase, 4 wire systems, there exist 3 possibilities of potential difference in practice.

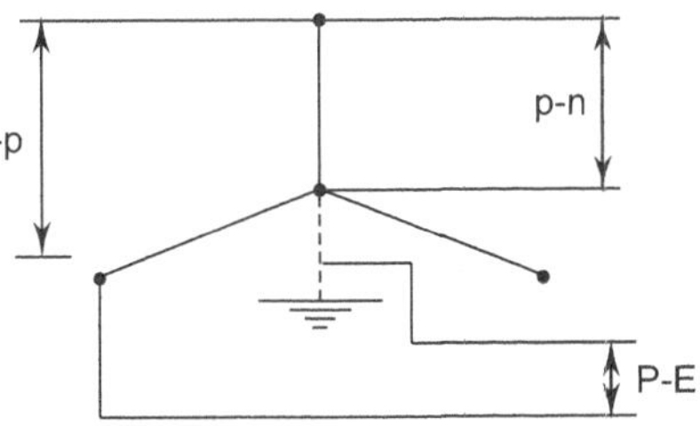

p-p — phase to phase
p-n — phase to neutral.
p-e — phase to earth.

ELECTRICAL HAZARDS

Is this electricity safe? Certainly it is safe, if normal precautions and safety measures are taken: Since electricity does not give any forewarning prior to any electrical accident and also the result of electrical accidents are of serious nature, one has to be more alert when using this power. Electrical hazards are the dangerous effects of electricity.

(*a*) Primary Hazards

1. Electric shock.
2. Electric burn.

3. Electric flash.
4. Electric fire and explosion.

(b) Secondary Hazards (Resulting In Voluntary Reflex Action)

1. Persons falling from height.
2. Dropping of tools and objects. Electrical hazards can also be broadly classified into two main categories:

(a) To living human beings.

(b) To property.

ELECTRIC SHOCK

Electric shock is the net effect of the direct passage of the current through a human body when it accidentally connects the gap between two potentially different points. It stimulates the body's nervous system by electric current. The current enters the body at one point and leaves at another.

It occurs when the body becomes part of the electric circuit. An electric circuit consists of a source of supply and two wires running to a load. When the switch Is closed, current will flow along one wire, through the lead and back to the source of supply. To maintain stability in the system and for other technical reasons, one wire is connected to earth at the source of supply. This earthed wire is called the "neutral" and the unearthed wire is called the "phase".

Shock may occur in one of the following ways:

1. By simultaneously touching the phase and neutral conductors, thus completing the circuit through the body.
2. By touching the phase conductor and standing on the ground.
3. By touching a metallic part that has become live by itself being in contact with an energised wire.

Calculation of Current through the Body

The amount of current through the body can be computed with the help of Ohm's law. In a simple equivalent circuit of an electrocution accident, current (I) through the victims body is given by.

where, r_2, r_3 = contact resistance at the tough points.

R_1 = any other resistance in the circuit.

rm = Body resistance (for the given touch points)

V = Potential difference.

Effects of Electric Currents on the Human Body

For a given potential difference or "V" Volts, the severity of the electric shock depends upon:

1. Magnitude of current.
2. Path of current.
3. Duration of current.

The magnitude of the current follows the electrical laws and is proportional to the Voltage of the circuit and inversely proportional to the resistance to the flow of current which exists in the circuit. The electric current passing through the human body affects the vital body system such as nervous system, respiratory system, cardiac system etc.

Magnitude of current	Effect
Upto-10mA	Tingling sensation.
10–30 mA	Pain is felt in the muscles.
30–80 mA	Pain increases, muscular contraction occurs, breathing becomes difficult.
10–200 mA	Uncontrolled switching of heart (Ventricular fibrillation), disruption of blood flow, death.
Above-200 mA	Instant stoppage of heart.

The body resistance rm is a function of various parameters including the path of the current. It also depends upon magnitude and frequency of the voltage, wetness of the skin, surface area of contact etc. This is evident from the table below.

Human Resistance to Electric Current

Body Area	Resistance (ohms)
Dry skin	100,000–600,000
Wet skin	1000
Internal body (Hand to foot)	400–600
Ear to ear	100

Electric Burns

Another hazard of electricity is the burning of body tissues. This burning can be because of two reasons:

(*a*) Joule's burns.

(*b*) Flashover burns.

Whenever electric current passes through an electrical resistance, heat is generated which is proportional to:

1^2rt where l = current in amps.

r = resistance of path in ohms.

t = time in seconds.

Similarly the current (I) through the resistance (rm) of the man's body, causes heat generation and subsequent burning of tissues in the path.

ELECTRIC FLASHOVER

This is generally the result of breaking the circuit of an electric current. The higher the current, the slower the rate of separation of the parts, the greater is the flash. The flash is directly the result of arcing.

Flashover burns are caused when the victim is in close proximity of a high current contact breaking, during which a severe arc is established in the air. Such burns are also caused when the earth accidentally goes close to the high voltage point soiling up an arc due to ionisation of the air between the body and live part.

The result of such an accident is the severe burning of the body. The currents are too high to cause the ventricular fibrillation. It also affects the eyes, if a person sees it with his naked eyes. The eyes feel the irritation and a terrible burning sensation. It may even lead to temporary or permanent blindness. The ultra violet rays of the flash are responsible for this.

ELECTRIC FIRE AND EXPLOSIONS

The electric current results into the generation of heat and if this heat is not dissipated properly or allowed to generate 'un control ably', it can cause the arc heating of the medium surrounding, thereby results into fire.

The fire hazard is also eminent when sparking takes place. This can be due to:

1. Loose contact
2. Phase to phase, phase to earth faults.
3. Lightning stroke.
4. Static electricity.
5. Sudden breaking of circuit causing an arc.

Similarly, ting spark in an highly flammable atmosphere like diesel/petrol/LPG installation may result in as an explosion.

Electrical Safety

1. Avoid inadvertent contacts.
2. If accidental contact establishes, see that the magnitude of current is minimum.
3. Do not allow high temperature spots, sparks.

   It is necessary that consistent efforts are made to follow above principles at every stage of an installation. These stages are:

   (*a*) Design and manufacture of the basic components.
   (*b*) Planning and design of tailor made system.
   (*c*) Installation and Commissioning.
   (*d*) Operation.
   (*e*) Maintenance.
4. Maintaining all the live parts inaccessible by using suitable barrier guards, interlocks etc.
5. Full knowledge about the work hazards involved.
6. Using hand gloves, rubber mats, safety shoes while working.
7. Ensuring the proper isolation from mains, and discharged to earth.
8. Permanent connections to be ensured.
9. Using hand lamps with wire guards.
10. Never deviate from the laid down procedures. Avoid over confidence and shortcuts.
11. Electricity and Fire are friends of none; thereby avoid electrical accidents with proper knowledge of electrical safety. Electrical accidents mainly happen due to many of the following reasons.
12. Unsafe equipment/ installation.

13. Unsafe work-place.
14. Unsafe work-practices.
15. Protect people from electrical shock caused by:
    (*a*) Improper insulation.
    (*b*) Improper guarding of electrical installation.
    (*c*) Improper grounding.
    (*d*) Improper safety devices viz fuze, auto shut off device, protective relays etc.
16. Display warning for any installation dealing with more than 220 volts.
17. Don't replace fuze in line circuit and without knowing the cause of blowing.
18. Don't attempt to remove a person who is in contact with the wire without insulating yourself and switching off the power supply.
19. While working in high voltage line, use high voltage safety shoes and gloves.
20. Don't throw water on live electrical item in case of fire.
21. Don't operate electrical ckts/switches when your hand and dress is wet.
22. In case of fire, fight fire after switching off the ckt.
23. Never remove earthing from electrical installations, it is an essential part.
24. Ensure non-availability of power supply in socket while connecting to it.
25. Don't allow unauthorized/untrained personnel to work on electrical installation.
26. Only authorized employees should issue clearance and permit to work. No work should be taken up without proper shut down.
27. Always be sure that there is no power, test for it before touching any electrical equipment conductor connected or disconnected.
28. Use the designated PPEs.

POWER PLANT AND POWER TRANSMISSION SAFETY

1. The first step in launching the plant for production of energy and transmission is to determine the load forecast for immediate future and long range plans. For this, one has to contemplate the voltage level, load capability and location of terminals for the transmission.
2. Line design :
    (*a*) Load flow and stability and cost of right of way to be considered.
    (*b*) Selection of line insulation for withstanding switch surges and impulses caused by lighting.
3. Line routing
4. Detours
5. Land survey
6. Aerial survey
7. Line Construction

Hazards in a Power Station

(*a*) Fire by short circuit fusion

(*b*) Non conforming to safety codes as under:

(*i*) Appropriate load flow

(*ii*) Proper selection of line insulation

(*iii*) Improper decision on number of insulations to be used or already used, phase clearances from conductors to ground and use of improper size and number of conductors.

(*iv*) Bad weather conditions, heavy rainfall, storm etc.,

(*v*) Telephone lines or other signal lines running parallel. It should not closer than 91.44 meters. For various power lines, minimum and maximum recommended distance are as under.

For	230 k.v.	minimum distance is	21-34 meter	Recommended 24-38 meter
	345 k.v	minimum distance is	38 meter	45-72 meter

(*vi*) Non providing protection and temporary guard structures as required for the protection of all personnel, equipments, existing structures, communication line, powerline etc.

(*vii*) Non-installation of danger signs

(*viii*) Disorder in tensioning equipments for all installations to prevent the cable touch the ground water or any above ground constructions including guard structures.

For generation, 11 k.v. step up with the help of step up transformers. String insulations are fitted on both sides of transformer tower. It is safe to use ACSR (Aluminium conductors steel reinforced) between tower to tower. Proper insulation must be provided between conductors to carry extra high voltage.

The power plant supervisor to see gas storage safety. It should be in proper size and maintain the gas pressure. Gas transmission pipe joining should be fool proof. Gas pressure to be maintained to run the turbine. Light system be fool proof. Lighting arrester be properly maintained. The generation frequency in our country is 50 cycles per second. There will be trip, if not maintained. The turbine may stop functioning. After production of required of power, periodical maintenance is a must. The main hazard is electrocution. In the event of this accident, the sources of power be disconnected/separated. Then give First Aid (Artificial breathing). Another cause of accident is short circuit. This happens due to weak insulation, overloading and over working.

Preventive measures

The workers to use PPEs as under:

(*a*) Insulatory gloves,

(*b*) Glass shield,

(*c*) Geoggles,

(*d*) Shoe (non conductive),

(*e*) oxygen mask,

(*f*) Helmet (Non conductive).

Remote control system

The computer would indicate the disorder if any.

Requirements:

(*a*) For distributor: 11 kv.

(*b*) For industries and house holds : 440 volt

(*c*) Big industry: 1000 volt.

Safety measures

1. Fixation of enclosures, coverings and barricading the power plant and transmission equipments and prime movers.
2. Indirect or remote lubrication, self lubricating bearings, travelling shaft polish rings, proper clearances, automatic or mechanical belt shifting, warning signals for starting and locks on starting devices.
3. Provision be there for centralised mechanical oiling equipment. It is a typical example of incidental safe guarding.
4. Sheet metal and wriemesh enclosures
5. Enclosures supplement guarding by location and incidentals to safety such as
   (*a*) Centralised lubrication,
   (*b*) Buffers and Bumpers,
   (*c*) Warning gongs,
   (*d*) over travel limit of the hoists, travelling cranes are used as additional means of preventing accidents.

SAFETY IN CHEMICAL INDUSTRY

Classification of Hazardous Chemicals (U.N)

The chemical, which causes an unreasonable risk to health and safety of a person or environment is termed as hazardous chemical. Chemical could be toxic, inflammable, explosive, corrosive, infectious, radio active or having other hazardous properties. Therefore rules and regulations have been formulated for storage, process and handling of hazardous chemicals. According to U.N. expert committee, the hazardous chemicals are classified as detailed below.

Class-I

(*a*) Explosives with mass explosion risk
(*b*) Explosives, which explode in mass but which has limited explosion and fire hazard.
(*c*) Explosive, having a mass fire hazard. There is an explosive rate i.e. GNNCFAL, which means 12845678

1. G—Gun powder
2. N—Nitrates
3. N—Nitro compounds
4. C—Chloratis
5. F—Florates
6. A—Ammoniation
7. F—Fire works
8. L—Liquid oxygen.

Class-II: Compressed gases, liquefied or dissolved gases under pressure

Class-III: Inflammable liquids *i.e.*,

(*a*) Liquid with flash points between 23°C
(*b*) Liquid with flash points between 23°C–65°C
(*c*) Liquid with flash points above 65°C.

N.F.P.A.		**Classification**
NFPA	:	National Fire Protection Association of America
ICSC	:	International Chemical Safety Cards
MSDS	:	Material, Safety Data Sheet

NFPA Classified 3 Categories

(i) Toxicity of the chemicals/Health of the chemicals
$H \rightarrow (0, 1, 2, 3, 4)$

(ii) Flammability of the chemicals
$F \rightarrow (0, 1, 2, 3, 4)$

(iii) Reactivity of the chemicals
$R \rightarrow (0, 1, 2, 3, 4)$

H_0	:	Like ordinary Material
H_1	:	Slightly hazardous
H_2	:	Hazardous-use breathing apparatus
H_3	:	Extremely dangerous useful protection (clothing)
H_4	:	Too dangerous to enter vapour or liquid
F_0	:	It does not burn
F_1	:	It must be preheated to burn
F_2	:	Ignites (Take fire) when moderately (limited) heated
F_3	:	Ignites at normal temperature (25°C)
F_4	:	Extremely Flammable
R_0	:	Normally Stable.
R_1	:	Un-stable if heated, use normal precautions
R_2	:	Violent chemical change.
R_3	:	Strong shock or heat may detonate, use monitors from behind explosion resistant barrier
R_4	:	May detonate (explode with a sound noise) Vacate area if material are exposed to fire.

Material Safety Data Sheet for Chemicals

Hydrochloric acid

Formula: HCI

Description

Colourless or slightly yellow, fuming liquid vapour, has a pungent odour.

Used in:

- ore reduction
- cleaning
- activation of petroleum wells
  Used as a chemical Intermediate.
- food processing
- boiler scale removal

Short-Term Exposure Effects

Inhalation

Vapour or mist

- Irritates the nose, throat and upper airways,

- causes coughing, choking, and nose and gum bleeding. Severe exposure results in— Corrosion of nose and throat,
- fluid accumulation in the lungs.

Eye contact

Extremely corrosive causes:

- Irritation
- severe burns
- blindness

Begin First Aid proceeds Immediately.

Skin contact

Concentrated solutions (greater than 38%) can cause severe burns.

Ingestion

Unlikely, but severe burns of the mouth, esophagus and stomach disorder would result. Can cause:

- nausea
- vomiting
- thirst
- diarrhea
- collapse
- death.

Long-Term Exposures Effects

Vapour causes

- tooth erosion
- persistent bronchitis
- nose and gum bleeding

Prolonged skin contact with dilute solutions causes inflammation.

Fire and Explosion

Does not burn.

Extinguish other materials, which are burning. Cool containers with water spray to help prevent rupture and reduce vapour.

Chemical Reactivity

Reacts violently with:

- strong bases such as sodium hydroxide.
- oxidizers such as peroxides. Reacts with the following to release flammable and toxic gases.

Corrosive

- sulfides
- phosphides
- cyanides
- acetylides.
- fluorides
- silicides
- carbides

Personal Protection

Inhalation: Wear suitable respiratory protection If vapour or mist concentration is unknown or exceeds exposure limits.

Skin: Wear, as needed

- gloves
- coveralls
- boots.

A suitable material is natural rubber. Wear acid resistant coveralls for high exposure situations. Have a safety shower/eyewash fountain available in the Immediate area.

Eyes: Wear a face mask, as needed.

Storage and Handling

Store hydrochloric acid:

- In tightly-closed, labelled containers
- In a cool area with adequate independent ventilation
- away from normal work areas and incompatible materials

Area should have:

- acid resistant floor
- approved drainage. When preparing solutions, slowly add acid to water to avoid splattering and boiling.

Clean-up and Disposal

- Only trained personnel should clean up. Ensure appropriate ventilation is provided.
- Use appropriate protective clothing and respirators.
- Stop spill or leak if possible.
- Contain spill using commercially available absorbent materials for hydrochloric acid.
- Neutralize traces with soda ash or lime.
- Dispose waste material in a designated landfill site.
- Comply with environmental regulations.

First Aid

Inhalation: Control source of hydrochloric acid or move victim to fresh air.

Begin artificial respiration immediately if required.

Avoid Mouth to Mouth Contact

Eye Contact: Immediately flush the affected eye with lukewarm, gently flowing water for at least 15 minutes, holding the eyelid open.

Skin contact: prevent contact

Immediately flush affected area with lukewarm, gently running water for at least 15 minutes.

Remove contaminated clothing and launder before reuse or discard.

Ingestion: Never give anything by mouth if victim is:

- Losing consciousness
- unconscious
- convulsing.

Rinse mouth thoroughly with water.

Do Not Induce Vomiting

If possible, have victim drink about 250 ml. (8 oz) of water.

If vomiting occurs naturally:

- Have victim lean forward
- repeat administration of water.

Note: Obtain medical attention IMMEDIATELY for all serious exposures. Consult a physician or the nearest Poison Control Centre.

HAZ CHEM CODE

The Haz Chem Code also called Emergency Action. It gives vital information to fire brigade and other emergency services on the action to be taken in case of spillage, leakage, fire in an emergency involving a hazardous substance.

Haz Chem Code for some selected chemicals

1. Hydrogen : 2SE
2. Nitrogen : 2T
3. Petroleum gases, liquefied : 2 WE
4. Hydrochloric acid (HCL) : 2R
5. Sulphuric acid H_2SO_4 : 2P
6. Carbon Tetra chloride : 2Z
7. Gas oil (Kerosene, Diesel) : 3Y
8. Hexane : 3YE
9. Naptha Petroleum : 3YE
10. Sodium Metal : 4 W

UKHIS: United Kingdom Hazard Information System

The Haz Chem Code consists of a number from 1 to 4 and any one of the letters P, R, S, T, W, X, Y, Z. Sometimes followed by the letter E:

E—Evacuation (Don't allow any person to the spot).

1. JET (you can apply water by jet)
2. FOG (finely dropped water)
3. Foam (A detergent powder mixed with water which uses as the blanketing action)
4. DryAgent/DCP (Dry Chemical Powder).
   - Compound of DCP: Sodium bicarbonate, Urea base Sodium bicarbonate, Potassium bicarbonate

The letters signify four aspects

(*i*) Method of controlling spillage

(*a*) by diluting

(*b*) by containing

(*ii*) What P.P.E. is to be used.

(*iii*) Reactive nature of chemicals

(*iv*) Need for evacuation or not.

Haz Chem Code For Fire/Spillage

P	V	Full	
R			
S		BA	
S	V	B.A for fire only	Dilute
T		B.A.	
T		B.A. for fire only	
W	V	Full	
X			
V	V	BA	
Y		BA for fire only	Contain
Z		BA	
Z		BA for fire only	
E		Consider Evacuation	

V — Can be violently or explosive, reactive.

Full — Full protective clothing with BA (Breathing appartus).

BA — Breathing apparatus plus protective gloves.

Dilute — May be washed to drain with large quantity of water.

Contain — Prevent by any means available, spillage from entering drains or water course.

Example: For Ethylene Oxide with U.N. No. -1040 containing not more than 0.2% Nitrogen. The Haz Chem Code is 2 P.E.

It could de-coded as

2 : For equipment is recommend

P : Can be violently reactive. Full body clothing with B.A. is necessary. This surface may be diluted and washed to drainout.

E : Consider evacuation as a first priority.

Example: 2 For motor spirit (petrol) with UN No.-1203.

The haz-chem code is 3YE and this could mean

3 : Use foam only'

Y : Can be violently reactive, use breathing apparatus + protective gloves is to be used in case of fire. The PPE is not required for spillage. Any spillage of the substance should be contained.

E : Consider evacuation as the first priority.

Preventive Maintenance of Pipelines and Tanks

Pipelines

(*i*) Blanking of chemical line should be done after making sure that the connecting valves, pumps etc. are closed/stopped.

(*ii*) Chemical should be drained out before blanking.

(*iii*) Safety appliances should be used for working on chemical lines. Apron, jacket, pants, hoods, handgloves, shoes can be of PVC materials.

(*iv*) Water should be made available near the site. In case of highly toxic chemicals, automatic water spray bath must be provided.

(*v*) A protocol to be prepared and signed by concerned agencies before the work is started.

(*vi*) Tools made of non-sparking materials (copper or brass) must be used for working in flammable chemical lines.

(*vii*) For replacement of pipelines flange packing or valves etc., even after fully draining, presence of entrapped chemical cannot be ruled out. It is therefore, necessary to open the flange by keeping a barricade in front of the person doing the job and after creating a gap, the job can be carried out by removing the barricade. Such barricading is a must while dechocking of jammed pipelines.

Tanks/Vessels

For carrying out maintenance work inside the vessels or tank, the following precautions are to be taken:

(*i*) The tank must be thoroughly dried and cleaned to wipe out the residual chemical.

(*ii*) Welding or gas cutting can be taken up only after testing explosive mixture and presence of traces of chemical/vapour.

Loading and Unloading of Chemicals

(*i*) For loading of chemicals into, tankers/wagons telescopic pipe must be used to avoid spillage during loading,

(*ii*) Wagons/tankers should be placed carefully just below the telescopic pipe while loading.

(*iii*) The operating persons must wear safety appliances such as PVC apron, goggles, hood, face shield, rubber/PVC hand gloves or PVC gumboots.

(*iv*) Ensure that the pump is stopped and valves are closed while opening the manhole cover.

(*v*) Insert the telescopic pipe well into the wagon.

(*vi*) After setting the valves, people should be removed from site.

(*vii*) Then the loading pumps to be started.

(*viii*) Level checking of chemical should be done only after stopping the pump.

(*xi*) The telescopic pipe can be removed from the tanker only alter the pump is stopped, valves closed and chemical is fully drained (Both suction and delivery valves to be closed).

(*x*) Loading of chemical must not be done during storm, rain and thunder.

(*xi*) If loading is done by hosepipe, it should be firmly tied, so that it does not come out under pressure of the chemical.

(*xii*) If chemical is fed into a bath or tank manually through portable containers, this should be done through a funnel having a platform to support the can. The lid of the can should be opened only after positioning the can on the platform. Chemical may spill out while positioning the can on the platform if the lid is opened before hand.

Training and Awareness

Persons working for handling of chemicals must be made fully aware of safe handling through:

(*i*) Organised training on SOP & SMP (Standard Operating and Standard Maintenance Practices), First Aid, Job Safety and Safety appliances.

(*ii*) Distribution and display of Material Safety Data Sheets (MSDS).

(*iii*) Organising workshop and mock drills for disaster management.

Considering the hazards associated with toxic and corrosive chemicals both towards persons and equipments, it is essential that all the above safety provisions are taken care of without fail. Although legislations have come up in a big way for control of toxic chemicals, its effective implementation will largely depend upon the co-operation between Industry, Government and Public. Much care is already being taken by Indian Industries in this direction but a lot more is required to be done to come to international level.

Permissible Limits of Exposure of Toxic Chemicals/Substances

Ref: Section 41 F and Second Schedule of Factories Act.

Substances	Permissible limits of exposure			
	Time weighted average concentration (8 hrs)		Short term exposure limit (15 mits)*	
	ppm	**mg/m³**	**ppm**	**mg/m³**
Ammonia	0.25	18	35	27
Benzene	10	20	25	75
Carbon Monoxide	50	40	400	440
Chlorine	1	3	3	9
Hydrogen Sulphide	10	14	15	21
Naphthalene	10	50	15	75
Nitric Acid	2	5	4	10
Nitric Oxide	25	30	35	45
Nitrogen Oxide	3	6	5	10
Phenol	5	19	10	38
Sulphur-dioxide	2	5	5	10
Sulphuric Acid	-	1	-	-
Toluene	100	375	150	650
Xylene	100	435	150	655

Note: 4 times a day with at least 60 min interval between succesive exposures.

Hazards

1. *Inhalation of monoxide*: Affinity of carbon monoxide is 300 times more than oxygen. It displaces oxygen rapidly and forms caboxy haemoglobins

2. If percentage of caboxy haemoglobin rises in the blood, the following complications may arise.
   CO in percentage CO% in haemoglobin

   (*a*) 35 1–15
   (*b*) 50 5–10
   (*c*) 100 10–20
   (*d*) 200 20–30
   (*e*) 500 30–50

   Symptoms

   (*a*) Headache, (*b*) Severe headache,
   (*c*) Nausea, (*d*) Unconsciousness,
   (*e*) Fatal.

3. In case of emergency, a plan is formulated on the site to minimise damage and evaluating help of some expertise in various fields. It helps to protect persons and property of the factory. It also helps to inform people and surroundings about happening, if it is likely to adversely affect them.

4. *Colour coding of pipeline*: In different industries, there are different hazardous chemicals are flowing through pipelines for achieving different types of operations. So, it is necessary to distinguish them by help of proper colour coding.

   (*a*) Flammable or water for fire protection : Red Colour
   (*b*) Dangerous Chemical for fire protection : Yellow Colour
   (*c*) Safe (water, air) for fire protection : Green
   (*d*) Protective material for fire protection : Blue

It is however different in different plants for their own rules and regulations.

Breaking of pipeline: Piping of liquid or gaseous chemicals flammable or toxic is a most complex task for any person. Because there is no direct vision contact over the material which is to be transported through the closed pipeline system. An operator has to rely on the flow meters, pressure gauges etc.

But when it comes to breaking of pipelines, containing flammable and toxic chemicals, it becomes more complex and also hazardous. It certainly is one of the most important causes of service injury.

Causes

(*i*) A portion of pipeline is cracked which required to be welded.
(*ii*) A portion of pipeline may be chocked due to deposition of dusty particles on inner side.
(*iii*) If part of pipeline required to be completely changed.
(*vi*) If leakaging of fluid occurs through flange joints or bend joints.
(*v*) If any extra equipments is to be maintained in pipeline.

It should be remembered that in every case we feel that a major hazard is associated with the breaking of pipeline then a safety protocol is prepared before attainment of the job.

HAZARD ASSOCIATED

Any kind of mishandling, risk taking or deviating the safety protocol may lead to serious hazard as described below:

(*i*) In case of flammable gasses or liquids in side the pipeline, and if pipeline is not flushed or cleaned properly,then fire or explosion may occur during cutting of pipelines by gas cutter.

(*ii*) In case of acids, person handling the job may be injured due to acid spray or acid contact.

(*iii*) If there is toxic liquid / gases inside the pipeline then it may lead to fatal or serious injury in case of leak or direct inhalation by the person attaining the job.

(*iv*) In case of flammable and explosive is transported through the pipeline, mishandling and inadequate safety precaution leads to serious accidents causing huge losses.

(*v*) In case of high pressure line, bursting of pipelines may happen during heating, gas cutting or welding.

Safety points to be considered for storing highly flammable materials.

(*i*) Solid material is generally stored in their own container in hopper's or bins.

(*ii*) Flammability of the material before designing storage equipment.

(*iii*) Combustible material are to be stored after dust spreading in the air kept lower to explosive limit.

(*iv*) Tight enclosure and dust collection system, to be provided.

(*vi*) Good plant house keeping will go far to prevent disaster.

(*vii*) All sources of ignition should be excluded from the area of potential explosive dust

(*viii*) Proper care against static electricity should be taken. Proper maintenance and inspection of the storage media, to avoid accident.

(*ix*) Portable container, hopper bins, packing bags etc., are the effective means of storage of solid flammable chemical depending upon their chemical stability.

(*x*) Fire extinguisher equipment should be properly maintained,

(*xi*) Proper maintenance and inspection of the storage media to avoid accident to a great extent.

(*xii*) Proper colour coding should be provided and maintained for storing flammable material.

(*xiii*) Flameproof lighting should be provided to avoid (ignition).

(*xiv*) Mutual reactive material shouldn't be stored side by side.

Guide Word Approach HAZOP

The HAZOP study is made to idehtify hazards in a process plant and operability problems, which could compromise the plant's ability to design intent. The approach taken is to form a must carry team that works to identify hazards by searching for deviation design intents. Following terms are used for the process for analysis

Intentions : It defines how the plant is expected to be operated.

Deviations : These are departures from intentions.

Causes : These are reasons why deviations might occur.

Consequences : Results of deviations should, they occur

The method uses guidewords, which are used to quantify or qualify the intention in order to guide and stimulate the hazard identification process. The guidewords are used to generate deviations from the design intent. The team then identifies cause and consequences of the deviations.

HAZOP guide words and their meaning

Guide words	Meaning
• NO	Negation of design intent
• Less	Quantitative decrease
• Part of	Quantitative decrease
• As will as	Quantitative increase
• More	Quantitative increase
• Reverse	Logical opposite to intent
• Other than	Complete substitution

Toxicity Index: How it is calculated?

Determination of Toxicity index 'T'

Toxicity Index 'T' is primarily based on the index figure for health hazard established by NFRA. The NAPA figures are translated in to a toxically factor 'Th' according to the table below.

NEPA Index figure Toxicity factor (Th)

0	→	0
1	→	50
2	→	125
3	→	250
4	→	325

In addition, the toxicity factor has to be corrected per MAC value of a toxic substance by adding to it a penalty Ts which is given in the table below.

MAC (PPM)		Ts
<5	→	125
5-50	→	75
>50	→	50

The toxicity index (T) is now calculated as follows :

$$T = Th + Ts/100\ (1 + \text{GPHtot} + \text{SPHtot})$$

In this, the material which gives the highest value of Th + Ts is determining factor.

GPHtot = The total of general process Hazard penalties.

SPHtot = The total of special process Hazard penalties

How DOW index for fire explosion risk calculated?

Determination of fire and explosion index 'F' and Toxicity index 'T':

For each separate plant element which contains flammable or toxic substance, a fire and explosion index 'F' or 'T' may be determined in a manner derived from that method for determining 'P' developed by DOW chemical company of united states. 'F' is calculated from = $F = Mf\ (1 + GPHt) \times (1 + SPHt)$.

Mf = Material factor measure for the potential energy of the dangerous substances present (according to NFPA Data).

GPHt = General process hazard a measure, for the hazards inherent in the process (From the nature and characteristics of process).

SPHt = Special process hazard, a measure for the hazards originating from the specific installation (Process condition, nature and size of installation etc.) The toxicity index 'T' is calculated from:

$$T = (Th + Ts)/100(1 + GPHt + SPHt)$$

Where Th = Toxicity factor (Obtained from NFPA data)

Ts = Supplement from MAC value.

Work Permit for Working at Height

1. Name of Agency
2. Exact Location
3. Nature of Job
4. Date
5. Time
6. Following safety precautions are taken while: working at height.
    - Sound Scaffolding/Working platform made, Safe approach made (Yes/No)
    - Barricade provided around Scaffolding/Working Platform (Yes/No)
    - Life line (One 16mm rope) fixed from one end to other and of the column (Yes/No)
    - Safety Net provided below scaffolding platform (Yes/No)
    - Explained the Hazard associated, Remedial measures and proper use of Safety appliances etc. to the workmen prior to execute the said job (Yes/No)
    - Adequate illumination system provided at job.(Yes/No)
    - Safety helmet/Safety Belt/Shoe/Goggles/Eye shield/Hand gloves issued and used (Yes/ No)
    - Number of workmen engaged: (Yes/No)

We hereby declare that above mentioned safety precautions have been taken to execute the job safely

Name and Signature of Supervisor:______

Name and Signature of Group Leader:______

Name and Signature of Engineer in Charge:______

What are the steps to be taken during large-scale release of Chlorine?

Properties of Chlorine

Colour-greenish yellow

Odour-pungent

Taste-bitter

It is heavier than air and slightly soluble in water forming a strong acid.

It attracts metals severely in presence of water corrugating it.

Health hazards: Cl is corrosive, it reacts with body moisture to form corrosive acid. It can cause respiratory injury ranging from orientation to death depending on the concentration and duration of exposure.

Permissible Limits

Threshold limit value of Cl_2 is as follows:

$$TLV = 1 \text{ ppm},$$
$$STEL = 3ppm$$
$$TWA = 0.5ppm$$

Emergency measures

In case of large scale leakage or spillage following steps be followed.

1. Any person taking chlorine should be equipped with emergency kits, self-breathing apparatus or masks and other appropriate items of personal protection,
2. In case of major leakage, all persons should be warned and take the leakage container to an isolated place.
3. Approach from the windward side so that escaping gas is carried down wind to a lower level under no circumstances, should be watered or other reactive liquid be directed to the leakage.
4. In case of large leaks, try to stop the flow of liquids by closing the valves.
5. After containing the liquid one can use fluoroprotien foam (FFFP) to prevent evaporation of liquids.
6. In case of large leakage, the alternating solution is to absorb the chlorine in all alkaline medium such as caustic soda, soda or lime slurry.
7. Organic solvent-should in no case be exposed to chlorine because in contact with organic liquids Cl_2 makes a highly poisonous solution.
8. A fire should be ignited or the arrangement should be made to provide fire source by which the chlorine leaked is used to react with burning material preventing it to vent to atmosphere.
9. Effective arrangement should be made to evacuate working personnel from the source of leakage to a safer place.
10. All the local authorities should be informed in order to provide safety to people living near by the factory.

Stable first-aid and medical treatment facilities should be provided in Case of severity of Cl_2 hazards. For the above accidents, management should always remain alert to take risk in well-planned and strategic manner.

How risk is calculated mathematically. Risk is likelyhood of a specific undesired event occurring within a specified period or under specified circumstances.

Assessment	-	For assessment of the risk, the 3 major parts to be followed first.
Identification	-	risk (Nature)
Estimation	-	Qualification or measurement
Evaluation	-	Acceptability or control

Consequences, E- Exposure, P- Probability

N	-	Consequence
Occurrence:		(*i*) Probability.
		(*ii*) Exposure.

Score Calculation:

Sequence × Exposure × Probability

Consequence	-	**Marks**
Numerous fatalities	-	100
Multiple fatalities	-	50
Fatalities	-	25
Serious injury	-	15
Casualty treatment	-	05
First aid treatment	-	01
Probability	-	**Marks**
Almost certain	-	10
Quite possible	-	06
Unusual but possible	-	03
Remotely possible	-	01
Very unlikely	-	0.5
Practically impossible	-	0.1
Risk score	-	**Marks**
Very high risk	-	Above-400
High risk - 200	-	400
Substantial Risk	-	80-200
Moderate Risk	-	20-80
Probably	-	Below-20

From these parameter risk score, it can be determined using a monograph, which provides it ranking N-risk and not absolute value.

Example. 1. Condition-1: Alive electrical naked wire hanged in 10 feet high in the room.

Risk—C × E × P,100 = consequences

(Numerous fatality, 01 = Exposure (rare),

0.5 = Probability (very unlikely)

Then, Risk = 100 × 1 × 0.5 = 500 (score)

2. Condition-2: A wire is projecting in the pathway in the working zone.

Then Risk—C × E × P, C = 01 (first Aid Treatment)

E = 10 (Continuous exposure)

P = 10 (Almost Certain),

Risk = 01 × 10 × 10= 100-Score

Comparing both the above conditions it is concluded that more risk is involved in condition-2.

WRITTEN SAFE WORK PROCEDURES

Sample Job Hazard Analysis

Cleaning Inside Surface of Chemical Tank - Top Manhole Entry

Step	Hazard	New procedure or protection
1. Select and train operators	Operator with respiratory or heart problem;	• Examination by industrial physician for suitability to work.
	Other physical limitations	• Train operators.
	Untrained operator-failure to perform task.	• Dry run
2. Determine what is in the tank, what process is going on in the tank, what hazards this can pose.	Explosive gas	Obtain work permit signed by safety, mainten-ance and supervisors.
	Improper oxygen level.	• test air by qualified person
	Chemical exposure: Gas, dust, vapor; irritant toxic	• Ventilates to 19–21 % oxygen & less than 25% LEL of any flammable gas. Steaming inside of tank, flushing and draining, then ventilating, as previously described, may be required.
	Liquid: irritant. Toxic	• Provide appropriate respiratory equipment - SCBA of air line respirator.
	Corrosive	• Provide protective clothing for head, eyes body and feet.
	Solid: irritant	• Tanks should be cleaned from outside, if possible
	Corrosive	• Provide parachute harness and lifeline.
3. Set up equipment.	Hoses, cord, equipment tripping hazards.	Arrange hoses, cords, lines and equipment in orderly fashion, with room to maneuover safety.
	Electrical-voltage too high, exposed conductors	• Use ground-fautcircuite interrupter
	Motors not locked out and tagged.	• Lockout and tag mixing motor, if present.
4. Stall ladder in tank.	Ladder slipping	• Secure to manhole top or rigid structure.
5. Prepare to enter tank	Gas or liquid in tank	• Empty tank. • Review emergency procedures • Open tank • Check of job site by industrial hygienist or safety professional. Install blanks in flanges in piping to tank (Isol-ate tank.) • Test atmosphere in tank by qualified person (long probe).

6. Place equipment at tank-entry poison.	Trip or fall	• Use mechanical-handling equipment • Provide guardrails around work positions at tank top
7. Enter tank.	Ladder - tripping hazard. Exposure to hazardous atmosphere.	• Provide personal protective equipment for conditions found. • Provide outside helper to watch, instruct and guide operator entering tank, with capability to lift operator from tank in emergency.
8. Cleaning tank.	Reaction of chemicals, causing mist or expulsion of air contaminant	• Provide protective clothing and equipment for all operators and helpers. • Provide lighting tank (Class I, Div. 1) • Provide exhaust ventilation. • Provide air supply to interior of tank. • Frequent monitoring of air in tank. • Replace operator or provide rest periods • Provide means of communication to get help, if needed. • Provide two-men standby for any emergency.
9. Cleaning up.	Handing of equipment, causing injury.	• Dry run. • Use material-handing equipment.

Planned Job Observation Report (Model)

1. Name = R.C. Sen
2. Check No = 6335
3. Department = Engineering
4. Occupation = Labourer
5. Job observed = Waxing Floor
6. Date = 2/4/2010.
7. Type of observation = ☐ Initial ☐ Follow up.
8. Time with the company = 5 Years
9. Time on present Job = 5 Years
10. Notification = Told in advance not told in advance

Job observation

Describe below any practices or conditions that deserve compliment or correction:

(*a*) Worker had equipment and supplies well organized, especially in their positioning at the work area. He was very systematic in clearing furniture and used prescribed dollies.

(*b*) He did not use color coded stickers required. I interrupted the worker since wax had an organic base and he did not ask me to check the ventilation as required on checklist.

(*c*) All other steps were according to the checklist; except for failure to use the non-slip pullovers on his shoes and the color code marking for furniture.

(*d*) Since he was aware of being observed, his conscientious manner of work during the observation appeared to be his natural way of doing that job.

(*e*) He did an excellent job of roping off area and caution and no smoking signs.

(*f*) The major need for correction was obviously the worker's failure to use the required check-list.

11. Have you properly complimented and/or reinstructed the worker on these observations?

☐ Yes ☐ No

12. Should a follow-up observation of this worker or task be made in the near future?

☐ Yes ☐ No

13. Describe any standard procedure method or equipment you observed that management should consider changing in the interest of safety, quality or productivity: (Details of description)
14. Supervisor/Observer
15. Check No. = 40586
16. Department = Labour.

FOLLOW-UP ACTION

A follow-up observation was made on 6/2010. Worker used the required checklist and followed all items on it. I thanked him for his good work and encouraged him to continue it.

Emergency Warning Alarming System

There should be a reliable system for informing the emergency services as soon as alarm is raised on site. The details of communication arrangement should be agreed locally and have a direct link to the fire brigade. Besides, good ventilaton system be ensured, which no doubt greatly contribute towards health of workers on the worksite. Since toxic materials are released as dust fumes, vapours and gases cause health hazard.

Safety in handling and storage of chlorine

- Handle all containers as though they are full.
- It is necessary to provide adequate unloading and storage facilities.
- All storage, handling and use areas should be properly designed, so that personnel can quickly escape in emergencies
- Storage location must be cool, well ventilated from heat.
- Chlorine must always be stored above ground because it is heavier than air.
- Ton container should be stored laying on trundles or wedged with rails, so they cannot roll freely.
- Full and empty cylinders should be marked and be stored separately to avoid confusion, which lead to an accident. .
- Protective covers for valves should always be securely in place even when ton cylinders are empty.
- Do not put water over leaking chlorine gas, as its solubility is very low. The leakage will increase due to high rate of corrosion.
- Never heat cylinder beyond 70 centigrade otherwise volumetric expansion of gas will cause the bursting of cylinder.

- Do not weld any piping or equipment until it is thoroughly purged with air.
- Make sure that piping is dry before admitting chlorine.
- Check leakage with ammonia solution, it gives white dense fumes.
- Workers should always carry escape type respirators.

Safety in handing and storage of hazardous chemicals

- To be stored in a dry and well ventilated room.
- While preparing solution, caustic soda should be added slowly and in small amounts to water to avoid violent boiling and splatter.
- Local exhaust ventilation controls should be provided for process such as transfer operations where caustic soda dust can contaminate work environment.
- Small spill and leaks can be washed away with large quantity of water. Drainage should be directed away form work area. Liquid spillages can be neutralized with weak acid.
- Steed in removing the chemical is of primary importance. In case of 15mins duration physician be consulted.
- If ingested, the chemicals may be diluted by large quantities of drinking water, contact with skin, safety slower should be used immediately and clothing can be removed later under the shower.
- Contaminated skin areas should be washed with very large quantities of water for as long as 1-2hrs.
- If it enters eyes, they should be copiously irrigated with water for at least 15mins.
- Contact lenses should not be used where caustic soda exposures exist wearing of chemical goggles of full face shield, mechanical filter respirator, rubber gloves aprons and boots are recommended.
- The mode of entry (caustic soda) into human system can be through inhalation, ingestion and skin contact. Its corrosive action on tissues causes burns and deep ulceration. Prolong contact with dilute solutions has a destructive effect on tissue.
- Hydrogen peroxide be stored in a cool, dry well ventilated area, away from heat and flame,
- Containers must be well covered.
- Route of entry is inhalation, ingestion, eye & skin.
- Contact with skin should be avoided but immediate flushing with water will prevent any reaction incase of accident. Contact with eyes, flush with plenty of water for 15mins.
- Remove the contaminated clothes and shoes. Flush the affected area with plenty of water.
- Inhalation, remove the victim to fresh air area ingestion have the victim drink water.

SAFETY IN EXPLOSIVE MANUFACTURING AND TESTING INDUSTRY

Fire is the main hazard in this Industry. It is therefore mandatory that while planning a new facility for processing or handling, storing explosives and hazardous substances, due care is taken very stringently to incorporate adequate safety measures to prevent on accident fire. All possible sources of fire are to be eliminated by judicious planning of the following main factors.

1. Selection of a Safe Site,
   (*a*) Away from any source of fire/fire prone area.
   (*b*) Soil not susceptible to erosion.

(*c*) Free from under-water courses.
(*d*) Should be well-drained and not prone to flooding.
(*e*) Should not be with continuous rock strata.
(*f*) Should be within easy reach of fire station, source of adequate water supply, railways, main roads, fuel, electricity, public transportation, sources of labour.

2. Adequate structural fire protection (using adequate fire rated materials)
   (*a*) Walls RCC/Brick of adequate thickness (Not less than 68 cms brick/45 cm RCC)
   (*b*) Roof shall not be less than 15 cms RCC.
   (*c*) Complete fire resisting separation between storeys, protected staircases and escape routes.
   (*d*) Floor should be paved/PCC or tiled and raised at least 30 cms above the ground level.
   (*e*) Buildings are with non-combustible structure including linings. The exclusion, as far as possible of combustible material within the explosive area (in an around an explosive building) and various operational measures intended to minimize/eliminate the risk of fire.
3. Orientation (so that effect of outside heat is minimised).
4. Adequate Safety Distances between buildings.
5. Protective features (Barricades/Traverses).
6. Appropriate electrical fittings (short circuiting, arcs, sparks and over-heating are eliminated).
7. Safe air-conditioning system with non-combustible duct fitted with fire dampers.
8. Proper lightning protection arrangement.
9. Protection to static electricity by adequately bonding and earthing each equipment/ instrument. (All equipment/instrument are bonded and earthed properly).
10. Minimising fire load to maximum possible extent.
11. Identified and eliminated source of fire/source of ignition.
12. Good House Keeping-Proper layout and proper equipment, correct material handling and storage, cleanliness and orderliness, place for everything and everything in its place.
13. Adequate water supply for fire fighting, (FHs & SWTs).
14. Fire Station with adequate number of Major fire fighting appliances and crews.
15. Fire points with adequate/appropriate first-aid fire fighting equipment.
16. In-built passive fire protection system. (Unit risk method by compartmentation with fire rated walls, doors).
17. Fire suppression measures, (Automatic drenchers/Sprinklers).
18. Fire breaks of PCC hard-standing with a radius of 5 M around the unbarricaded/untraversed buildings.
    (*i*) The above planning are duly implemented and/ followed up by organising and controlling to ensure no deviation. For this, a written down fire prevention plan of the unit is made and implemented sincerely.
    (*ii*) With the above actions followed by good house keeping the chances of accidental fire is remote. Therefore, the provision of fire detection cum alarm system may be over emphasised. It is always better to prevent a fire occurring than to detect it once it has occurred. Therefore, eliminating/minimizing fire hazard around the facilities is the key

element of fire safety. The automatic fire detectors, particularly the conventional smoke sensors and heat sensors are NOT suitable inside explosive buildings for the following reasons:

(*a*) Even a well designed and properly installed automatic fire alarm system will not be able to render reliable and trouble-free service unless high standards of maintenance and supervision are ensured during the entire service period of the system. Regular inspection and scheduled preventive maintenance are critical and should include all the components of the system.

(*b*) The smoke and heat detectors have to be fitted against the ceilings/false ceilings inside the room/ building under protection.

(*c*) The detector fittings do not conform to the electrical fittings of FLP/DT /TE types.

(*d*) The detector circuit inside the building is maintained live round the clock during their total serviceable period.

(*e*) Deposition of explosive vapours/explosive dust inside the chamber of the detector device can not be ruled out.

(*f*) The detector owing to the reason at (*e*)above, need to be cleaned thoroughly every day after the working hours which is not a practical proposition.

(*g*) The live detector circuit may be a source of spark/fire, although we have FRLS cables now.

(*h*) The detector units and the insulating cables of conductors used for the fire detection-cum-alarm system shall increase the fire load which adds fuel to the fire.

(*i*) The installation of fire detection-cum-alarm system may render the occupations careless to Good Housekeeping.

(*iii*) Therefore, the smoke type detectors and heat detectors are not suitable inside an explosive building. However in an extra hazardous explosive process operation (deflagrating but not detonating explosive), if the situation demands, automatic flame detectors coupled with automatic extinguishing system may be provided for detecting any mishap at the working machines/instruments and simultaneously pouring the extinguishments. The above datectors are grouted in the walls/floors (inside a narrow window opening) so that only the viewing lens of the detector unit shall be protruded inside the building, the area under protection. The cables of conductors and all other units of the detector circuit shall be laid outside the building/room under protection. The system may also be laid underground and through non-combustible conducts up to the working machine.

(*iv*) The flame detector such as UV/IR detectors are used where the need of automatic fire and explosion suppression system can not be overemphasized, An automatic detection system without having been coupled with automatic extinguishing system is of little value for fire involving explosives. The details on UV detection and suppression system with diagram is given below.

(*v*) In view of the above for an explosive building, with good house keeping, there is little requirement of an automatic fire detection-cum-alarm system. However, in case of any fire around the building/inside the building owing to inadvertent carelessness in maintaining good house keeping, manually operated electric fire alarm located outside the building (near the entrances/exits) may be provided to call the emergency services/fire fighting services.

Ultraviolet detection and suppression system

AIM: Suppression of initiation of a fire, preceding to deflagaration explosion.

1. Detection of a fire within 10 milliseconds of its inception (Ultra Fast Detection).
2. The detection system shall operate the deluge system to deluge the affected area within 50 milliseconds of inception of a fire in its very incipient stage (A deluge valve operating ultra fast).
3. The system shall sound an available as well as visual alarm simultaneously with the flow.
4. The system consists of:
   (*a*) UV-Detector Geiger Mueller Sensor
      (*i*) Responds to UV-Radiation of wavelength 1850 to 2450 angstrom (0.1850 to 0.2450 Micron)
      (*ii*) Effective cone of vision 90
      (*iii*) Effective distance of vision/to 3 metre
      (*iv*) Shall not respond to normal ambient light. Eliminate false actuation ,
      (*v*) Least Troublesome
      (*vi*) Shall not be affected by wind, rain, humidity, Change/extreme of temperature and pressure.
      (*vii*) Detect flame/spark/ARC in 10 Milliseconds.
   (*b*) Explosive valve (Squib Type)
   (*c*) Wet-Line System
   (*d*) Discharge Nozzles to flow at 7 kg/cm^2 pressure
   (*e*) Pressure tank, prepressed at 7 kg/cm^2
   (*f*) Pump
   (*g*) Compressor
   (*h*) Water/Reservior.

X-rays	Cosmic rays	x-rays	UV	Visible 3968-A	IR 7594A	Micro Radio
Lower ← 4000				7600 → Higher		

SAFETY IN HOSPITAL INDUSTRY (RADIATION SAFETY)

Ionizing Radiation and Radioactive material has been present from geological times in the environment. The discovery of Ionizing Radiation and Radioactive material from Pitchblende by Rontgen and Becquerel respectively, made man more conscious. Furthermore, with the advent of newer techniques for measurements and with establishing the law enforcing agencies for Radiation protection, human beings are very much protected. The law protecting agencies, equipments for its exposure, measurement and surveillance, made man more and more educated. By the year 2000, the protection from Ionizing Radiation will be an important concern. The importance of radiation has been felt more and more from the beginning of the century due to advances made in our understanding of newer concept of a risk versus benefit. The medical use of ionizing radiations, specialty X-rays and those emittetion emitted has been established in the first three decades and is being utilized presently for the detection of diseases and evaluation of physiological and pathological behavior not detectable in any other way.

Radiation and its effects on health have been studied by various expert bodies over half a century and more is known today about radiation risk than about those of particularly any other physical or chemical agent in our environment. The effect of large doses of Radiation on human health is well understood and such doses are clearly hazardous. Various models have been proposed to predict how the health effects of low level radiation related to the radiation dose received. The differences among these predictions are so small that they make it very difficult to verify anyone theory conclusively. Some kind of radiation can penetrate the human body and produce highly reactive electrically charged atoms and molecules and free electrons in the material making up individual cells. This ionizing radiation, when received in very small amounts has been termed low level radiation. Examples of low-level radiation abound: We receive it from the natural environment (*e.g.*, earth, building material & our food) and from man made sources (certain medical diagnosis, treatment and some consumer products.)

This discourse endeavors to show how radiation worker using X-ray/linear accelerators machines of radioactive materials (sealed or unsealed) for diagnosis or treatment, can avoid exposing themselves, patients and the general public visiting hospital to undue hazard. Our main aim is to protect life, creatures and environment from Radiation and if it is for diagnosis or for treatment, then the Radiation dose should be As Low As Reasonably Achievable (ALARA). So far there is no threshold of Radiation dose that can induce any harmful biological effect. However, there is evidence that high doses of Radiation, there is an increase in rate of Incidence. Sometimes, it is observed that even small doses of radiation may increase the cancer risk marginally. Low doses may also induce mutations, effects the reproductive organs *etc.*, and these mutations may bring out changes in genetic disorder.

In any hospital premises, the main sources of Radiation are Instruments/Radiopharmaceutical, used mainly for Radiotherapy, Radiodiagnosis and Nuclear Medicine. These sources may be of open or closed type. Medical applications like Radioimmunoassays, Invitro assays and other Radio-chemicals used for Research also contribute to the population dose.

Mainly, any hospital has three categories of personnel, who has the chances of the radiation exposure. They are:

(*a*) Radiation workers

(*b*) the patients attending Radiotherapy, Radiodiagnosis and Nuclear Medicine clinics for therapy and diagnosis

(*c*) the General Public/Attendants of patients. In order to develop safety culture, responsibility of the operating 'person may be fixed by the regulatory control authority to provide safety and security from radiation, to various categories of people.

The concept of radiation dose reduction has originated with the understanding of physics of Ionizing Radiation, where strength to a given radiation source can be correlated with simple function of time, distance and shielding attenuation. Therefore, we can reduce the dose by increasing the distance between the source and the individual, and interposing shielding for photon, X-rays, high energy beta emitters and by using smallest amount of radioactivity needed for investigation or therapy. With the advent of radiopharmaceuticals, the fourth concept originated at housekeeping, Radiation dose resulting by inhalation, ingestion or absorption, requires an additional measure of care beyond that necessary for X-ray units and sealed sources in any hospital premises.

The radiation safety concept is to ensure the protection of an individual (patient attendant), the environment and the succeeding generations. The radiation safety and security in any hospital premises may be site specific, state wide, nation wide or universal that should rest on the following general principles of ICRP. They are

1. Justification
2. Optimization
3. Limitation.

(a) *Justification:* The use of Ionizing radiation should be used only when it is absolutely necessary, and benefit to society.

(b) *Optimization:* It should follow the ALARA by taking social factors into consideration, as low levels of exposure of ionizing radiation may have effects on population.

(c) *Limitation:* Radiation exposure should be kept minimum (below the permissible limit) without any risk.

The main sources of ionizing radiation are machine induced X-rays, electrons from electrical devices and the alpha, beta and gamma rays emitted from radioactive substances. X-ray is most frequently used Ionizing Radiation, classified on the basis of voltage.use to accelerate electrons between cathode and anode of the X-ray tube. The higher the voltage the greater the penetration and the tube voltage varies from kv to Mv or maximum photon energy. There applications are as follows.

Diagnosis	:	20–150 kv.
Low Voltage (Skin therapy)	:	150 kv to 1 Mv.
Ortho Voltage (Deep Therapy)	:	150 kv to 1 Mv.
High Energy Therapy	:	1 Mv and above.

Of all X-ray procedures, fluoroscopy, if misused can involve the highest risk to the patient and staff. The preparation and manipulation of the patient should be done with the beam switched off. A spring loaded fluoroscopy switch with remote control will definitely avoid the danger of the equipment being left unnecessarily or accidentally on. Moreover, if sufficient information can be achieved by radiography, fluoroscopy should not be attempted. The beam size should be kept as small as possible. Operating procedure should be followed strictly by the staff. The machine should have long focus skin distance (40 cm).

Risk association with various hospital radiological modalities are as follows:

X-rays diagnosis and the associated risk: The risk related to diagnostic radiology can be minimized or can be completely eliminated by the use of proper techniques as well as trained personnel. There is some evidence of an increased incidence of leukemia and of breast or liver cancer among patients frequently undergoing fluoroscopy of the chest during the treatment of tuberculosis and of an increased incidence of leukemia and cancer among the children exposed in utero.

Radiotherapy and the associated risk: The patient undergoing radiotherapy for malignancy of the evident side effects due to be accepted in order to achieve the therapeutic dose effect. The benefit obtained is definitely better than the stochastic somatic consequences. However, increased risk of leukemia has been reported after radiation treatment of spondlytis.

Nuclear medicine associated risk: Health risk is minimum to the patients as compared with diagnostic radiology. However, risk associated with therapeutic use of unsealed radionuclide remains the same as other forms of radiotherapy.

Natural background of radiation and associated risk:

(*a*) Cosmic Radiation entering the earth's atmosphere from outer space, the intensity at places depends mainly on altitude above the sea level.

(*b*) Terrestrial radiation emitted naturally occurring radionuclides in the earth's crust. The concentration depends on the geological structure in the particular locality.

(*c*) Internal natural radiation emitted by naturally occurring radionuclides within the human body, the concentration depends mainly on their concentration in foodstuff.

Radiological safety	Nuclear safety
Justification of practice	Establish and maintain effective defense to protect individual, public and environment against radiation hazard.
Optimization of Protection:	To ensure
(dose constraints and risk constraints	1. Exposure during normal operation are below the limits and ALARA 2. Mitigation of extent of Exposures during accidents.
Individual dose and risk limits	1. To prevent with high confidence accidents to NPP. 2. Radiological consequences of any of DBA and minor and likelihood of severe accidents are extremely small

The dose limitation for occupation workers and to general public are as follows:

Occupational exposure

Eff. dose 20 mSv/year averaged over 5 years.

Eff. dose 50 mSv/year averaged over 5 years.

Eqult. dose 150 mSv/year to Lens-of eye in a year.

Eqult. dose 500 mSv/year to extremities and skin in a year.

(In special circumstances of interventions, a worker may be exposed upto 100 mSv in a year).

Member of public

Eff. dose of 1 mSv in a year

In special circumstances with Reg. Auth. approval upto 5th Sv in a year but averaged in 5 years not to exceed 1 mSv/year.

Eqlt. dose to lens of the eye 15 mSv in a year.

Eqlt. dose to skin 50 mSv in a year.

Applies to

(*a*) Sum of exposures from all manmade sources but excluding medical exposures.

(*b*) Sum of external exposure in a year and committed exposure from in take of radionuclide during the year.

Chronic exposures

222 Rn in houses -20- 600 Bqm3 annual average.

222 Rn in work place 100 Bqm 3 annual average.

In addition, inhalation of air borne material causes serious internal hazard. The absorption, retention and elimination of material taken by lungs depends on the size, solubility of the particles and the rate of respiration of an individual.

In following radiographic procedure, film should be supported mechanically and exposure tends to be very short in order to minimizing the blurring of the image. Locally, radiography machine should be fixed, mobile machines should have adequate output so that exposure time can be kept short. The accurate optical localizer should be kept small. The room layout must be conducive with maximum working efficiency. The protective apron should be provided to cover the patient from neck downwards during exposure. Adequate film processing unit is an essential part of radiation protection, poor quality of image may force a patient for repeated exposure, thus doubling the quantity of radiation exposure. The principle of Radiation safety helps in creating Radiation Security in any hospital premises while practicing it in hospital, using strict law enforcement, is the duty of the administration to protect the Radiation worker, the Patient and the Environment of the hospital from each other. The principles are as follows:

1. *Management responsibility:* Safety Culture, Fixing of Responsibility, Regulatory Control.
2. *Strategy of defence:* Successive Barriers between Worker, Machine and the Hospital Environment.

   Prevention of Accident

   Accident Mitigation
3. *General principles (Technical):*

   Proven Controlled Practice

   Quality Assurance

   Human Factor
4. *Safety assessment and verification:*

   Radiation Protection

   Operating Experience and Safety Research

   Training Course and Symposia.

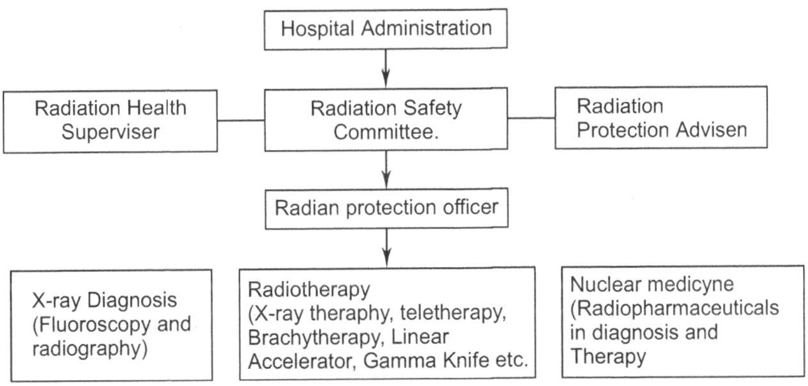

Planning of Radiation Facilities

The allocation of floor space and Radiation Protection requirements of Radiation facilities should be considered in the early planning stages of a new Hospital. The planning of new Radiological facilities requires the close collaboration of the architect, radiologist, radiation protection advisor and the hospital administrator. The location of radiotherapy facilities determine, to a large extent, the amount of shielding required, designing and the construction. The shielding \ protection factor should be considered in the architectural design. In the larger hospital it has been found advantageous to have separate department of Radiology, Radiotherapy and Nuclear Medicine.

The patient with radioactive sources, whether receiving intracavitary, or interstitial therapy, contribute hazard to staff in the department. Warning alarm should be fixed and warning notices must be pasted on bed side working site indicating safe time and the minimum safe distance. All the working staff, the patient bearing sources must have the Radiation Exposure monitoring devices and should be monitored. Care should be taken to exclude pregnant women from such work. The urine excrete from the body should be through the toilet directly that has been set aside.

Posting Requirements

Condition	Posting
5 mrem in 1 hour at 30 cm from the source or shield surface	Caution, Radiation Area
100 mrem in 1 hour at 30 cm from the source or shield surface	Caution, High Radiation Area
100 rads in 1 hour at 1 m from the source or shield	Grave Danger, Very High radiation area.
Air Concentrations exceeding the DAC area.	Caution, Air borne Radio-activity area.
Use or storage of 10 times the quantity	Caution. Radiation Material

Radisisotopes and other radiation sources made so many advances in understanding biology and medicine, at the same time they have created unique problems associated with its increasing use. This may produce harmful biological effect. The balancing of benefits versus risks is difficult when the ranges at doses concerned are those for which biologic effects have not yet been observed. Toxicity of raw material, quality control of Radiological equipment radiopharmaceutical, open corridor for public/attendant, proper direct drainage for radioactive waste, their disposal through ducts connected directly to sewage, air-conditioning of the working area, shielding of sources, proper inventory control are to be verified with delegation of power and responsibility may be fixed by law enforcing radiation regulatory board. The health survey is very much essential from time to time. This is only the way where radiation safety will assure radiation security to workers/ patient and the environment.

Some suggested safety measures

1. Avoid direct exposure and direct look from a shorter distance for a long time
2. Power level density level > 10 mw/cm^2 normally are hazardous due to long exposure
3. Use main limit criteria where RF entry permissible limit decalred as hazardous ,
4. Never work on "Wave guide" or Antena with relation on.
5. Use shielding of RF generating source.

Safety Guidelines in Handling, Storage and Transportation of Explosives

1. Discharge accumulated static electricity from your body before approach to explosive.
2. Wear cotton dress and safety conducting shoes.
3. Earth the explosive device/container properly.
4. Use conducting mat for standing on it while working on explosive device.
5. Use Non sparking tools.
6. Keep EED always shorted when not required.
7. Keep EED away from Radio transmission source or vice versa.
8. While working with explosive device, keep always ambulance, Nursing Assistant with first Aid kit as stand by.
9. Keep suitable fire extinguishers Tender as stand by.
10. It is preferable not to handle explosive when there is lightning/ storm/weather warning.
11. Ensure that the operation building is with flame proof electrical wiring and fitments, No naked wire or loose connections are there.
12. Work site lighting must be flame proof.
13. Before use of EOT crane, check for crane earthing.
14. No petrol vehicle is allowed near to the place of operation/handling.
15. Never Carry any match box/lighter/cigarettes etc to the building.
16. Do not store flammable material and fuels, paints, grease etc in the same building/place.
17. Do not fiddle with electric connections. Let the trained and designated person do the job.
18. Keep the work area always clean.
19. Restrict Man Limit at the operation site. Follow check list to progress the work.
20. Follow the SOP - Whenever available.
21. Do not work with explosive device if you are tired or over-worked.
22. Check that the building earthings and lightning arrester resistances are within limit. Measure earthing resistance every three months and keep a record.
23. Before using the EOT crane, check that the crane has been tested periodically and test certificate to the effect is available.
24. Do not hurry and keep your head cool while working with explosives.
25. If you are not conversant or have a slightest doubt stop work with explosive device.
26. When there is an out break of fire with explosive device, it is better to vacate the place immediately and proceed to safe shelter rather than trying to fight the fire.

Storage Transportation

1. Store the explosives in an approved magazine only and category wise.
2. It is always mandatory to store different category of explosives in separate magazines/ rooms.
3. If warranted and there is paucity of space, different categories must be separated by a sand-bag barrier wall.
4. During storage, the explosives must be properly packed/erated and stacked.
5. Keep the magazine ventilated and floors clean.
6. Wooden covered flooring with nobes is always desirable in a magazine.

7. Keep the magazine surroundings clean (2 meters), empty boxes, paper crates, over growths etc to prevent fire hazard.
8. See that there is no water logging around the magazine
9. Check the earthing and lightning arrester continuously.
10. Record temperature (Dry and Wet) periodically and maintain a record.
11. Ensure that the magazine is guarded 24 hours for security.
12. Allow no unauthorised person to enter the magazine area.
13. Matches, lighters, cigarettes must not be taken to the magazine area.
14. Ensure that fire class symbol board and hazard level board are well displayed near the magazine.
15. Check that fire hydrants are functional and there is adequate water in the static tank.
16. Check availability, serviceability and expiry for First Aid Fire Fighting appliances.
17. Check for fire buckets with sand and water respectively at the work site
18. The explosive van must be equipped with First aid fire fighting equipment.
19. Hazard level symbol must be painted on the vehicle, while transporting
20. Instructions for handling of emergency with important phone Nos must be available with the explosive vehicle driver.
21. Do not smoke inside or near the explosive vehicle.
22. Keep safe distance between your vehicle and preceding vehicle:
23. Do not consume alcohol before or during driving.
24. Never leave the van unguarded.
25. Enroute, if there is a need to stop, park the vehicle to the extreme side of the road and at a desolate place.
26. Try best that the van is always with an escort.
27. In a prepared missile convoy, pilot van, Ambulance with nursing assistant and Fire Tender must form a part.
28. If there is an accidental fire in the van, but not with the explosive, first try to put out and control. If not possible run away to a safe distance. Do not allow others to go near and seek help.
29. If it is an explosive fire do not try to control, and save yourself. It can not be controlled and you can salvage nothing. Only try to protect the surroundings from behind a cover.

Safety During Convoy Movement

- While transporting explosives from one place to other place in the range, it should be transported in a convoy formation as follows.
- Pilot Vehicle - Explosive Vehicle - Fire Tender-Ambulance
- Stand by Vehicle.
- The pilot vehicle and explosive vehicle must have RED FLAG display on right side of the vehicle.
- The minimum distance from explosive vehicle to pilot vehicle and fire tender shall be 50 meters.

- In case the explosive in explosive vehicle catches fire, all the personnel should evacuate upto min 30 m.
- A positive communication (VHF) between pilot vehicle and stand by vehicle should be maintained throughout the convoy movement.

Safety Precautions During Storage, Handling and Transportation of Liquid Propellants

(a) *RFNA (O-Fuel):*

- Highly reactive with metals like Iron, Copper and their alloy.
- May produce FIRE with wood and cellulose product.

Toxic effects

- Highly corrosive.
- Causes yellow skin and severe burn when in contact with skin.
- Contact with eye can cause severe damage to eye.
- Repeated exposure may cause ulcer in the nose, mouth and irritation of respiratory tract. First Aid Medical Treatment is a must.

(b) *RFNA affecting eyes:*

- Flush affected eye with water for 15 minutes.
- In case of severe exposure—Hospitalise.

(c) *RFNA spilling or contact with body:*

- In case of spillage, neutralise with sodium bicarbonate.
- In case of contact with body; take an emergency shower and then consult a Doctor.

Fire and explosion hazard: The RFNA has moderate explosion hazard since it can react explosively with many reducing agents.

It can cause a dangerous disaster, if RFNA is heated to decompose, (since on decomposition highly toxic fumes of NO x will react with water vapour/steam to produce heat and toxic, corrosive and flammable vapours).

Prevention of fire and explosion hazard: Keep storage area clean of trees and vegetation.

- NO smoking in RFNA storage bldg.
- NO naked light.
- In case of fire, fight with large amount of water.

Personal protection:

- Storage area should be well ventilated.
- Avoid body contact.
- Ensure availability of serviceable showers and eye wash fountains.
- Wear protective dress and face mask.

G-FUEL

Composition

- An equal proportion blend of Isomeric Xyledine and Triethyl amine.
- Yellow (dark) looking oily liquid of characteristic smell of fatty amines.

Reactivity

- Reacts vigorously with acids and oxidising agents.
- Hypergolic in nature (*i.e.* self igniting when in contact with oxygen).

Health hazard

- Affects by ingestion, inhalation and skin contact.
- Causes mild skin irritation and severe eye irritation.
- May damage kidney and liver.

Emergency treatment

- Irrigate eyes with water.
- Wash contaminated areas of body with soap and water.
- Treat skin bums in normal manner.
- Artificial respiratory support to be kept standby.

Preventive measures

- Operation in well ventilated area.
- Use full protective rubber clothing and goggles.
- No physical damage to container.
- Keep away oxidising material and flame.
  First aid: The working space and laboratories must have first aid kits containing boric acid solution, iodine cotton and bandages.

In case G-fuel splashed on skin

- Wash affected area with water and soap.
- In case of large/severe affection take shower and consult Doctor. G-Fuel Splashed in the eyes.
- Wash eye thoroughly with water followed by 0.5% boric acid solution.
- Consult doctor.
- If G-fuel has been confined to mouth, wash mouth a number of times. Never swallow mouth-wash.
- If swallowed, give copious amount of water or milk to drink.

UNSYMMETRICAL DIMETHYDRAZINE (UDMH)

- Highly flammable and hypergolic.
- Reacts with strong oxidouts and porous materials (*e.g.,* wood) to ignite.
- Decomposes to produce irritating nitrogen oxides.

Health hazard

- Enters body through skin, inhalation and ingestion
- Irritates skin, eyes and respiratory tract.
- Inhalation may affect central nervous system, may cause liver damage blood disorder and death.

Emergency treatment: In case of inhalation, supply fresh air and keep the patient in half upright position. If necessary, provide artificial respiration.

Preventive measures

- Persons handling should be examined periodically.
- Must use protective clothing, face shield & breathing apparatus.
- Choose well ventilated area for storage/handling.
- Availability of serviceable fire fighting equipments, hydrants and alarm system.
- Flame proof electrical fittings.

First aid measures

Inhalation : Supply fresh air & artificial respiration.

Splash : Remove contaminated cloth.

- Wash affected part with copious water.
- Consult doctor.

Waste disposal methods: Water spray should be used to reduce vapours, collect leaking and spilled liquid in sealable containers as far as possible, absorb remaining liquid in sand.

Don't wash away.

Don't let it enter envimoment.

Safety in Storage and Handling of Acids

1. Acid should be stored in container, which is compatible to that particular type of acid.
2. The acids are of highly corrosive in nature and shall be stored in a well ventilated room.
3. Protective clothings viz. PVC hand gloves, PVC apron, shoes and full face masks having antiacid canister should be used while handling the acid in bulk.
4. Fumes of highly toxic acids are dangerous in nature. The canister of such kind of acids should not be opened without wearing safety dresses like mask, gloves, apron etc.
5. In case of spillage, wash the area with a large quantity of water. Suitable neutralising agent (alkaline solution) for a particular acid should be kept near handling area.
6. The storage/handling area should have suitable drainage system to clear the neutralised acid.
7. A first Aid box containing medicines should be always available at work site.

SAFETY IN HOTEL INDUSTRY

Hotel industry is always affiliated with three basic hazards.

(*a*) Health, sanitation and environmental hazard.

(*b*) Plan Layout of swimming pool and drawing.

(*c*) Fire hazard

1. While dealing with health and hygienic aspect, we must look to the following good housekeeping.
2. Cleaning and disinfecting.

3. Maintenance, repair and replacement of air conditioning equipments, and machineries.
4. Priority be given for electrical safety.
5. Inorder to keep the environment free from gas, toxic elements, infection and diseases regular spraying of disinfectants is a must.
6. Congestion and crowd control activities be undertaken.
7. Ingress and egress system in case of fire hazard should be planned.
8. Safety in transport of in mates.

(*d*) Swimming pool tragedies are occasional occurrences, which ends in loss of human lives. The accident happen to faulty Layout and Construction of the Swimming Pools inside the Hotel Premises:

There is a standard regulation for swimming pool erection. Tourists and guests die due to head injuries caused while diving from the high platforms, due to dash against the pools bottom floor. The hotel management must adhere to the building regulation and avoid accident on the score. Besides, qualified trainers be engaged to rescue guests incase of drowning ppes for guests and rescue personnel should be strictly ensured.

(*e*) The most important aspect of safety hazard in the hotel industry is Fire. The causes of building fire, which has been elaborately discussed earlier under Building Industry is also most similar here. Therefore, fire safety measures are to be provided right from the stage of drawing, construction and electrical installation. In most cases, fire havoc is caused due to electrical short circuit. Regular, maintenance, and repair must be undertaken. On the event of fire that occurs beyond 2nd floor and above, means of escape should be followed as discussed earlier.

SAFETY IN OIL, GAS INDUSTRY (TANKERS)

These are three types of oil and gas safety.

(*a*) Crude Oil,
(*b*) Product Tanker (oil)
(*c*) L.P.G. Tanker, natural gas and chemical gas.

Approved Design of Tankers

Equipment of a design that has been tested and approved by the appropriate authority should be adopted. The authority should have certified the equipment as safe for use in a specified hazardous atmosphere. Improper commissioning of Industries can seriously affect environmental features, such as air, water, land, vegetation, animals, peoples etc. Proper care should be needed at this time of sitting rather than going in for measures at the later stage. The petroleum and gas industries shall maintain the following distance from the civil area.

(*a*) 25 km away from ecologically and sensitive areas.
(*b*) 0.5 km from high tide lines in coastal areas.
(*c*) 0.5 km away from flood prone rever system.
(*d*) Major settlements.

If it is within 50km, the industry should be atleast 25km, the projected growth boundary of the settlements. As per G.O. MS No.1 dated 6.3.84, no Industry causing intensive water pollution shall be located within 1km radius from the banks of revers, streams, lakes and dams.

Common petroleum products in tanker

(*a*) Crude Oil
(*b*) Aviation gasoline
(*c*) Jet fuel
(*d*) Kerosene
(*e*) Motors spirit
(*f*) Highspeed diesel,
(*g*) Ordinary diesel,
(*h*) Furnace oil
(*i*) Lubricating oil.

Hazards

1. Toxicity (Petroleum Gas) :
   (*a*) Dulls your sense of smell.
   (*b*) Make you dizzy.
   (*c*) Gives headache.
   (*d*) Produce diminished sense of responsibility.
   (*e*) Irritate eyes.
   (*f*) Speech becomes blurred.
   (*g*) Staggering.
   (*h*) Confusion.
   (*i*) Loss of consciousness.
   (*j*) Arrest of breathing (large quantity inhalation).
   (*k*) Arrest of heart.
   (*l*) Death.

(*a*) Petroleum (liquid) :
   (*a*) After skin contact causes irritation.
   (*b*) Blocks skin pores and cause rash and allergy.
   (*c*) Longer period of contact causes cancer.
   (*d*) Smoke emanating from oil fire pollutes the atmosphere upto a distance of 3km from the site. The inhabitants are affilicted with itching of the eyes, breathing problems, respiratory problems, fever and vomiting, skin allergies etc.

Fire Fighting in Oil Tanker

Tankers are subjected to strict safety regulations. The oil tanks are normally replaced with inert gas, when oil is discharged from tank as fire precaution arrangement. The fire risk varies on tankers on the basis of types of cargo and the size of the ship. Cargo of heavy refined oil present, relatively causes less fire risk then crude oil and petroleum having low flash point. The danger is less when tanks are full and properly sealed. The risk is greater when tanks are partially empty or fully emptied but tanks contain gas, the problem is normally tackled by following inserting procedure properly. The fire in tankers may be caused due to collision which may rupture the tanks when flammable mixture are formed and air reaches the tanks. The fire caused from tanks may spread on decks and reach superstructure or may also spread on water surface.

When oil is burning in a tank, large quantity of foam will be required. The supply of foam concentrate and water should be continuous to be successful in combating such fire. The attack should be from wind ward side from the deck of the ship or from the fire boat or a tug from nearby. Thick layer of foam should be maintained for some hours even after the fire has been extinguished and cooling operation of hull should also continue to eliminate chances of reignition of oil. The officer-in-charge of fire service must maintain close liaison with ship's master and Chief Engineer for necessary information and guidance.

To distinguish a gas from flammable liquid vapor, a substance that has a vapor pressure greater than 40 psi at 38 °C can be considered as a gas. A gas normally means a substance that exist in a gaseous state at normal temperature and pressure (21 °C at 14.7 psi).

According to chemical properties, gases can be classified as flammable, non flammable, reactive or toxic. A flammable gas is simply any gas that will burn in normal condition of O_2 in air. Like flammable liquid vapor, flammable gas has flammable, range and finite ignition temperature.

Non-flammable gases = O_2, N_2, Ar, He

Reactive gases = F_2, Acetylene, vinyl chloride

Toxic gases = Cl_2, H_2S, CO

Compressed Gases = O_2, Ethylene, H_2 Acetylene

Liquefied Gases = LPG. MPS, Methylacetylene propadience (stabilised)

Cryogenic Gases = liquefied gas at temperature and pressure *e.g.*, O_2 (– 183 °C) methane (– 160 °C)

FIRE PROTECTION CONSIDERATIONS

Cylinders and small tanks of flammable Gas should be protected by:

1. **Segregation:** Separating the hazard by distance, at least 10 feet from other operations or a non-combustible barrier (as high as the container) of at least 1/2 hour fire resistance rating can be used.

   Hazardous materials should not be stored next to oxidizing gases. Combustibles should be eliminated in the gas container area. Storage rooms should be of non-combustible construction.
2. **Confinement:** To prevent the escape of a flammable gas and the consequent forming of a flammable mixture in the air, the gas must be confined in a container or piping system of a design that is adequate to withstand pressure and temp, to which it may be subjected to be leak free and to be located so as to avoid physical damage.
3. **Ventilation:** All indoor storage areas should be ventilated. If there is strictly storage and no filling of container involved, the ventilation rate may be low (around one air change per hour). Normally the rate will be adequate when the con. of the gas in the gas-air mixture does not exceeds 25% of the lower flammable limit. LPG being heavier than air, will settle in low areas such as pits, presenting a problem and thus require low level ventilation for control.
4. **Pressure relief:** Excessive pressure can result from exposure to sunlight and fires or from warming of cryogenic gas valves. To help alleviate this pressure so that the tank will not fail, pressure relief valves are installed on compressed and liquefied gas tanks. These devices can be activated by pressure, heat, gas flow or a combination of these. Containers of highly toxic gases do not have pressure relief devices, since a prematurely opening or leakage device would present a greater hazard. Pressure actuated devices use the principle of suture disc, spring controlled relief valve, or hydraulic valves. As the pressure quickly rises, these devices actuates so that the container can handle the pressure. Temperature actuated devices use fusible metal plugs or bimetal springs. These devices operate well at high temperatures or in fires. They are not effective for over filling pressure increases.

5. **Excess flow:** Excess flow actuated valves are designed to close when equipment failure releases a large volume of gas. The valve operates in a narrow pressure drop range when the leaking gas exceeds the predetermined flow rate of the valves.
6. **Emergency shut-off:** The emergency shut off valve is normally located where the swivel type or hose piping is connected to the fixed piping of the system. Temp,sensing devices at the valve automatically shut the valve in the event of a fire. The valve can also be operated manually installed at remote but safe location.
7. **Source of ignition:** Ignition sources should be eliminated. Cutting and welding operations should be prohibited where flammable gases are being stored. Some other sources of ignition are friction sparks from metal to metal, metal to hard material contact, flames, hot surfaces, smoking materials and static electricity.

   The major container hazard for LPG is the boiling liquid expanding vapor explosion (BLEVE). This is a type of pressure release explosion where the container is heated by an outside source such as fire impingement. Even though container weakening is more often caused by fire, it can also occur when the container is punctured, bent or damaged for other reason. The liquid vaporizes with a large liquid to vapor expansion. This, combined with the weakening of the tank, cause tank failure. Failure occurs before the pressure relief valve can relieve the pressure. The fireball from a BLEVE can be quite spectacular due to the rapid mixing of vapor and air and the subsequent ignition by the fire that caused the BLEVE. The atomized droplets of liquid burn as they are propelled through the air. The magnitude of the BLEVE depends Upon the weight of the container pieces and the amount of liquid that vaporizes. Insulated tank designed for fire exposure conditions can help in protecting against BLEVE. Water spray from hose streams or a fixed system entrained on the portion of the tank that is not in contact with the internal liquid will provide protection.
8. **Safe handling:** The following practices should be observed when handling cylinders of compressed gases:
   (*i*) Avoid accumulation of spare cylinders than you really need.
   (*ii*) All cylinders should be properly secured in a vertical position to an immovable object *e.g.*, bench or wall.
   (*iii*) Never drop cylinders or permit them to strike to each other violently.
   (*iv*) Valve protection caps should also be screwed into position of top of tank in storage.
   (*v*) Cylinders containing flammable gasses should be electrically grounded before attaching equipment or opening valves.
   (*vi*) Never open a gas cylinder valve that has no pressure regulator attached.
   (*vii*) Pressure regulator fittings are standardized to fit only cylinders containing gasses with which they are compatible.
   (*viii*) Never tamper with safety of cylinders or valves.

Filling Density

Filling density is expressed as a percentage of the total water capacity, which is the amount of water required to fill a container. For LPG, this density can range from 48% to 57%, depending on the specific gravity of the LPG, the size of the container and the container location. Appropriate filling density is necessary to ensure a vapor space above the liquid to allow for sufficient expansion if the temperature increases.

Liquefied Petroleum Gases

LPG, a volatile substance are liquids when under pressure but vaporise upon the release of pressure; consequently stored and transported as liquids but used in their gaseous states. These gases include propane, propylene, butanes and butylenes.

Except when used as raw material in a chemical reaction, LPG are provided with an odorant (usually mercaptans) as an aid to indicate the pressure or leakage as it has a distinctive odor.

Store container of LPG will rupture from heating and softening of the unwet crown, and while modern welded tanks may not rocket, this rupture results in the release of the liquid contents with flashing of the material into the vapor form, immediately creating an high energy release and resultant fire and explosion. Under such condition, the operation of the relief valves will probably not be sufficient to prevent rupturing, although under light exposure conditions, they may be expected to operate satisfactorily.

The potential failure source that could occur at each of the six areas include employee operating errors, inadequate equipment design or specifications, valve failure and other equipment failures.

Fire Fighting

Dry chemical powder is a very effective extinguishing medicine for LPG fires. A DCP extinguisher of adequate capacity should be provided at each LPG installation. The fire should not be extinguished unless it is possible to shut off the flow of gas. Extinguishing the fire and permitting the continued flow of gas with the possibility of vapors reaching a source of ignition will generally result in a much greater area of involvement and loss. Thus, controlled burning can be desirable in many cases.

In multiple container installations, adequate water in the form of hose streams from a nearby hydrant or hydrants should be available to play upon the unwet crown portion of the tank not in contact with internal liquid of the storage containers to prevent rupture and involvement of all containers. Over pressurization and rupture of containers, as attested by loss records, can and do result in disasters of catastrophic proportions.

The need for the use of hose streams will be greatly reduced, but not eliminated by:

(*i*) Insulating the tanks

(*ii*) Molding or burying the tanks

(*iii*) Providing automatic fixed water spray systems.

Insulation of the tanks, using a non-combustible insulation to limit the temperature of the metal in the unwet crown to 427 °C after 2 hours of exposure to an average temperature of 982 °C is recommended. The insulation will require no maintenance, as compared to the maintenance required for the water spray system and single failure point possibilities of fixed automatic equipment. Burying and mounding of the tank will require corrosion protection and present problems in subsequent tank inspections.

The location of LPG containers with reference to important buildings, electrical substations, boiler houses, exterior process units, vital cooling tower, etc. should provide for maximum separation to avoid exposure damage to important facilities. In addition to adequate separation, it is also highly desirable to orient the axis of LPG containers so that the tanks are parallel to and not pointed at important facilities.

Current energy problems and specifically the shortage of natural gas are compelling many industries to add LPG storage installations. In case the minimum containers distances can not be met, either a free standing fire wall should be constructed. Fire wall should extend slightly above the height of the container, about one feet but where tanks are exposing important buildings, particularly if multistorey, this height may be increased. Instead, it should be of reinforced concrete/fire walls or the other means of protection should also be used where minimum distance of 25 feet can not be provided between groups of tanks. Above 1000 liters capacity LPG installations as a rule, are required to comply with certain statutes. Guidelines on such matters as location of tanks, spacing and location of non-flame proof electrical equipment as well as distances between plants and boundary are given in the petroleum rules 1976.

(*iv*) Fire protection should conform to the scale laid down in Oil Industry Safety Directorate (0ISD)-144Vol. IV para 10.5.

(*v*) Most codes of practice agree on recommended minimum distances which increases with the size Of storage. The 'Gas cylinder Rules' safety distances as applicable in India is stated below.

The following distances shall be kept clear at all times between any building, public place, public road or any adjoining property which may be built upon and the storage shed used for the storage of LPG cylinders:

Qty. of LPG in …… …. cylinder (KG)	Minimum Distance to……. … … …., be kept clear (meters)
0-100	NIL
101-1000	3m
1001-4000	5m
4001-8000	7m
8001-12000	9m
1 2001-30000	12 m
Over 30000	15 m

Water Spray System

Early cooling of the vessels with water is essential to contain the damage to a greater extent. Small vessels can usually be kept cool by water from a hose reel installation and certainly by the fire Brigade hoses, but larger vessels becomes progressively more difficult to handle because of the shell are involved and thus rather more sophisticated means may be required such as monitors. Above approx. 50–60 tones aggregate capacity, even the monitors may not be adequate and certainly a fixed medium velocity water spray system becomes essential.

The application of water spray to a vessel enveloped by fire will reduce the heat input rate to a value of the order of 6000 BTU/ft. 2/hr of exposed surface wetted by the content when the unit rate of water application is 10 LPM/m^2 of exposed surface. Wind effects must also be taken into consideration when designing water spray systems and it is usually desirable to project water on to a tank from sprayers placed not farther than 600mm away from the surface. The location of sprayers should be carefully considered to cover the entire surface uniformly and in particular the vicinity of tank fittings such as safety valves, vents, etc.

Sprayer pipe work should be galvanised which not only provides corrosion protection but also ensures that the pipes are internally descaled. The sprayers should be of simple design, having a single orifice to produce as large a waterway as is feasible for the flow rate required. Strainers should be fitted at the nozzles where these are less than 10mm dia. The detection system should be robust and reliable. The general practice is to provide a 25mm pipe work system charged with compressed air and fixed at various levels around the protected vessel and fitted at intervals with quartzoid bulb sprinkler heads to serve as fixed temp, heat detectors. The function of these is to release the air pressure or operation which in turn causes the deluge valve on the water supply line to open sprayers to operate. Detector sprinklers are usually spaced not more than 2.5m apart horizontally. Additional detectors are fitted close to vents and other tank fittings.

EARTHQUAKE MANAGEMENT AND SAFETY

What is there common in Latur and Los Angeles, Columbia and California, Uttar Kashi and Kansai in Central Japan? Even a kindergarten kid will now promptly reply — It's earthquake (EQ).

The sudden unwarranted rattling of the ground, vertical or lateral, commonly known in Indian context as "Bhuchal" or "Bhucomp", depending upon their intensity (generally expressed in terms of Richter Scale), speed of propagation and distance from the epicenter (the point from which the ground movement originates), invites immense miseries to the inhabitants of the tremorised zone. Collapse of high rise buildings and thus trapping and burying of persons under the rubbles causing death and injuries, ripping of railway lines, bridges and lines of communication, damage to underground water mains, trunk gas supply lines and electrical circuit, destruction of road and express highways are very common sight in any earthquake striken area.

Against the odds of earthquake, the planners and designers of civil construction are since long, are after some suitable antiseismic devices. Each earthquake prone nation, based on their experience, research and wisdom, developed their own norm and standards of protection against the shock-waves resulting from earthquake. "But what is surprising" commented Dr. Sergey Smirnov. Professor of Structural Mechanics at the prestigious Moscow Building Institute, "after every major earthquake, if the destructions are too great, the scientists review and correct the existing national standards of seismic calculations as a routine measure".

It is now known to many that the present situation in the sphere of seismic defense and seismic stability is quite unsatisfactory because of limited knowledge readily available in the field. That is the reason; every strong earthquake exhibits new and newer types of unexpected damage pattern of buildings and other structure inspite of existing seismic defense measures adopted on pre-fixed calculated norms. Kobe in Kansai region Japan is the latest victim in this regard, when in a pre-dawn feat on 7.2 Richter Scale on 17th January, 1995, a devastating earthquake had completely stalled the life of the entire industrialised port city of Western Japan Kobe, killing 4340 persons, injuring 22.514 persons, demolishing about 12.000 buildings and thus making about 3,00,000 people homeless, completely ripping the much glorified stilted railway lines for superfast "Bullet", damaging bridges and lines of communication, disrupting water, gas and electric supply and above all, twisting and destructing several sections of the once Japan's pride column mounted elevated express highways connecting Kobe with other important cities of Japan. The anticipated cost of damages, a very conservative estimate reveal, is to the tune of $80 billion US dollars. This cost may have liked many times by this time after, Kobe received 1304 after-shocks till 30th January, 1995.

Elevated super-expressways of Japan had so long been considered as the excellence of Japanese engineering skill with superior high-tech designs and constructions and meeting all the standards and norms for anti-seismic measures prescribed by their nation. As the old time, no once thought that the glamour ship "Titanic" can not sink, Japanese designers and engineers were of firm belief these expressways were impregnable to any strong earthquake. Now, that vanity and pride had completely been shattered down to the earth after the wrath of the current earthquake. Questions had risen from many corners amongst Japanese experts doubting the validity of the present national anti-seismic standards. "We had firm belief that these Japanese highways are safe", admitted Yasuo Hashimoto, the Director of Emergency Service Department of Tokyo, a city that is proud of its anti-earthquake protective measures, "but now, we have seen many scenes of twisted and collapsed highways, it is difficult to regard damages having no connection with us". He further added, "We had conducted research on basic assumption that our recently built urban structures are completely impregnable by big earthquakes, but now, we may have to revise our assumption".

A Tokyo University seismologist Tadao Minami commented "*It is similar to what happened in San Francisco or Los Angeles, even though the Japanese expressways were believed to be stronger*". Prof. Motohiko Hakuno, a Tokyo University Professor of Engineering, lamented saying "We could have done better, the structures could have been stronger".

Like their US counterpart, Japan's elevated expressways and railway lines and bridges are supported by column with dozen of vertical rods and fortified with "Hoop Bandings" for additional protections against swaying. Osamu Yoshizaki, an official of the Japan's Construction Ministry's concerned section dealing with the express highways, already stated that a part of Hanshin Highway had been beefed up with additional bandings to meet the latest earthquake resistance requirements. It is not yet known whether the collapsed and damaged sections of the super-highways had adequate bandings or not. Local Report already raised the question of insufficient bandings in expressways. But according to the Director of Japan's Public Work Research Institute, Yasayku Koga, what happened was beyond anyone's control and had happened, but all expressways should strictly adhere to the earthquake resistance requirements prescribed for in the national standards. According to Horoyuki Fuzikawa, the Director for Highways of Japan's Ministry for Construction, an enquiry committee has already been constituted to look into the earthquake proofing norm and an investigator had already been dispatched to disaster site for making an on the spot assessment of situation. Depending upon the assessment of the investigator, the newly constituted committee may consider a thorough review of the existing national standards for earthquake resistance.

Some experts blamed the Government for placing too much emphasis on strength over flexibility for contributing to the degrees of damages in the highways and other structures. But the in-charge of Architecture Division of Japan's Construction Ministry Mr. Tatsua Azuhato differed with the above view stating that Japan's building and highway byelaws are not that rigid and stiff not to withstand any strong earthquake as claimed by many.

Anti-seismic designs remained a chronic headache to the planner because of inherent complicacies involved therein. For better understanding, the diagrammatic representation at Fig. l requires careful study. The shock-wave of earthquake causing movement of the ground first passes through the foundation to the walls and travel one floor at a time. The first floor walls then transfer the stress to the second floor and second floor to the third floor and so on till the top floor is reached. From the top floor, the stress is reflected back towards the ground. The shock-waves of earthquakes are generally recorded by various instruments like seismograph, accelerometer, velocity seismograph etc. To determine the intensity of the seismic disturbance, the speed at which the shock-wave propagates to other directions, commonly knowing science parlour as velocity, and also to locate the epicenter of the earthquake *i.e.*, the point from which the original shock-wave initially originated. Dr Sergey Smirnov, Professor of Structural Mechanics at the Moscow Building Institute is a veteran in the field of earthquake engineering and also a pioneer in the studies of the seismic defense and seismic stability, who is of the opinion that the absence of reliable seismic defense measures and reliable seismic stability to buildings/structures, even those of special seismo-defensive constructions prove ineffective during crisis. The scientist are either partly mistaken or are unimaginative concerning the nature of seismic influences and the reasons for seismic destructions, which quite often causes major catastrophies in case of any moderate or high intensity earthquake. Dr. Smirnov is also very critical about the sensitivity of the presently available seismographic instruments which in his opinion are blind or inert to major disturbance of seismic activities. Thus, according to Dr. Smirnov, the existing theory of seismic defense, seismic stability and seismic calculations are not quite correct because of lack of proper data recorded by available instruments, which does not reflect the real nature of earthquake.

Dr. Smirnov's caustic comments above are mainly based on his intensive research and experimentations. He emphasised the studies of peculiarities of pattern of seismic destruction in buildings and other civil constructions and then co-relating them with the seismic movement of the ground during earthquake recorded by various instruments clarified above. During analysis of the facts and figures thus obtained, certain peculiar findings surfaced which Dr. Smirnov explained in a simple way in his paper under caption "Discordance between real seismic destruction and present calculation" published in Vol. VII No. 1/94 of International Civil Defense Organisation (ICDO) Journal published from Geneva, Switzerland. A brief gift of his findings and explanation with pictorial representation at Fig. 2 are discussed below:

(*i*) There are many inexplicable and curious shear cracks and shear destructions observed in the columns and walls of the buildings after most of the earthquakes, which contradicts the soil movement recorded by seismographs. (Fig. 2D)

(*ii*) These unexplainable destructions can never be caused by the rocking which correspond to the seismic movements of the ground registered by seismic instruments like seismographs, accelerometers and velocity seismograph etc. (Fig. 2 C&D).

(*iii*) The above facts indicates the possibility of existence of some powerful unknown influences operating during earthquake, which cannot be recorded by the available seismic instruments. Perhaps, these mysterious influences are the major causes of seismic damages which invariably accompanies with the appearances of strange shear destruction and shear cracks.

(*iv*) So, it is logical to suppose that an earthquake must include some non-fixed quasi shock waves in the ground which have an almost vertical front speed. Only such waves can cause very big shear stresses and such relevant destructions. These unknown shear stress wave can be manifold greater than their counterparts appearing in the building during rocking, but remains undetectable by available seismic instruments due to their lack of sensitivity or inertness to such undefined shock waves.

The absence of co-relation between the real shear destructions of the columns and walls and seismic ground movements recorded by various seismic instruments and their contradictions have further been explained by Dr Smirnov stating that soil movements recorded and described by seismic accelerograms cause sideways swaying (horizontal) and rocking by massive part of the buildings such as reinforced concrete Dallas or floor platform causing mutual displacement of the ends of columns and walls. Such displacement may in turn cause the curving or bending deformations in the columns and in the high walls. Bending deformations cause the curving and crooking cracks and destructions which are known to many (Fig. 2A& B).

There are differences between bending and shear cracks which can be distinguished very simply by studying the locations, pattern etc. Curving or bending cracks are horizontal and normal to the axes of the column and beam where stress is concentrated. They are not "draft", but only appear in the tensile zone and can not cross the pressed zone of the concerned section of the column. They are wide open near the edge in the tensile zone (Fig. 2A&B).

In contrast, strange shear cracks appear in earthquakes are inclined and "draft". Their most incomprehensible and mysterious characteristics is the absence of bending cracks supposed to be usually associated with. It is not very easy to explain as to why very often they appear in the middle section of the column where the shear force supposed to be caused by bending movement, is almost absent (Fig. 2 C & D).

Further, analysis of any photographs of seismic destructions, would reveal such shear crack which contradicts the seismograms and accelerograms. According to Dr. Smirnov, it is quite likely they are caused by so called ground waves strange shear stress which are expected to be manifold stronger than the shear stresses caused by rocking.

Dr. Smirnov concludes that all seismic calculations and seismic defense measures reflect soil movements registered by accelerometers and because of it, the scientists do not take into account the danger of shear destructions. So, the shear destructions, having a wave but not of rocking nature, cannot be excluded by traditional seismic defense measures directed against rocking. They can defend buildings only against reasonable rockings of upper floor Dallas and against curving deformations and destructions of the column by swayings and rockings. But according to Dr. Smirnov, these seismic defense measures prove futile against the strange quasi shock waves and against their shear destructions.

In the post-log, it may also be mentioned in the above context that Dr. Smirnov, based on his above hypothesis, had also already designed some models of reinforce concrete, which — is stated to have withstood the strong shock wave impulses and other allied effects of earthquake when subjected to artificially produced simulated conditions achieved by underground explosions.

It is perhaps for the betterment of the humanity, if the Japanese experts and their counterparts in India relook in to the basis of calculations for seismodefensive devices in search for greater seismostability. Predictions are already there in the air that in near or distant future, a massive earthquake to the tune of 8 to 9 Richter scale may strike India in the central Himalyan region bordering Nepal involving wide spread areas and thus making approximate 200 million people as destitute.

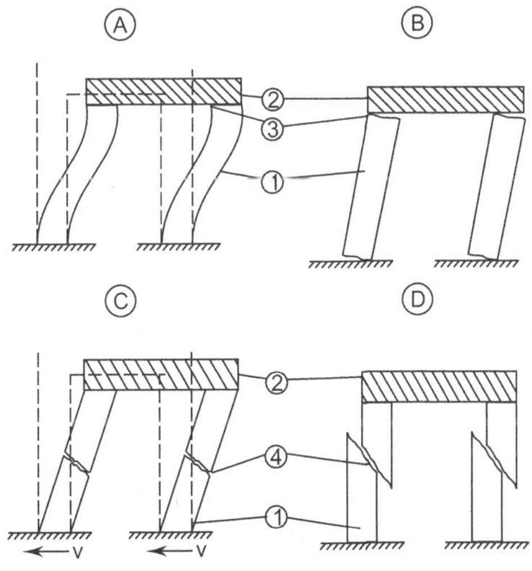

Illustrations showing expectable verus real seismic destructions of columns

A. Expectable bending deformations
B. Expectable bending destructions
C. Real shear deformation
D. Real shear destruction of columns

1. Pinched r.c. column
2. Massive upper floor slab or dalle
3. Bending cracks
4. Shear cracks

I - inert forces: V - velocity

Drawings are schematic Earthquake effects on buildings

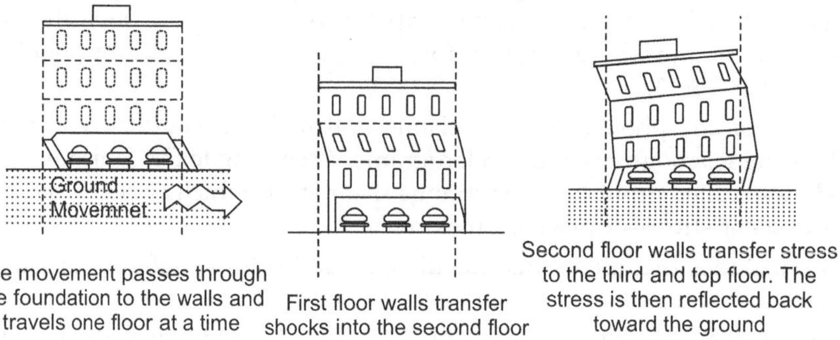

Source: The principles of earthquake resistance in buildings.

RAIL SAFETY

Historical Background

1. The first railway line in India was opened to traffic on 16th April, 1853 between Mumbai and Thane. In 1883, the Consulting Engineers, who were entrusted with the line construction and called earlier as Inspectors, were officially recognised. The Railway Board was established in 1905, under the Indian Railway Board Act 1905. The railway Board became the safety controlling authority for the working and operation of the railways in India.
2. The Govt, of India Act 1935 provided the functions for securing the safety both of the travelling public and of the persons, operating the railways, including holding of enquiries into the causes of accidents.
3. In 1939, Pacific Locomotive Committee was constituted, who suggested its independence from the Railway Board.
4. In 1940, the Central Govt, endorsed the suggestion.
5. The railway inspectorate was thus separated from the railway board and was attached with the Deptt of Communication on 12.05.1941.
6. The railway inspectorate was redesignated as the commission of railway safety on 01.11.1961. Subsequently, the govt, inspectors and chief govt, inspectors were also redesignated as commissions of Railway Safety and Chief Commissioner of Railway Safety. The Commission, as an independent statutory body has completed 69 years of its existence on 12.05.2009
7. There were only six zones by the year 1951–52 which increased to around 16.
8. The post of Commissioner of Railway Safety for each zone was created.
9. At present, the commission of Railway Safety headed by Chief Commissioner of Railway Safety (CCRS) with his headquarter at Lucknow and nine commissions reporting to him. Each Commission is being assisted by Seven Deputy Commissions.

10. The Commission of Railway Safety has also been active member of International Transportation Safety Association (ITSA) for the last 8 years. The objective of the Association is to share the good practices of independent investigations in different countries to disseminate knowledge and to explore the possibilities of mutual assistance in the investigation, depending upon the availability of expertise and resources.

Duties and Functions

The Railway Board generally refers to the Commission the matters relating to modification of trains, track, locomotive, rolling stock and revision of Rules embodied in general rules, Rules for the opening of new lines, schedule of dimensions etc.

The duties and functions of the Commissions are laid down in the Railways Act 1989 and rules framed there under and the executive instructions issued from time to time. These are briefly-

(*a*) Inspection of new lines and certify for their fitness.

(*b*) Conduct investigations on serious accidents,

(*c*) Sanction, execution of new works and installation of running track, affecting the safety of the travelling public such as building bridges, remodelling of station yards, line capacity works and resignaling works etc.

(*d*) Conduct periodical inspections of railways and report to the central govt.

(*e*) Examine technical aspects of new rolling stock and tender advice on their introduction in the existing net work and to sanction their running at higher speeds.

(*f*) Authorise carriage of oversized consignments, stipulating the conditions for their movements.

(*g*) Recommend and sanction infringements to the schedule of dimensions prescribed by the govt, of India.

(*h*) To oversee the accident prevention efficacy of zonal railway admn. by reviewing reports of departmental enquiries into less serious accidents.

(*i*) Tender general advice to railways on matters concerning safety of train operation.

Types of Accidents

Railway accidents can be broadly classified under the following.

1. Derailments
2. Collisions
3. Level Crossings Accidents
4. Fires and miscellaneous other categories.

Broad causes of accident

(*a*) Due to fault of drivers.

(*b*) Due to wrong signalling.

(*c*) Due to failure of Electrical communication.

(*d*) Vehicles escaping from stations.

(*e*) Portion of train, left in a block section.

(*f*) Train in a block section without authority to proceed.

(*g*) Failure to report on conditions likely to affect running of trains, to controller or centralised traffic control operation.

(*h*) Train parting.
(*i*) Track, including unsafe bridges.
(*j*) Normal damage of tracks.
(*k*) Sabotage.
(*l*) Work of the terrorists in exploding a portion.
(*m*) moving on highspeed other than the approved one.
(*n*) Removal of fishplates and fittings by miscreants.
(*o*) Sleepers damaged .
(*p*) Infringement of traction bond.
(*q*) Removal of traction bond
(*r*) Damage or non-functioning of signal wires causing obstruction to train.
(*s*) Excess loading.
(*t*) Due to natural calamities.
(*u*) Public travelling on roof and foot board.
(*v*) Rail fracture.

5. There are manned and unmanned level crossings, Major and maximum level crossings. Lives of men and cattle are lost due to negligence of general public, who cross these crossings without sufficient care and caution. In respect of manned level crossings, the accidents occur due to negligence of gateman, who fails to act (closing the gate) in time.

   Train fire is a major hazard that occurs due to the following reasons.
   (*a*) Smoking by the passengers.
   (*b*) Carrying stoves, gas cylinders, inflammable items.
   (*c*) Unauthorised cooking with kerosene stove and LPG gas stoves in the compartment.
   (*d*) Throwing burning match stick.
   (*e*) All types of electric fires.
   (*f*) Mechanical fires emanating from trucks by friction.
   (*g*) Self ignited fires from the carbonised materials.
   (*h*) Fire from Air conditioners.
   (*i*) Fire from coal and other minerals being transported without thick cover.

Accident prevention

1. Good upkeep and maintenance of track, signal system.
2. Providing train actuated warming device (TAWD) or anticollision device.
3. Providing train protection and warning system (TPWS), Auxiliary warnings system (AWS) and Anti-collision device (ACD).
4. Track circulating of points, zones in yards,
5. Checking unauthorised carriage of inflammable materials and banning smoking while parcels in parcel vans.
6. Rail, wheel and axle fractures need to be prevented by better diagnostic systems and timely action for their replacements.
7. Checking overloading of wagons.
8. Train doors to be closed during the run.

9. Improve design of toilets in coaches for reducing corrosions in rails.
10. Repair, maintenance and replacement of bridges.
11. Provision of appropriate fencing of tracks before introduction of highspeed passengers carrying trains in order to effectively control the trespassing of the same.
12. Streamlining the working of level crossings by
    (*a*) Interlocking of level crossing with signals.
    (*b*) Keeping the normal position of level crossings open to road traffic.
    (*c*) Providing train actuated warming device (TAWD) or Anti -collision device (ACD)
    (*d*) Replacing the level crossings with Road overbridges or Road under bridges, where ever feasible.
    (*e*) Not providing voice recorders at the stations connected to manned level crossings.
13. The driver, conductor, T.T.E.S. and all railway personnel must have sufficient rest hours. They fell asleep during their duty hours due to insufficient rest. So also consumption of alcohol and drug etc. made them inefficient in taking right decisions. Most of the conductors in passenger trains, fleece the passengers in allotment of berths and with this ill-gotten money, procure alcohol, go to the pantry and merrily enjoy the drink and sleep in A/c sleeper berths, (remaining vacant for some reasons) without allotting the same to deserving commuters as per rule. This results, theft, trespass, unauthorised entry of hooligans into the compartment.
14. Providing emergency escape breathing apparatus for all crew members on freight trains. carrying hazardous materials.
15. The railway personnel sometimes are afflicted with arrogance and avoid observation of Rule book. They pay lip service to the safety, codified safety measures, but function as per their own shortcut. They are quite sceptic on safety instructions. They never acknowledge their fault.
16. They are to be trained in ERGONOMICS, and soft skill management to make them good managers, leaders and develop their skill. They are to be effectively trained in safety management and fire fighting.
17. So far expecting efficiency, it is incumbent on the part of Govt, to provide adequate health measures for the employees, specially for the employees, who are constantly on the move.
18. Upgradation of technology,

In case of an accident to a passenger train in which death and/or grievous injury has (or is reported to have) occurred to other than railway staff, officers required to attend the site are:

1. AGM or GM,
2. CMD,
3. CSO,
4. Territorial HOD of Civil Engg. Deptt.
5. SAG officer, of Mechanic branch as nominated by CME,
6. SAG officer, electrical Deptt, nominated by CEE,
7. SAG officer of signal,
8. Chief Security Commissioner.

Incase of other accidents the GM/AGM may instruct officers to proceed to the site.

Comments

Something is obviously very wrong with our railway system. It has expanded enormously and its inadequacies and shortcomings are manifesting themselves in the same proportion as evidenced by an alarming increase of disasters and loss of life.

The Indian railway system is spread over the length and breadth of the country and constitutes a stupendous network for transport of goods and passengers-66,000 km of rail lines, 1.6 million workers, an annual budget of Rs 28,000 crore, and 1300 trains carrying 13 million passengers and 1.2 million tonnes of goods every day.

However, it's mounting record of accidents and failures has become a source of serious alarm. The most recent accident in July, the Gaisal rail disaster, accounted for nearly 400 deaths and 600 injured. The last November's accident at Khanna in Punjab, where two trains rammed into each other claimed 210 lives and 400 injured. These major' accidents sent shock waves across the country, but as time passes, these are merely becoming statistical data. Let us take a look at other gruesome statistics.

In five months, from July to November 1998 there were as many as 211 accidents other than the Khanna accident, resulting in 157 deaths and 403 injured.

These accidents took place on all the railways, the Central Railway, Eastern Railway, Northern Railway, North Eastern Railway, Southern Railway, South Central Railway and Western Railway. In the five preceding years, there were 165 collisions, 1600 derailments (practically one every day) and 318 level crossing accidents, involving 728 deaths. In one year alone in 1996–97, there were 2,965 rail failures, 12,580 engine failures, 37,150 cases of defective wagons, 3,140 cases of poor brake power and the astounding 1,25,259 signal failures.

These figures are indicative of the faults and defaults in the railway system. The disturbing fact is that these figures may not be even be accurate as it is well known that figure of accidents and mishaps are "filtered" by the subordinate staff and even at higher levels. They are kept under wraps. The disturbing fact about the mishaps is that at least 70 per cent of these are eventually traced to human failure.

A major cause of the high incidence of accidents and mishaps is the unfortunate and continuing dependence on the human factor in the present era of modernisation. Other causes include: existence of over aged assets; lack of reliability of installed equipment; and inadequacy of safety devices which could cancel a wrong operation cause by human failure. Another important cause is that of political pressure for carrying out "populist" programmes of expansions and additions to the network which normally may not be feasible or necessary. In the last few years, as many as 900 new trains have been introduced, a large number of them because the political masters considered these additions necessary for the objective of expanding their "vote banks".

While expanding resources on all these expansions and additions, the normal maintenance of, existing equipment and the vital considerations of safety fail to receive the necessary and sufficient attention. It is reported that when railways spend nearly Rs 14,000 crore annually on disbursing salaries, only about Rs 250 crore is set aside for the requirements of signaling and communication which are vital for ensuring safety.

Accidents are obviously a matter of very serious consequence to the country and the reputation of our railway system. A very disquieting fact is that while enquiries are ordered for examining the causes of accidents, reports of past enquiries are hardly ever made public. They are treated as private and confidential reports as though railway property is the private property of the Railway Board. No information is available in the normal course about whether in relation to any serious accident any severe penalty has ever been imposed on the defaulting staff and cases brought to court for trial. In fact, there are reports that whenever any action is taken against any official, it is written off in due course. An Inspector of South Eastern Railway who was held responsible for a serious accident was merely suspended for six months and reinstated thereafter without any cut in his emoluments.

Technological up gradation in relation to every aspect of the functioning of the railway system is of paramount importance in the context of safety requirements and the enormous expansion of the railways. Priorities of investments for this purpose have to be given special importance simultaneously while curbing the demand for expansions on populist and political considerations. Expansion will inevitably be required for creating third and fourth corridors for passengers and goods where the existing corridors are intensively used. Maintenance of tracks and the rolling stock must receive maximum possible attention from the viewpoint of safety. Areas of technological backwardness, obsolescence and absence of effective management practices must be expeditiously overcome. There must be constant monitoring, training and retraining at all levels of the railway administration.

While looking at this totality and while considering the steps that need to be taken to avoid railway accidents in future, one cannot help taking account of the grave tragedies and the harrowing particulars of the gory scenes of these accidents. The Gaisal accident is one such incident which shakes one's insides. It was in this accident where according to a report: "Victims in the front carriages were torn to pieces by the mangled metal; limbs ripped from bodies of passengers hung from gaps in the wreckage; torn arms, hands and feet littered carriages, and broken bodies had been tossed like scraps of meat across the wreckage. Railway coaches pilled up like multi-storeyed buildings. One on top was burning, people were screaming and there were bodies all over the places. These are the scenes that should drive us to ensure that these do not ever get repeated.

From the platform of common cause, we have taken the entire matter to the Supreme Court urging it to direct the Railway ministry and the Railway Board to immediately set up a group of known experts from within and outside the Government to examine the various aspects related to the functioning of the entire railway system; that these expert groups should complete their task and submit their recommendations within three months, and that the railway administration should implement these recommendations within a period of one year. Lot of hue and cry but little action.

Each major train accident generates a lot of hue and cry. Apart from the customary "high-level" inquiry, the minister of the day announces grandiose plans for installation of various safety devices.

The announcements, however, often remain on paper. Commenting on the status of four important safety aids, identified by various committees over the past 20 years, the Comptroller and Auditor General (CAG) has said none of these has been fully implemented. The four devices are

(*a*) track circuiting, which eliminates the chance of a train arriving at an occupied line at a station,

(*b*) auxiliary warning system (AWS), which can prevent collision of trains by automatic application of break if the driver overshoots the signal,

(*c*) radio communication between train and control centres and

(*d*) remote control train activated automatic warning device at unmanned railroad crossings.

After the Ferozabad accident in 1995, the government set a one year time-frame for complete track circuiting of all stations on main routes. However, as both the Khana Committee and the CAG point out a lot of ground on the front still remains to be covered. The same is true of other safety devices. In some cases, the projects have even been frozen. According to the CAG report finalised a few months ago, the safety aid plans, to be installed at unmanned railroad crossings, have not been installed at any of the 707 such crossings on main routes. During 1993–98,581 people died in accidents at these crossings. Plans for installation of this device have been going cm for more than 20 years.

Both the Khanna Committee and the CAG have pointed out that large number of coaches and wagons, not fully fit from the safety view point, are sent out to traffic, ignoring their rejection by the independent neutral control organisation (NCO).

According to the CAG report, 64 per cent of the coaches and wagons turned out of workshops and sick line and rejected by NCO were passed out to traffic during 1993–98. A large number of these rejections were on safety grounds.

The Khanna Committee has found that there are 1,560 stations with overaged signalling gear — in some cases even older than 35 years. The committee has recommended replacement of the overaged gear within a five year time-frame.

According to the committee, of the 1,19,724 bridges that straddle the railway network, there are 51,340 bridges of 19th century vintage. Applying the thumb rule of a structural engineer that the average life of a bridge is 100 years, this is a grave safety threat. The committee has recommended that all bridges, more than 100 years old or declared "distressed" by the railway, should be subjected to "super" checks and remedial action should be taken immediately.

AVIATION SAFETY

Aviation Safety and Security is discussed at airport, aircraft and airlines levels. The present challengs are examined in the light of past experiences of the safety and security programmes. Utmost vulnerability exist in passanger cabin and cargo hold. The essential features of aviation safety are:

1. Pre departure screening
2. Screening of check in baggage
3. Cargo safety.
4. Non observance of model code of conduct enshrined in the Aircraft Rule 1937, Federal Aviation Regulations of U.S. and The Air Transport Safety bill of U.K. by the pilots/ commanders, so far consumption of alcoholis concerened.
5. Disobeyance of Safety rules like flying when the atmosphere is foggy, stormy and heavy rain lashed the routes. In these situations, the visibility factor becomes a problem and aircrashes generally happen.
6. The biggest threat today is hijacking the designated flight by terrorists and placing various demands, by torturing even killing passengers on board.
7. Sometimes, flight hazard takes place due to negligence of pilots on the following matters.
   (*a*) The pilots donot act in conformity with the A.T.C. directives like maintaining the height, the take off and landing time schedule, the direction on landing on a particular runway to avoid congestion, accommodating extra passengers in the cockpit, pilots'cabin etc.

(b) Lack of regular upkeep and maintenance by the technical personnel of the air crafts before flights. Instances of deflation of wheel tyres, opening of doors, malfunction of air conditioners etc. are there which causes serious hazards to the air traffickers.

8. In the present days, violence, ill treatment to the air hostesses, quarrel, assault and infighting among the crew members, disobedience of directives of safety code of conduct by the passengers in the mid air etc. poses threat to smooth air travel.

Remedies

1. Pre departure screening of passengers and cargo should be very strict and fool proof.
2. To avoid the potential terrorists and hijackers not to sneak in to the air craft, the immigration officials must check the passenger' passports, visas in the entry points. Forged passports, visas are the common sight today. Proposal for mandatory passagner profiling is being contemplated, Which would form an important regimen in pre screening schedule so also their cargo. Instances of loss of bagges in the flight, is a common happening today.
3. Pre flight maintenance and repair of planes should be ensured.
4. Drunken pilots/commanders/crew members should be obstructed entry into the plane and departmental action be taken against them.
5. Non observance of safety rules and norms explained by the crew members to the passengers through P.A. system/live demonstration etc. in the mid air, should be strictly dealt with. In case of violence, the concerned passenger should be handed over to the airport security/police when the plan lands in the next destination.
6. The Pilots/Commanders be punished, if they entertain extra passengers in their cabin.
7. The entry door to the Pilot's cabin must be secured from the inside and in no case ingress to the cabin be allowed under any circumstances.
8. Lighting of the runways before landing of aircraft must be ensured by the grounding staff.
9. Air Traffic Control Staff should remain alert in matters of exit and entry of planes in the airport. They must ensure that the pilots follow their directions and stick to their designated lane, runway and maintain required height and path.
10. The planes should not be allowed to take of at the time of dense fog, heavy rain and storm, during which the visibility of the pilots become extremely poor.
11. Regular medical checks, particularly the sight of the pilots be done.
12. Drunken, erring pilots threat a serious flight hazard. Breath analyser tests of the pilots, which is mandatory under rule 24 of the Aircraft, Rules 1937 should be strictly followed. According to Indian Society of Aerospace Medicine, an in-flight study of pilots with blood alcohol levels of 0%, 0.04%, 0.08% and 0.12% suggest that even quite low blood concentration of alcohol cause significant performance decrements. Blood alcohol concentrations of 0.04% are associated with substantial increases in number and potential seriousness of procedural errors. In U.K., the Transport Safety Bill, puts the alcohol limit for its pilots at 20mg in 100mg of blood. The bill gives wide powers to the police, to enter into the aircraft and other places, if they suspect an offence has been or being committed. Many pilots have been arrested by airport security officials in U.K. Pilots face suspension and a habitual offender can lose his licence. The bill also empowers severe punishments to the pilots found with unacceptable alcohol level in their blood while doing even "ancillary to aviation function" like reporting to the operation department for other sundry administrative work. Likewise, under section 91.17 of Federal Aviation Regulations in U.S., the permissible blood-alcohol content is 0.04%.

Their "Company Handbook" also says, employees, who fail the blood-alcohol test are liable for instant dismissal.

In India, airlines conduct random pre flight breath analysis tests on pilots reporting for duty on rare occasions, for international flights into India, those checks are carried out post flight. Most scheduled carriers have a doctor appointed by the operations department for this job. At times D.G.C. A's air safety officials undertake surprise visits to the airports to monitor the checks conducted by air lines.

Rules 24 of Aircraft Rules 1937 prescribes that the pilot, copilot, engineer, cabin crew, members shall not have detectable alcohol in their blood, breath or urine, while operating the flight. It means that permissible alcohol level should be zero. The rule further says that all air lines shall establish a system of preflight medical checks for their crew to ensure compliance of this requirement and the procedure so established shall form a part of the operation manual. But no where, the rule spells out the nature or kind of punishment, while some airlines ground their pilots for a straight three months, others have arbitary rules. A first time offender may be let off with a warming, while a serial offender could also be suspended.

The Indian Rule is silent about punishment to be metedout to offender pilots and crew, while in U.K. and U.S., their rules framed speaks the lapses on this score as criminal offence, when negligent road drivers are punished under the provisions of Motor Vehicle Act, pilots should also be similarly punished and penalised for their lapses and negligence when lives of hundres are involved.

13. Fuel shortage in the mid air occasionally pose a problem for which the pilots usually land in alien territories and cause innumerable political and security problems. On the other hand, the passangers suffer from severe mental torture and harassment for delay in release of impounded air craft after completion of a set of parleys at the political level.
14. Bird menace is the biggest air safety hazard, acknowledged in the history of flight hazard. To ward off this menace, the airport environment must be conducive for the flight regimen to be smoothly observed.
15. Anoter airport safety is to keep strict vigil in guarding its premises and boundary. Somtimes tresspassers pose a problem in Safe movement of planes. At times stray cattles used to enter in to the air field through the openings made in the compound wall. The security guards should be alert enough to stop any type of tresspasses. The compound walls also be properly maintained and timely repair affected.
16. Preventive measures like, technology of imaging system, trace detection, non-imaging electromagnetic technology, explosive detection system (EDS) are taken to make safe air travel.
17. Since air crashes result total loss of human lives on board of the plane, post accident measures are confined to retrieval of dead bodies, cargo and blackbox to make accident analysis and preparation of insurance claims of the legal heirs of the deceased and other compensatory benefits along with return of the cargo etc. are some of the responsibilities the Airline authorities are expected to accomplish.

SAFETY IN STONE CRUSHING INDUSTRY

In stone crushing units, the following hazards are very common:

(*a*) Severe air and water pollution.

(*b*) Workers are affiliated with ashthama, silicosis and other respiratory diseases, due to inhalation and ingestion of injurious substances like harmful dust.

Preventions Measures

(*a*) The workers should be provided with masks and respirators of all types.

(*b*) The accumulated water and water sources nearby get polluted and become unsafe for drinking, by the men and cattle. So continuous disinfection, purification, water treatment plant process, become a mandatory duty of the crushing unit owner.

(*c*) Steps may be taken to raise quick growing trees in and around the unit, to check spread of dust by wind.

(*d*) Water sprayers/sprinklers need be commissioned to spray water in and around the unit at intervals in order to check further air and water pollution.

(*e*) Workers to be provided with soap and washing powder for cleaning their body and clothes.

(*f*) They should be medically checked at intervals and medicines supplied for asthama and respiratory ailments

(*g*) Residential colonies of workers should be located far away from the site. Women workers are prohibited to bring infants with them to the site.

(*h*) The company should start welfare measures in villages situated near the units under corporate social responsibility (C.S.R.). Provision of drinking water preferably deep tubewells raising forest cover by large scale plantation and crop rescue operation.

(*i*) Workers handling boulders (cutting, lifting, removing, placing etc.) are required to wear appropriate ppes while handling blasting operations, use and storages. Dynamites explosives should be as per approved procedure (discussed in earlier chapters), wire netting be ensured to protect damage by splinters.

SAFETY OF SHIP VOYAGE

Safe Access to the Ship

1. Safe access to the ship in port should where possible be ensured by an accommodation ladder or gangway that can be adjusted to cope with any movement of the ship, such as one caused by the rise or fall of the tide or by ranging along the quay.
2. The gangway should give safe access to a clear safe part of the quay.
3. Seafarers should take care at all times when boarding or leaving a ship. They should not use unauthorised means of access and should never attempt to board or leave the ship by stepping or jumping from the quay to the deck or from the deck to the quay.
4. Any means of access and the deck leading to the accommodation should be well lit during hours of darkness and at other times as necessary.
5. A lifebuoy should be provided near the gangway with a lifeline attached.
6. Where practicable a suitable net should always be rigged to prevent people from railing onto the quay or between the quay and shipside.
7. Seafarers should take special care in using the gangway after they have imbibed alcoholic beverages.

Manual Lifting and Carrying of Loads

1. No seafarer should, as a regular practice, attempt to lift any load over 55 kg unassisted. This load should be reduced as required in light of physical limitation, advancing age or in young persons.

2. Before lifting any load, attention is to be paid to
   (*a*) sharp edges, nails, splinters, projections and
   (*b*) greasy or slippery surface that may affect the grip.
3. The deck area over which the load is to be moved is to be free of obstacles and not slippery.
4. Unless the load is light or fitted with a proper handle, it should be firmly gripped with the palms of the hands and not only with fingers.
5. A person preparing to lift a load should:
   (*a*) not bend from the waist but should take a firm balanced stance with feet not too wide apart and the body close to the load
   (*b*) bend the knees
   (*c*) keep a straight upright back with the chin tucked in and
   (*d*) get a good grip on the load.
   (*e*) try the weight with a short trial lift and get help if unable to manage alone
   (*f*) once sure of lifting the load, raise it by straightening the knees, keeping the load close to the body.
6. Where possible or when lifting to a high level the person should—
   (*a*) use a bench or support to make the lift in two stages and
   (*b*) adjust his or her grip as necessary for carrying or lifting to a second level.
7. To put the load down the lifting procedure should be reversed, preferably in two stages, letting the knees do the work of lowering, not bending from the waist, but keeping the load close to the body.
8. Persons carrying load should ensure that:
   (*a*) the load does not obscure vision.
   (*b*) they do not trip and
   (*c*) they do not injure their arms by catching them or the load, on any obstruction.

Anchoring and Mooring Operations

1. The windlass anchors, cables and chains should be carefully maintained and regularly inspected for damage and defects.
2. Any anchor, cable or chain showing defects should be withdrawn from service.
3. When not required, anchors and cables should be properly secured by stoppers etc. and such securings should be removed only when it is intended to raise or lower the anchor.
4. Raising and lowering of the anchor should be under the supervision of a Responsible officer.
5. Before the anchors are raised or lowered the Responsible officer should ensure that there are:
   (*a*) No persons in a dangerous position *e.g.*, between the windlass and the hawse pipe and
   (*b*) No small craft or obstructions under the bow of the vessel.
6. The person operating the brake should wear safety goggles or face mask when letting go the anchors.
7. The person incharge of mooring operations should ascertain that there are no persons in a dangerous position before any heaving or letting go operations is commenced.
8. Any winch or capstan under use in mooring operations should be under the control of an experienced seafarer at all times.

9. Seafarers should be alert and ensure that they are in a safe position when in attendance at any mooring operation.
10. They should not stand in the bight of a rope or wire, nor where practicable in the bight formed the mooring winch and the coiled rope.
11. They should never stand or move across a rope or wire that is under strain or being worked, when possible they should stand behind the winch or capstan when any rope or wire is under strain.
12. When drums ends are being used to haul mooring ropes the following precautions should be taken:
    (*a*) the person handling the rope should have a firm footing
    (*b*) the deck behind the drum ends should be preferably provided with anti-slip surfaces or treated with anti-slip paint
    (*c*) the drum ends should be kept free from grease and paint
    (*d*) any rollers or fairleads used should be seen to turn smoothly
    (*e*) the person heaving on the rope should stand on the opposite side of the drum to the lead of the rope and
    (*f*) a second person should be employed to take up and coil the slack as it is taken in.

Work on Deck or in Under Deck Spaces

1. Persons working on deck shall wear:
    (*a*) well fitting safety shoes, with reinforced toe caps and a good gripping sole at all times,
    (*b*) protective head gear when and there is a danger that the head may come into contact with swinging or falling objects and
    (*c*) clean well fitting, full sleeved overalls or boiler suits, buttoned up securely.
2. Persons should wear Appropriate Hand Gloves when:
    (*a*) handling or picking up sharp edged objects or hot pipes or equipment
    (*b*) there is a danger of fingers being nipped or
    (*c*) handing ropes or
    (*d*) handling chemicals
    (*e*) when carrying out hotwork.
3. Personnel should take care to avoid obstructions, which could cause them trip, or to hit their heads or limbs on low or projecting obstructions. All such hazards should be painted by bright conspicuous colour. Where practicable, notices should be posted reading "Watch the Step" or "Watch your head".
4. Loose gear on deck should be properly secured against movement.
5. No equipment should be used or operated unless the prescribed certificates of tests and examination are on the ship and are current and valid.
6. The safe working Loads of all lifting equipment should be clearly marked and they should never be exceeded.
7. Seafarers should take care:
    (*a*) never to stand or pass under the path of any load being moved, nor under a suspended load
    (*b*) never to leave loads suspended longer than necessary, nor leave equipment where it would create hazard

(*c*) never to enter a cargo compartment or enclosed space without informing a responsible officer

(*d*) never to walk on hatch beams or coaming edges when a hatch is open

(*e*) to use appropriate safety line and harness when any work is to be performed that introduces the danger of falling.

8. Work should not take place over side whilst the ship is underway.
9. When working over side, the personnel concerned should wear safety harness and lifelines in addition to life jackets or other suitable flotation protection.
10. When working near the propeller, the Engineer officer must be informed and the propeller must not be turned.
11. Persons working on a stage should wear Safety Harness belt. This belt should not be secured to the Stage but elsewhere with a separate line.
12. The hauling and lowering of personnel should be by hand and not by mechanical means.
13. Persons working aloft, overboard or ion staging should keep any tools in suitable containers properly fastened to prevent danger to persons working below or passing below.
14. They should arrange their work so that one hand is always free for holding on.
15. When work is to be carried out on the funnel or near the siren, the responsible officer should ensure that:

    (*a*) the duty bridge and engineer/officers are aware of the fact that such work is proposed;

    (*b*) the siren cannot be operated, by effectively shutting off the source of power;

    (*c*) boiler steam valves are not lifted;

    (*d*) the emission of steam, gases or fumes such as would be likely to endanger personnel is prevented; and

    (*e*) appropriate notices are posted on deck and in the engine room.

Work in Boiler and Machinery Spaces

1. Personnel should wear anti-slip footwear with reinforced toecaps as well as suitable head-gear.
2. Spaces in which hearing protection needs to be worn should be so indicated by warning notices at the doors into such spaces and the equipment should be donned before entry to the space.
3. Machinery spaces should never be used as a drying room for clothing or wet equipment.
4. All moving parts of machinery should be provided with permanent guards or railings or fencing.
5. All hot pipes such as steam pipes, exhaust pipes and compressor discharges could constitute a safety hazard and should be lagged or shielded and warning notices posted.
6. The Machinery spaces must be well lit at all times.
7. The Walkways should be free of oil or grease. Anti slip coating may be given as required.
8. Ventilation should be provided to maintain a comfortable atmosphere.
9. All stores and tools should be properly slowed and adequate arrangements should be made, particularly with heavy stores, to secure each item in rough weather.
10. Any tools being used above a lower platform level should be in a suitable bag or box or secured so as to prevent them from falling.

11. Personnel should wear suitable eye protection, protective clothing and dust masks as required when grinding or chipping or when engaged in dirty or dusty jobs such as scaling of boilers or hand cleaning of boiler tubes.
12. All trace of oil or cleaning fluids should be thoroughly washed from the skin after exposure.
13. Personnel should never
    (*a*) Start, stop or operate any machinery unless authorised to do so: nor
    (*b*) Work or clean in the vicinity of moving machinery unless it is adequately guarded; nor
    (*c*) remove swarf or turnings from the vicinity of moving machines.
14. No wood planks or other combustible material should be stored in the machinery spaces, particularly around boiler casings or engine exhausts and uptakes or in the funnel.

Electrical Equipment

1. Protective equipment such as rubber gloves and rubber boots should be used whenever there is a risk of electric shock, but should not be regarded as providing fully adequate protection against such risks.
2. Protection against contact with live equipment should be afforded by
    (*a*) placing the live parts out of reach;
    (*b*) effective enclosure of the live part; and
    (*c*) adequate insulation.
3. No work should be done on any equipment until the equipment has been made electrically discontinuous.
4. Precautions such as locking out and tagging switches and fuses should be taken out to prevent the current from being switched on while the work is in progress.
5. before the current is restored, a responsible person should ensure that no personnel remain in a dangerous position.
6. Flexible cables should be:
    (*a*) not be laid on surfaces that are oily or wet with corrosive liquids;
    (*b*) be kept clear of moving loads, running gear and moving equipment;
    (*c*) no to be used to lift the portable lamp or portable light to which they are connected;
    (*d*) be of robust construction; and
    (*e*) have additional protection where they are likely to be subjected to rough usage or moisture.

Catering Services

1. Refrigerator chamber doors should be fitted with -
    (*a*) a device of adequate strength to hold them open in a seaway: and
    (*b*) a means of opening them from inside.
2. Refrigerator rooms should have a means of sounding the alarm both inside and outside, preferably in the galley.
3. The arc immediately outside should have a non-slip surface.
4. If leakage of refrigerant in a reefer space is suspected:
    (*a*) a warning notice to that effect should be posted on the door; and
    (*b*) personnel should not enter.

5. Do not store Dry ice or Solid CO_2 which at times arrives with provisions in the Reefer chamber. It evaporates and leaves a Oxygen free atmosphere that can be fatal.
6. When food stuffs are being chopped or cut with a knife, the fingers should be bent inwards the palm of the hand with the thumb overlapped by the forefinger.
7. The knife should be angled away from the work so that the blade is slopped away from the fingers as the work is fed to the knife.
8. Care should be taken never to leave pans or fat unattended on an oven or heating stove, nor to allow water to come into contact with hot fat.
9. Water should never be used to attempt to extinguish burning fat. It is preferable to smother it.
10. Absolute cleanliness should be maintained in respect of food, crockery, cutlery, cooking utensils and stores.
11. Smoking should be prohibited in kitchens; galleys, pantries, storerooms or other places where food is stored, handled or prepared and notices to that effect should be displayed

Safety in Accommodation

1. Care should be taken to keep all accommodation whether individually or collectively used, in a clean and tidy condition.
2. Clothing should be hung in drying rooms to dry and if that is not possible care should be taken to keep clear of light bulbs, radiators and any equipment where a source of radiant heat, could lead to afire.
3. When personal electrical equipment is installed, all the equipment, including the connection plugs, should be checked by a competent person.
4. Multi outlet plugs should never be used.
5. After smoking, all butts and ashes are to be properly extinguished and placed in a non-spill ashtray.
6. For your own safety and that of your mates, never smoke in bed.
7. Personal portable electrical space heating device should not be permitted in living accommodation,
8. A non-slip bath mat should be provided to avoid slipping or falling in bathrooms.
9. Personnel should not smoke when using aerosols.
10. When hand irons are used, precautions should be taken to avoid burns and scalds. Portable hand irons and ironing boards should always be returned to the stowed position whenever the user leaves the laundry room.
11. All electrical equipments should be switched off or electrically disconnected when the user leaves the room.
12. Personnel should behave in a way to serve the best interests of the community, to avoid giving offense, *e.g.*, by holding noisy or unruly parties or playing radios or music loudly when others are trying to sleep and by doing there in, keeping the accommodation clean and tidy.

DRUG AND ALCOHOL HAZARD

Drug and Alcohol abuse and its adverse effects on safety is one of the most significant problems of our time. Recognizing the potential serious impact of marine incidents, the marine industry have over the years developed guidance aimed at encouraging safe ship operation and protection of the environment. Most companies are now operating with strict policies related to drug and alcohol use on board their ships. Most companies have a clearly written Drug and Alcohol policy that is easily understood by the seafarers. In orders to enforce their policy they have rules of conduct and control in place, with the objective that no Seafarer will navigate a ship or operate it's on board equipment whilst impaired by drugs and alcohol. Seafarers are now subjected to testing and screening for drugs and alcohol abuse by means of a combined program of unannounced testing and routine medical examination.

The misuse of legitimate drugs, or the use, possession, distribution or sale of illicit or un prescribed controlled drugs on board ship cannot be condoned and should be prohibited. In addition any use of prescribed controlled drug which causes, or contributes to unacceptable job performance or unusual job behavior should require the seafarer to be excused from duty until such times as he is repatriated, or treatment and its after effects cease.

Recognising that all seafarers must be able to respond at any time to emergency situation, the International Maritime Organisation (IMO) is issuing guidance to administrations on maximum permissible blood alcohol content and abstinence and abstinence periods in the Part B of the revised STCW convention.

Officers and Ratings should be aware that local regulations may be in place and where this is the case, it is recommended that these be strictly adhered to.

PERSONAL LIFE—SAVING APPLIANCES

Similar to FFA, each ship is well provided with appliances and survival craft for saving of all life's onboard in case the ship has to be abandoned. The same principles as in the FFA applies here also.

Survival craft

1. Life boats
2. One may come across different types of life — boats on different ships as:
    - (*a*) open
    - (*b*) partially enclosed
    - (*c*) self-righting partially enclosed
    - (*d*) totally enclosed
    - (*e*) totally enclosed with air support system
    - (*f*) totally enclosed with fire protection system.
3. For passenger ships, the capacity of the life boat is generally sufficient for every person on board. For cargo ships, the capacity is generally twice the number of persons onboard. Life boats are either launched by davits or by free-fall method. Embarkation of the boats may be done either at the embarkation deck or through ladders when the boat is in the water. Nowadays, it is possible for all personnel, on a cargo ship to embark a life boat and leave the ship. This means that there need not be any person left on the ship to lower a life boat.

    Life rafts: Life rafts are inflatable. They can either be manually launched or launched by davits or they can also automatically float-free from a sinking ship.

    Rescue boats: Depending on the types of ship (passenger/cargo) a minimum number of rescue boats have to be carried. A life boat may be classed as a rescue boat also, if it conforms to requirements.

4. Personal Life Saving Appliances
   (*a*) *Life buoys:* According to its size each ship should carry a specific number of life buoys distributed on the deck. Some of these life buoys should have life lines, selfigniting lifts or smoke floats.
   (*b*) *Life jackets:* Life jackets as available and approved may be rigid or inflatable. For every ship, the number of life jackets to be carried will depend on the maximum number of persons onboard that ship at any particular time. A whistle is to be firmly secured to each life jacket. Each life jacket will require to' have a fixed or flashing light also.
   (*c*) *Immersion suits and thermal protective aids:* These are provided for at least 3 persons in each non-enclosed life boat on both passenger and cargo ships. If immersion suits are provided there is no need of thermal protective aids.

Fire Prevention and Fire Fighting in the Board of Ship

Areas of fire hazard

(*a*) Machinery Spaces (Engine Room)
(*b*) Accommodation Areas (Galley)
(*c*) Galleys
(*d*) Radio room, battery room and electric equipments.
(*e*) Cargo area
(*f*) store Room.
(*a*) **Causes:**
   (*i*) Combustible liquid
   (*ii*) Oil soaked insulation
   (*iii*) Hot surfaces
   (*iv*) Hot work
(*b*) **Accommodation Areas**
   (*i*) Combustible Liquid
   (*ii*) Hot Surface
   (*iii*) Electrical equipment
(*c*) **Galley (same as above)**
(*d*) **Cargoes**
   (*i*) Spontaneous Combustion
   (*ii*) Oxidising Cargoes
   (*iii*) Compressed Flammable Gas
   (*iv*) Reactive Cargoes
   (*v*) Explosives.

Adoption of Safety Procedures

Safe Practices: General Safety Procedures.

- Shipboard fires are often caused by carelessness and ignorance.
- Do not smoke in "No Smoking Zones". Use only Safety matches.

- Ensure matches are extinguished and cigarette ends properly stubbed out Prior disposal.
- Conspicuous warning notices should be displayed in any part of the ship where smoking is forbidden.
- Hot work permit to obtain prior welding and gas cutting. The vicinity or the place should be free from combustible materials: A bucket of water and portable fire extinguisher to keep stand by for use, one man to be on stand by duty to keep water on sparks of welding/gas cutting and take necessary action.
- In case of fire; attempt to control the emergency by means available at hand. Raise fire alarm, shout "Fire Fire, Fire". Report to the Bridge.

Measures for Reducing Fire Hazards in Engine Room

- Paints, spirits, Oil, Wood should not be kept in Boiler room and machinery spaces.
- Engine exhaust/hot pipe lagging to be kept in place, if gets damaged – to repair/renew immediately.
- Suitable metal containers should be provided for the storage of cotton waste, cleaning rags in the engine room. It should be emptied at frequent intervals safely.
- The precaution should be strictly followed to avoid fire due to welding, gas cutting or other hot work in the engine room.
- Any oil leaks in the Engine room should be rectified immediately. Spillage of oil/paint solvent to be wiped immediately.

Measures for Reducing Fire Hazards in Galley

- Care should be taken in particular to avoid overheating and spilling of Cooking oil on the hot plate.
- Extraction — fan filters should be kept clean at all times.
- Means to smoother fat or cooking oil fires, such as Fire blanket be readily available inside the Galley.
- All Electrical wiring inside the galley should be well maintained and kept clean and dry.

Measures for Reducing Fire Hazards in Accomodation Areas

- No smoking in Bed. Ashtrays to be provided and used only at authorized area.
- Cleaning gears to stow in lockers. All bins to be emptied after days work.
- All fixed electric heaters are to be fitted with suitable guards. Dry clothing on the heater should not be permitted.
- All the electrical fittings in the accommodation should be firmly secured. Authorized persons only interfere with electrical fittings. Personal electric appliances should be connected to the ship's supply only with the approval of Electrical officer.

Measures for Reducing Fire Hazards in Cargo Spaces

- Hatches should be cleaned properly after every discharging.
- Attention is to be given to deck area near upper hatches when hazardous cargo is being loaded/discharged.
- Ensure cargo is stowed and ventilated in accordance with rule.

- The atmosphere in cargo compartment is to be inerted as required.
- Fire doors of compartments/deck to keep close.
- Smoking on deck should be strictly prohibited during the cargo work.
- All paints and painting materials to be stowed properly and all containers to be covered tightly only in paint store.

Fire detection and fire alarming system

1. Heat detection
2. Smoke detection
3. Flame detection

Fire extinguishing system

1. Carbon dioxide system
2. Foam (low and high expansion)
3. Sprinkler system and pressure water spray
4. Chemical powder applicants
5. Emergency fire pump

Fire fighting equipments

1. Fire hoses and nozzles
2. Portable fire extinguishers
   (*a*) water type,
   (*b*) Foam type,
   (*c*) Dry powder type,
   (*d*) Carbon dioxide type
3. Fire blanket.

Fireman's PPEs

Fireman's outfit: The fireman's outfits enable fire fighting team to approach the seat of fire and to carryout rescue duties and provide adequate protection against the effect of smoke and heat.

Every ship should carry fireman's outfit, each consisting of:

- Breathing apparatus.
- Protective clothing to protect the body from heat radiation and burns from the fire.
- Boots and gloves of rubber or other electrically non-conducting material
- Rigid helmet.
- An electric safe hand lamp with a minimum 3 hrs. duration.
- An axe with a short insulated handle.
- Strong fire proof life line (36 meters length each minimum 2 nos.)
- A belt for carrying the ancillary equipment.

The breathing apparatus should be of self contained apparatus of approved type. The self contained apparatus should consist of a face mask attached by a flexible hose to cylinder containing air and supported on a frame and harness. The capacity of the apparatus should not be less than 1200 liters to give normal functioning time of 30 minutes.

Ship Fire Fighting Organisation

General emergency alarm: General emergency alarm is the signal which summoning crew and passengers to their muster station whenever any emergency occur. Generally emergency alarm signal consists of seven or more short blast followed by one long blast on the ship's whistle or siren and additionally on an electrically operated bell or other equivalent warning system. The ship's main power supply or the emergency source of electrical power shall power the general emergency alarm in case the ship's power supply fails. The system shall be capable of operating from the navigating bridge and also from the other strategic points.

It shall be audible throughout the entire working places, engine room, and all the accommodation. On hearing the emergency signal, all persons must go to their muster station as per the muster list. They must put on long sleeved boiler suit and donning their life jackets and swing in to action to save the ship by way of controlling the situation and preparation of life saving appliances.

FIRE CONTROL PLANS AND MUSTER LIST

Fire Control Plans

1. In all ships the Fire control plans shall be permanently exhibited for the guidance of ship's personnel, showing clearly the following:
    - For each deck the control stations.
    - The various section enclosed by "A" class division, the section enclosed by "B" class divisions.
    - The particulars of the fire detection and fire alarm systems.
    - Means of access to the different compartments, decks, etc.
    - Ventilating system including particulars of the fan control positions, the position of dampers and identification numbers of the ventilating fans serving each section.
2. One copy of such plans shall at all times be available on board in an accessible position.
3. Fire control plans shall be kept up to date, any alterations being recorded thereon as soon as practicable.
4. Description in such plans shall be in the official language of the flag state.
5. The instructions concerning the maintenance and operation of all the equipments and installation on board for the fighting and containment of fire shall be kept under one cover, readily available in an accessible position.
6. In all ships, a duplicate set of fire control Plan shall be permanently stored in a prominently marked weather tight enclosure outside the deckhouse near gangway areas for the assistance of shore side fire fighting personnel.

Muster List

The master of the vessel shall prepare the muster list before the ship proceeds to sea. The format of the muster list for passenger ship must be approved by the Government. Muster list must be kept updated at all times. The muster list shall specify:

1. All the special duties assigned to the different members of the crew against their name rank.
2. All the special duties for fighting the fire, having regard to the ship's fire control plans, such as manning of fire parties, fire fighting equipment etc.
3. Carrying of emergency radio, EPIRB; SART as detection equipment.
4. Ship's pyrotechnics (12 numbers from the bridge).
5. Sounding of an emergency alarm.
6. The order to abandon ship by the master of the vessel.

The copies of the muster lists must be posted in conspicuous places including the navigating bridge, crew accommodation and engine room, In passenger ships, the muster list shall show the location of passenger muster stations. Usually the muster stations will be public rooms, so as to protect passengers from the weather.

Communications

There are various means and type of internal communication systems provided on board a ship for both normal and, emergency situation.

The following are a few examples of them:

- Intrinsically safe magneto telephone systems;
- Voice powered telephone system;
- Voice pipes;
- Normal intercommunication telephone system;
- Fixed Intercoms between strategic points;
- Public announcement system;
- Hand held radios.

These systems are so positioned, that they provide communication between strategic positions, such as bridge, engine control room, radio room, cargo control room, machinery spaces, generator room, steering gear compartment, galley, signal deck, boat stations etc.

Personnel Safety Procedures

Most of the accident occurs onboard the ship due to the ignorance and negligence of seafarers to wear the appropriate protection gear.

In the past, during emergencies on board a ship, most of the lives that had been lost are due to the fact they have not followed the safety procedure or get trapped within the ship, being unaware of the escape routes to the nearest muster/boat stations. Therefore, it is mandatory for every person joining a ship, must read the training Manual and learn the ship's emergency escape plans, at the first available opportunity. Persons using safety equipment should know its limitation and check them each time before use. Special protective equipment should be provided and worn by seafarers who may be exposed to particular corrosive or chemical substances. The manufacturer's instructions to be consulted for use and maintenance of any equipment.

Periodic Shipboard Drills

The main objective of shipboard drills and training is whenever the ship's life and property is threatened by fire, explosion, flooding, collision, stranding, etc., to save the life and property of ship. The ships personnel are trained in organized manner to respond to combat such types of emotion to bring them under control so that the life of people and the ship's property can be saved out at sea. The shipboard drills are required to be carried out regularly it accordance with merchant shipping regulations.

- On passenger ship the abandon ship and fire drill to be conducted weekly.
- On cargo ships (other than passenger Ship) the drill to be conducted at least once in fortnight.
- The drills of the crew shall take Place, within 24 hours of the ship leaving a port if more than 25% of the crew has not participated in abandon ship and fire drills in the previous month.
- Each member of the crew shall participate in at least one abandon ship and one fire drill every month.
- The timing of every drill should vary so that personnel who have not participated in the previous drill may take part in the next drill.

It is important that the drill should be carried out realistically, approaching as closely as possible to emergency conditions.

Any person detects any defect or deficiency during the drill should be inspected and made good without delay.

Onboard training in the use of ship's Life saving and Fire fighting appliances should be given as soon as possible but not later than two weeks after a crew member join the ship. Details of abandon ship and fire drills and onboard training should be recorded in the log book.

The Fire patrol system is such that all ships shall at all times when at sea, or in port (except when out of service), be so manned or equipped so as to ensure that any initial fire alarm is immediately received by a responsible member of the crew.

Fire Fighting Methods

Fire: The basic fire handling technique can be summarized as follows:

Find: The fire, learn of its size, location and combustibles involved

Inform: the bridge and sound the afarm immediately—even if the fire appears small.

Restrict: the fire by closing doors, isolating fuel and electrical supplies and closing ventilation.

Use Boundary Cooling

Extinguish: the fire by using the correct type and quantity of Fire Fighting media. Uses CABA and protective clothing. Good organisation and speed are essential.

Fire fighting procedures ship voyaging in sea

1. Fire Alarm initiated
2. Crew assembled at fire stations
3. Fire parties assembled and prepare for action
4. Ship's course and speed altered
5. Fire pumps started
6. Fire fighting action initiated

Additional procedures at port

1. Call the port fire brigade
2. Inform port authority
3. Confirm that the master is in charge
4. Confirm that port fire brigade will take charge
5. Port authority to inform of hazards to the dock installations
6. Always use public address system

Responsibilities in the port and docks

The master of merchant navy ship is absolutely responsible for safety of men and ship on behalf of the ship owners. When the ship is in a port inside the docks or in a shipyard, the master is responsible for his ship and it's safety. He can ask fire service to stop fire fighting operation on his ship and may decide to leave Jetty or moorings, if he thinks necessary to do so for the safety. However, harbour master can also take decision to cease fire fighting operation and direct any ship to move out to safety distance or he can have it towed to the safer area. He has also the right to refuse entry of a ship into harbour if he considers that vessel constitutes danger to the port installation and its surroundings.

Fire fighting operation in ports

Ships in port generally do not maintain as strict a water tight integrity as they do at sea. Hatches and doors to unmanned or empty spaces some times kept open for various operations like cargo being loaded and unloaded, tanks/holds being cleaned and inspected and provisions/personnel coming aboard. This relatively "open" condition also contributes to spread of fire on the vessel. The fire department arrives and begins putting water on the fire. If the water is not removed or drained off, it begins to fill the space. This cause water level rise to a certain level, above the deck before it begins flowing through the hatch to the deck below. This reduces ship's stability in two ways *i.e.*, "high weight" and "free surface".

Preventive Measures Against Capsizing

This hazard is likely to happen on the following reasons.

(*a*) During fire fighting operation (water logging)

(*b*) Due to over/loading of cargo on the port.

(*a*) An important task, fighting a ship board fire is to control excessive water collection at various places. This includes two operations *i.e.*, maintaining/establishing water tight integrity and removing water from the ship after it has done its work on fire extinction.

The following points should be given careful attention.

1. Keeping the fire area sealed off from the rest of the vessel for non-spread of fire to unaffected areas.
2. Water does not spread to other areas and create flooding.
3. Water does not have to move to lower decks to create free-surface problems.

   In some other cases, counter flooding, filling of a tank or a compartment on the side of the ship opposite to the heal may be required to control the heal. This however should not be attempted by a fire officer, which is left for best judgement of ship master and ship operators.

If dangerous condition exist on a vessel, the numbers of personnel on board should be kept to the minimum. All passengers should be evacuated, and the main fire-cranes kept in readiness on the quay side or as near as is considered safe. However the master of a vessel carrying certain dangerous substances must have his vessel in a constant state of readiness to move, provided tidal conditions permitting.

(*b*) The ship may capsize due to overloading:

The ship must be loaded with the cargo as per approved weightage. While loading, other ancillary aspects like moisture contents, wet materials, drainage of water from wet materials, causing floor flooding and excess increase of weightage etc. be meticulously tanken into consideration. Example. Black rose ship capsized in the anchorage area of paradeep port, orissa.

Piracy on the High Seas

"Sea Piracy" was considered as an old and outdated term which was prevalent in the medieval period. After development of technology, watch and ward system, sophisticated arms and ammunitions, Sea Pirates ceased to operate. At present, the activities of the pirates have gained momentum and they are now active in the Somalian Coast, in the Gulf of Aden. Their modus operandi is to hijack the ships and capture the captain and crew members and demand hefty ransom for their release. After the sum is paid, they release the ships to move to their destination. In order to ward off this hazard, some nations including India have deployed their warships to guard the coast and overpower pirates using arms. Besides, the ships, passing through this designated channel, equip with fighting forces, arms and ammunitions in order to combat with the Somali Sea Pirateds, inspite of patrolling by multinational naval forces in the strategic shipping lanes that link Europe to Asia through the busy Gulf. They are trying to extend their range and strike deeper into the Indian Ocean.

Collusion Hazard

Due to invention of sophisticated machineries, obstruction detection systems and signaling systems have largely improved. Collusion with huge bergs are the events of past. In very rare circumstances, collusion of ships with each other on high seas takes place at present.

Training

Statistics compiled by the Institute of Indian underwriters (ILU) shows increasing trend in the total loss of ships and life at sea. This definitely demand higher priority and greater attention for upgrading safety standards, need for proper and regular training programmes for proper maintenance, operational fitness and equipment and system. Training is the best possible way of ensuring effective fire fighting and facing other sea hazards described in detail above. Maritime training institutes should provide classroom and simulator training which may include study in fire risk, fire prevention, pre-planning and liaison for operational command and control of fire on ships and docks. The training regiemen should include class room lectures with audio-visual aids, model rooms, practical drill grounds and concrete/metal ship models for practical make-up with all features common to most ships. Situations should be created by opening and closing number of valves, hatches, and doors to simulate conditions of damage, fire spread and floodings etc. Besides, ensuring safety and quick response, to attend to fire outbreak, or any other hazard, it is very important to make ship staff and shore personnel, aware of the fire hazard and prevention which is the chief danger for the ships.

MINES SAFETY

The conditions of mines in India today is intolerable and strict implementation of all safety laws by companies under coal India and other mine owners, operating under M.L. and P.L. are most essential. There is drastic fall in under ground production and serious accidents have gone up. The following are the types of accidents in mines operations.

1. Miners are forced to work on seven days in a week without rest and this has led to deterioration of safety and gross violation of mines Safety Act.
2. In many mines, there are no pit safety committees, which is mandatory under mines safety act. These committees are watch dogs in relation to safety of miners.
3. Only fatal mine accidents and about 1 % of serious accidents are investigated by officers of Director General of mines Safety and that too, the D.G.M. is under staffed.
4. In well type mines, miners get trapped due to collapse of overhead earth. No scaffolds or very weak ones are fixed in most of the lanes and by lanes which causes hazard.
5. The well type mines are not filled with sand to its surface, after operation and extraction of coal and minerals. They are left with makeshift protection, The rain water enters/per colates into those erst while mines. At the time of operation in the neighbouring area/mine, in the process of operation, the said flood water suddenly submerges the newly dugmines and entraps the miners, who die due to drowning/suffocation inside deep gorges before reaching to the hands of rescuers.
6. Land slides are a common sight in these mines and the miners are trapped for days together due to closure of path ways. They die due to lack of oxygen, food and drink before arrival of rescue parties.
7. Sometimes the miners die, inhaling poisonous gasses in the well type mines when they move deeper and deeper belts for exploitation.
8. The miners are some times, scantily attired with safety equipment and forced to work under hazardous circumstances.
9. The adverse work environment sometimes causes serious and minor accidents.
10. Sometimes due to communication failure, the miners fail to board the elevators in time, When the supervisor in the well signals for their return. At times, instances of fatigued miners resting in the tunnels are left, when others leave after working hours. The supervisors on the pit's mouth do not follow meticulously the safety rules in lifting/evacuating the number of miners he had sent underground in his shift.
11. In open cast mines, the biggest danger is the fire, which occurs at the drop of a hat, coal, specially is a self ignited material. Due to heat in summer days, the coal catches fire, inside and on the mouth of the mine, when dumped after extraction and fire havoc arises.
12. During rain, the open cast mines get flooded and after draining operation, the miners enter into the mines. The working place, becomes wet, slippery, the dumpers/carriers while carrying the load, face accidents on their return journey to the surface mouth. This results injury and death of drivers/operators.
13. The land slide, power failure, communication failure, failure of P.A. systems, inhalation of toxic and poisonous gases, electrical hazards, mechanical and machine failures, absence of protective barriers, barricades, scaffolding supports also contributes largely for creation of hazards in the mines.

14. Insecure blasting operation, non clearance of tunnel passages, common to both the above type of operations cause safety hazard.
15. Another most important health hazard that prevails in and around the mines is the air pollution. The toxic gases emanated from the coal/mineral beds when cut, affect the health and hygiene of the workers.
16. In respect of mines other than coal, the environment becomes intolerable, when operation starts. The garbage/trash, collected on the surface/mouth of the mines, some times catches fire along with self ignited jungle fire or human negligence fire and create fire hazard. Poor garbage disposal, denudation of forest and vegetation growth accelerates the environmental pollution, which ultimately affect the miners/workers in respect of their health.
17. The M.L./P.L. (mining lease/prospective licence) holders hardly equip the workers with P.P.E.S. They are forced to work under hazardous conditions, compelled to work for more hours, contrary to Mines Safety Act.
18. The miners/workers are paid less than the approved wage and free medical service, treatments is given to them when afflicted by serious injury/diseases.
19. The companies/mine owners/operators, generally operate in the remote jungles, mountains and uninhabited areas. They generally engage the local labourers on daily wage basis (both men and women) and largely exploit them. No records of their engagement are maintained. They do not extend the benefits prescribed under various statutes nor the daily wage workers are a ware of labour welfare provisions in the various labour laws.
20. The workers in this sector are unorganised except coal mine workers. Therefore, their voice does not reach the ears of administration. The company officials in collusion with the government officials like mine, labour, police, revenue and heavily bribing them, indiscriminately exploit these mute souls and reap rich profits from this business. Even if some trade union leaders flock on these places, they too become worst exploiters in collusion with govt, officials. Their purpose is to reap benefit from both the sides (workers and employers) and try to create vote bank of their own.
21. Another most glaring example of environmental hazard caused by the companies /mine owners is rampart cutting of forest growth beyond the designated area (licenced). They deposit the licence fee, lease amount for the area, they are allowed to operate (in acres including the cost of forest growth standing on the land. But the unscrupulous mine owners heavily bribe the revenue/mine/forest officials and encroach upon vast areas having succulent timber growth, beyond their designated area and lift the valuable minerals. Not only, the state becomes losser but the valuable forest growth is annihilated and create ecological imbalance.
22. Their action too evokes public resentment for their evacuation, air and water pollution, exploitation of national wealth without paying adequate compensation.

Hazard prevention and mitigation

1. In respect of well type mines, the entry point personnel like safety officer/supervisor should ensure that the miners have put on P.P.Es and carrying the required apparatus.
2. The well sufficiently illuminated,
3. The communication system be fool proof. The persons/workers posted at watch points at different depths are alert and regularly exchanging with the supervisors at different levels and on the mouth of mine at the surface level, uninterrupted power supply must be ensured.

4. The blasting operation below the mine must be as per approved procedure. Before evacuation at the end of the day, dynamites/explosives be fixed and be fused by remote control system from the surface.
5. To protect from toxic gases and fumes, masks, breathing apparatus be used.
6. Work should be started after inspection in detail by the supervisor/foreman of the explosion site. He should be satisfied that no living dynamite of explosive is there on the site of explosion.
7. The tunnels be properly lighted and the rails are all right, and the coal carriage cart/wheel barrows could smoothly run to the elevator point.
8. The loading operation should be fool proof.
9. Emptied tunnels be supported by strong scaffolding systems preferably steel, scaffoldings be used.
10. Waters be drained out as early as possible. The exhaust system of the machineries used inside the mine, should be properly sealed. All the machines, running by power, oil etc. be operated from the surface. The electrical installation inside the well should also be fool proof.
11. In case of water leaking from the tunnel walls it should be reported to the management for its stoppage. It may be the water accumulated in the previously used mine, has not been filled with sand and the water is percolating through the wall of the operating new mine. Example, Chasnala coal mine tragedy in Dhanbad district of Bihar, years back.
12. The environmental safety must be ensured at the place of work.
13. In respect of open cast mines, the garbage /trash disposal must be ensured in order to prevent fire havoc.
14. The coal dumping yards, should be well protected and water sprayed at intervals to prevent self ignition of fire due to excessive heat, during summer days.
15. Draining of water, putting cofferdams etc., be undertaken to prevent water logging and continuous of smooth extracting operation.
16. The dumpers used for carriage of coal/mineral to the surface should move carefully in the zigzag path ways as per traffic rules and regulations. Their exhaust system should be fool proof so as to stop sparking.
17. The miners/workers must put on dust preventing mask and other required P.P.E.S. In the tunnels they should use gas masks, SCBA etc., to escape from toxic gases.
18. The same usual blasting operation, cutting, digging, chipping, loading, transporting of coal/minerals must be as per approved norms, prescribed in the Mines Safety Act.
19. To stop/prevent pollution hazards, water tenders be commissioned to spray water at intervals, the nearby water sources/ponds/tanks etc. are to be regularly disinfected. Alternative green belt be created by afforestation in and around the mine site.
20. Corporate social responsibilities like provision of drinking water supply, building primary health centers. Health check up camps, educational establishments, road communications, agriculture production, putting irrigation canals be provided in the locality, by the companies operating in the belt
21. Exploitation of workers including contract labourers/casual be stopped. Welfare measures as per labour laws be taken up with earnest zeal and enthusiasm.
22. The activities of unscrupulous companies/mine owners in wanton destruction of forest growth, theft of mineral from unlicensed areas should be stopped and the offenders need be brought to book by Govt.
23. The first aid system on the site of mining operation should be introduced and treatment must be available to injured workers.

6

Occupational Health Hazards (Industrywise)

HEALTH

The 14th World Congress on occupational safety and health which was held in Madrid, Spain from 22nd to 26th April, 1996 reaffirmed the importance of protecting working people against injury and diseases at work. It also reaffirmed that humanization of work is a basic requirement for the achievement of social justice.

According to a report from the International Social Security Association (ISSA), in every three minutes, some where in the world, a person dies in an occupational accident or from an occupational disease. Each second, at least four persons are injured in the work place. Out of every 100 fat accidents, 7 are occupational accidents, 37 are traffic accidents and 56 occur outside working hours.

All the nations in the world have expressed serious concern for occupational safety and health. In this connection, every one has felt the need for health protection at the work place. To reap the economic benefits of new technology and foreign investment, countries must insist that occupational safety and health is an added value within a new concept of productivity, which would include the quality of production, its social benefits as well as its impact on workers' safety and health, the environment and the quality of life. There is also a need to strengthen the occupational safety and health capacity to handle technology transfer issues, while simultaneously ensuring safety and health in existing industrial sectors and enterprises. It is admitted that small and medium enterprises still encounter accidents in large numbers. As such, all are urged to pay greater attention to health measures for SMES with emphasis on promoting awareness, education and training. Chemical hazard also play a significant role and its hazards under the title "living safety with chemical risks." We should have a policy of fund management of hazardous chemicals through acquisition of knowledge and capacities to manage chemical risks.

The world health organisation set a goal "occupational health for all' which will provide a tool to create awareness and political commitment at all levels by all countries.

Safety and health at work places will remain a major component in sustainable development programmes. Comprehensive and coherent national policies on occupational safety and health need to be formulated and effectively implemented to contribute to sustainable development.

The social, ethical and health values, that have been translated into rights and obligations as I.L.O. standards, have to be implemented involving government, employers and employees.

In the globalisation of our economy, we must insist that human dignity and well-being are sacred values not to be sacrificed in the name of economic efficiency. The development of a safe health culture in corporating policies of responsible care, total quality management and continuous quality improvement will certainly go a long way to bring valuable benefits for the future, to all working people.

While discussing "Health" issues, it may be borne in mind that all Safety Measures undertaken are meant for Safety of Health. A worker should work under safe environment and avoid health hazards, that are likely to happen. This is the prime motto. "Health" is the focal point, round which everything revolves. Hence, both safety and health are co-related and move hand in hand. In the foregoing chapters "Health" aspects have been very elaborately discussed along with the Industrial hazards. Industrial accidents, health and environmental hazards have been enumerated in detail and remedial measures already suggested. The occupational diseases, that affect the workers at their work places have also been covered. Because, we cannot separate these two, while discussing the hazards and safe work procedure.

However, a brief list of occupational diseases in respect of each and every occupation is given below for better appreciation and information of the readers. Again, I have placed First Aid under "Health" which is an inalienable part of "Industrial Safety and Health."

Occupational diseases can be curbed/controlled by adopting following methods.

(*a*) Preventive and control method.

(*b*) Determination of necessary facts.

(*c*) Corrective action based on facts.

There is also necessity of co-operation between, physicians, indsutrial hygienists and engineers, when occupational disease problems arise.

For example: Health hazard in stone crushing units, where workers are prone to affliction of diseases like asthma and silicosis etc. A list is given below.

Industry	**Diseases**
1. Chemical industry	(*a*) Poisoning
	(*b*) Dermatitis
	(*c*) Respiratory Problems
	(*d*) Itching of eyes
	(*e*) Suffocation
	(*f*) Un-consciousness
	(*g*) Headache, Severe headache, Nausea
	(*h*) Ulcer in the nose, mouth and irritation of respiratory throat.
	(*i*) Damage of kidney and lever
	(*j*) Nose and gum bleeding. Corrosion of nose and throat. Fluid accumulation in the lung.
	(*k*) Severe skin burn, Burns in mouth

	(*l*)	Vomit, diarrhoea, stomach burn
	(*m*)	Tooth erosion, bronchitis, convulsing.
2. Caustic making, Dermatitis, Asthma, T.B, Cement Industry, Lime burning		
3. Insecticide Industry		Skin diseases, irritation of eyes, ulcer
4. Solvent Industry		Poisoning and dermatitis
5. Rayon, Rubber Industry		Acute chromic poisoning
6. Garages, Gas manufacturing units		Carbon monoxide poisoning
7. Electroplating, Paint Industries		Chrome sores, ulcer
8. Plastic		Skin irritation, liver disorder
9. Granite Quarry		Silicosis
10. Petroleum (Inhalation) (Toxicity hazard)	(*a*)	Dulls sense of smell, dizzy, headache, irritation of eyes. Blurring of speech, staggering, confusion, loss of consciousness. Arrest of breathing. Arrest of heart. Death.
11. Inert Gas		Asphyxiation, tiredness in the face, confusion, unconsciousness and quick death, if not treated immediately
12. Construction Industry		Body injuries of all types, breaking a bones. Damage to the organs. Disablement, death due to fall from height, Head injuries etc. Electrocution, death, electric burn, shock.
13. Hospital Industry	(*a*)	Diseases due to radiation'
	(*b*)	Drowning in the defective swimming pools
	(*c*)	Head injury in diving.

Developing an on-the-job training programme. A sample check list is attached.

Sample

Checklist for developing an

On-the-job training program

1. **Prioritize jobs needing on-the-job training:**
   - list all jobs for each worksite
   - identify most hazardous jobs
   - identify most important jobs
   - finalize list of jobs needing on-the-job training.
2. **Develop standards for each job:**
   - determine regulatory requirements
   - check manufacturers' standards/industry practices
   - set timeframe to perform job under normal conditions
   - List special condition.

3. **Develop procedures:**
    - include job standards and critical tasks
    - List tools required
    - List special abilities required
    - List job steps
    - identify hazard points
    - choose critical checkpoints
    - schedule testing of critical checkpoints.
4. **Demonstrate job to worker:**
    - Layout procedure
    - Layout tools
    - explain procedure
    - explain hazards
    - explain as you perform work
    - respond to questions
    - test worker
    - prepare report.
5. **Observe worker doing the job for the first time:**
    - review procedure
    - review tools
    - review critical tasks
    - review hazards
    - observe and coach
    - discuss observations with worker.
6. **Coach and train as needed:**
    - test progress with questions
    - perform critical-point checks
    - conduct review.
7. **Observe worker doing the job independently:**
    - allow minor errors
    - stop only if critical
    - review, reinforce, evaluate correct minor errors
    - question understanding of procedures.
8. **Set schedule for review:**
    - follow-up commitment
    - spot check
    - set performance objectives.
9. **Continue communication:**
    - communicate effectively.

Purpose for Which Health and Safety Traning is Required

Training Requirement	Executive	Line and Staff Managers	Foreman and Supervisors	Employee	Safety Committee Members	Health and Safety Representatives
Knowledge of company Occupational Health and Safety Policy and objectives of 5 Star Program.	Develop// review policy and objectives.	Interpret policy for the development of operational objectives Reporting accidents. Briefing reviewing performance.	Create understanding amongst own work force.	Understand objectives and policy and how they affect him.	Understand objectives and policy and how they affect him.	Understand objectives and policy and how they affect him.
Knowledge of principles and philosophy of accident prevention.	To design and develop a policy and program objectives as keep it up to date.	Development of operation objectives to be consistent with the Occupational Health Safety Program.	Implement Occupational Health and Safety Programme to understand and adopt safe practices and procedures for accident.	Understand: Why accidents happen. - need for program - what role they play in that program.	Understand: Why accidents happen. - need for program - what role they play in that program.	Understand: Why accidents happen. - need for program - what role they play in that program.
Appreciation of relationship between accident prevention and business operations.	Allocation of resources.	Development of operation objectives. Design standards of plant, etc.	To understand the need for safe and efficient operation.	Understand the need for safe and efficient operation.	Understand the need for safe and efficient operation.	Understand the need for safe and efficient operation.
Appreciation of legislative requirements for health and safety.	To design and implement policy.	Tender and contract. Develop operation objectives. Design standards of plant, etc.	Implementation of all legal requirements.	Understand their role and responsibilities and knowledge of regulations as they apply to the workforce.	Understand their role and responsibilities and knowledge of regulations as they apply to	Understand their role and responsibilities and knowledge of regulations.

Communications skills: - group basis - individual basis - oral/written skills.	Make policy known and gain effective implement-ation.	Provide information and liaison.	the workforce. Briefing of workforce regular contacts, discussions, etc.	Understand: what internal communication system exists.	Recommend to the employer and employees with respect to health and safety at work.	Recommend to the employer, employees and Trade Union (s). Recommend to the employer and employees improvement for health and safety of the workplace.
Knowledge of plant processes/products human factors and recognition of potential hazards in key result areas of own department/function.	Allocation of resources and agreement on objectives.	Development of operational objectives. Provide information as basis for policy, maintenance and monitoring of plant etc.	Supervision of subordinates. Inspection to identify hazards.	Ensure that his employer and supervisor are aware of any hazards of which he knows that may endanger himself or another employee.	Identify hazards and to make recommendations to the employer and employees for the improvement of health and safety of the workplace.	Inspections to identify hazards.
Knowledge of plant processes/products human factors and recognition of poten-tial hazards in key result areas of own department/function.	Allocation of resources and agreement on objectives.	Development of operational objectives. Provide information as basis for policy, maintenance and monitoring of plant etc.	Supervision of subordinates Inspection to identify hazards.	Ensure that his employer and supervisor are aware of any hazard of which he knows that may endanger himself or another employee.	Identify hazards and to make recommendation s to the employer and employees for the improvement of health and safety of the workplace.	Inspections to identify hazards.

Under the general Safety and Loss Control training element, we believe that the only positive way to change negative attitudes is through effective training and motivation. The following check-list for developing an on-the-job training needs analysis, should help develop effective training for all levels within the company.

To identify and apply the specific communication and motivation activities which will ensure that all employees, at all levels, have the required knowledge, skills and attitudes. There are a whole selection of simple but effective motivational activities which can be used to achieve this basic goal, the majority evolve around the quality management concept or the winners vs losers. However, my favorites and most frequently used for management/ slide is as follows:

ARE YOU A WINNER, OR A LOSER?

- A SAFETY WINNER says — Let's find out.
  A LOSER says — Nobody knows the answer.
- A SAFETY WINNER says — There ought to be a better way to do it.
  A LOSER says — That's the way it's always been done.
- A SAFETY WINNER goes through a problem.
  A LOSER goes around it and never gets past it.
- A SAFETY WINNER makes commitments.
  A LOSER makes promises.
- A SAFETY WINNER thinks, I'm good, but not as good as I could be.
  A LOSER thinks, I'm not as bad as a lot of other people.
- A SAFETY WINNER tries to learn from those who are above him.
  A LOSER tries to find fault with them.
- A SAFETY WINNER who makes a mistake says, I was wrong.
  A LOSER who makes a mistake says it wasn't my fault.

The 5 most important words are: YOU DID AN EXCELLENT JOB.

The 4 most important words are	: WHAT IS YOUR OPINION?
The 3 most important words are	: PLEASE HELP ME.
The two most important words are	: THANK YOU
The least important word is	: I

To systematically identify the most critical jobs/tasks that have been or could be involved in 80% of the behaviours related to major or catastrophic loss of people, property or process and develop effective procedures or practices guiding loss free performance.

The vast majority of Safety and Security programs are not successful because they concentrate on the physical aspects of loss prevention and thus identify only the unsafe or unsecured conditions, they overlook the unsafe practices of people which give rise to the greatest number of loss producing incidents by a ratio of 4 to 1.

A system therefore, which systematically identifies those jobs which in the past have given rise to injuries, or which have the highest potential for incidents and which establishes set procedures for the correct, safe and efficient performance of work can go a long way in eliminating many of the basic causes of injuries and losses in the workplace.

Purpose for which Health and Safety Training is Required

Training requirement	Executive	Line and Staff managers	Foreman and supervisors	Employee	Safety Committee members	Health and Safety representatives
Knowledge of rules, regulations, standards of operating procedures.	Allocation of resources and also the design of a policy and agreement on objectives.	Development of rules, regulations and standard operating procedures. Development of Disciplinary policy.	Ensure correct operation. Adoption of safe working practices. Conformance to disciplinary Policy	• Comply with Rules, Regulations and Standard Procedures. • Understand Disciplinary Policy	Advise and consult with respect to Rules, Regulations and Standard Operating Procedures.	Advise and consult with respect to Rules, Regulations and procedures. Monitor the adherence to the company discipline-any policy.
Analytical skills for job analysis, job safety including investigation techniques, job procedures.		Development of proper job procedure. Development of Accident investigation procedures.	Accident and incident investigation. Control and monitoring of performance.	Ensure that the works in a manner that is consistent with proper job procedures.	Accident Investigation. Advise and consult on the proper performance of work.	Accident/Incident investigation. Recommend to all parties improvement to the proper performance of work.
Selection/appraisal techniques, job instruction and coaching skills.	Developing of Policy.	Development of effective selection procedures with emphasis on matching the man to the job and appraisal systems. Briefing/training of subordinates.	Training of workforce Supervision and retraining. Department induction and training.			
Reporting procedures.		Reporting accidents.	Reporting accidents/incidents-deviations.	To report accidents/Incidents-deviations.	To report accidents/incidents-deviations.	To report accident/incidents deviations.
Interpretation of information.	For updating policy.	Use of information of preventive measures. Accident investigation and follow-up action.	Use of info, for preventive measures. Monitoring of performance. Plant inspecttions and follow-up action.	Understand information and what it means for his/her job Legal aspect of 'right to know'.	Proper utilization of information for preventive action.	Proper utilization of information for preventive action.

Human relations/ supervisiting requirements.	Conditions of employment.	Supervision/briefing of subordinates.	Attitude of workforce supervision and the development of effective shop-floor discipline.			Understand personal and motivational factors affecting people at work.
Emergency planning.	Development of Policy Allocation of resources.	Development of procedures for emergency planning. To include fire/security control as well as serious accident contingency plan.	Training to meet requirements of plan. Ensure trial runs of plan.	Understand requirement of plan.	Make recommendations for plan. Understand requirements of plan.	Make recommendations for plan. Understand requirements of plan.
First Aid	Development of Policy carry out First Aid Allocation of resources.	Establish procedures to Training. Take First Aid Training.	Arrange for First Aid. Take First Aid Training.	Take First Aid Training.	Take First Aid Training.	Take First Aid Training.
Personal Protective Equipment.	Development of Policy. Allocation of resources.	Establish procedure for use of personal protection equipment. Ensure that Health and Safety considerations are addressed at the design stage.	Ensure of personal protective equipment. Ensure of maintenance of personal protective equipment.	Understand need for use of personal protective equipment as equipment.	Understand need for - Use of P.P.E. where required. - Advise on the effective use of P.P.E.	Advise on the effective use of P.P.E.

These jobs must be identified, analyzed and set out in easy to understand format and made available to the supervisor so that safe job instruction and proper job observation can be performed.

The name of the procedures may vary from company to company but the essence of the safe work procedure should be that it provides a method for performing a job in the safest way possible. Due regard should be given to the production and environmental constraints exist at the workplace. It is important to note that these procedures are not job descriptions but are task specific.

Another critical aspect which I incorporate under this is the Safe Work Permit System and of course adequate lockout facilities. Both of which are designed to reduce risks by incorporating strict precautionary measures.

Tips for being healthy at work:

(*a*) Eat balanced meals every day, including breakfast.
(*b*) Avoid co-workers, who are sick (without being rude, if it is contagious in nature.)
(*c*) Drink at least 8 glasses of water a day.
(*d*) Take frequent breaks through out the day.
(*e*) Fully use vacation days.
(*f*) Quit smoking and drinking.

First Aid

First Aid is the immediate treatment given to a sick/accident victim before medical aid to avoid worsening of the condition of the patient.

Treatment of illness in work places requires some understanding of Anatomy and Physiology of human body. A preliminary training must be imparted to the personnel on functioning of the following systems.

(*a*) Circulatory system. (Heart and blood vessels)
(*b*) The respiratory system
(*c*) The digestive system
(*d*) The urinary system
(*e*) The nervous system
(*f*) The skin
(*g*) The skeletal system
(*h*) The muscular system.

We must start with common ailments that occur, in order to give First Aid before removing the patients to the nearest hospital for further treatment.

Fractures

A fracture is the partial or complete breakage of a bone.

Causes

1. *Direct force:* The bone breaks at the spot of application of force: *e.g.*,
   (*a*) Bullet passing into bones.
   (*b*) Severe fall on a projecting stone.
   (*c*) Wheel passing over the body etc.

2. *Indirect force:* The bone breaks away from the spot of application of force *e.g.,* Fall on outstretched hand.
3. *Force of muscle action:* Occurs when there is a violent contraction of group of muscles. This happens very rarely. *e.g.,* Fracture of ribs due to violent cough.

Types of Fractures

1. *Simple (closed) fractures:* The broken ends of the bone do not cut open the skin and show on the outside.
2. *Compound (open) fractures:* The broken end of the bone is shown outside, thus causing infection later,
3. *Complicated fracture:* It may be simple or compound causing injury to an important organ like spinal cord, brain, liver, spleen, major blood vessels and lungs.

Signs and symptoms

1. Pain at the spot of fracture or around it.
2. Tenderness *i.e.,* pain on gentle pressure over the injured part.
3. Swelling of the area and discolouration.
4. Loss of normal movements.
5. Deformity of the limb. The limb may loss its normal shape. Sometimes, the muscle will pull up the lower free ends, causing apparent shortening of the limb.
6. Irregularity of bone. If as in the leg bone, the break is under the skin, the irregular outline of the bone can be felt easily.
7. Crepitus (Grating): When one end of the bone moves against the other, a crackling is heard which is known as crepitus.
8. Unnatural movement at the spot of fracture can be felt.
   N.B. : The last two signs should never be tried by the First-Aider.

Management

The aims are

(*a*) to prevent further damage
(*b*) to reduce pain
(*c*) to get medical aid as soon as possible.

Fracture often occur in major accidents. Therefore, it is common to find other injuries also. There may be more than one fracture in the same patient or even in the same limb. And the first aider should act accordingly depending upon the condition of the patient.

1. If the casualty is bleeding heavily and wounded severely, that should be treated first.
2. If there is no immediate damage to life, temporary attention to fracture is enough.
3. Handle gently: avoid all unnecessary movements of the injured part. Simpler the first aid, the better.
4. Treat the fracture on the spot, so that the fractured ends stabilized and patient is ready for transport.
5. Treat for shock.

6. If the broken ends of the bone show out, do not wash the wound or apply antiseptic to the ends of the bone.
7. Never attempt to bring the bone to normal position or reduce the fracture.
8. Steady and support the injured part immediately, so that no movement is possible. This stops further injury and helps to stop bleeding. And also the broken ends of the bone damaging arteries, nerves and muscle is prevented.
9. Immobilise the fracture area and joints above and below fracture side by using bandages splints where available and where a recovery is confident of their use.

Infections

(*a*) *Using bandages:* Usually it is enough to use the other limb or the body of the patient as the splints. The upper limb can be supported by the body, the lower limb by the other limb (provided that also is not fractured). Most fractures (except forearm) can be immobilized there.

1. Do not apply bandage over the area of fracture.
2. The bandaging should be fairly firm so that there is no movement of the fractured ends. But not too tight in which case the circulation of blood to the area will be stopped.
3. Always place padding material between the ankles and knees and other hollows, if they have to be tied together, so that when limbs are bound together they are comfortable and steady.

(*b*) *Using splints:*

1. A Splint is a rigid piece of wood or plastic material or metal applied to a fractured limb, to support it and to prevent movement of the broken bone.
2. Reasonably wide splints are better than narrow ones.
3. They should be long enough so that the joints above and below the fractured bone can be made immobile.
4. Splints should be well padded with cotton or cloth so as to fit softly on the injured limb.
5. Splints are best applied over the clothing.
6. In an emergency splints can be improvised with a walking stick, an umbrella, a piece of wood, a book or even firmly folded newspaper.
7. Use of splint; becomes obligatory only when *e.g.*, both legs or both thigh bones are broken.

WOUNDS

When any tissue of the body is cut or torn by injury, a wound is caused. There will be bleeding from the injured part and it also form an opening through which germs can get into the wound. Wounds vary in extent and depth, depending on how they are caused, but any wound is usually one of the following kinds:

1. **A cut (incised) wound:** This is a cut or clean split of the skin and flesh beneath. Usually caused by sharp instruments like a knife or a broken glass, but any blow with an instrument, may split skin, particularly where it is relatively thin and tight as on the scalp. Since blood vessels are clean cut wounds bleed very much.
2. **A torn (lacerated) wound:** This has raggedly torn edges, and a portion of skin may be missing. This is caused by a running machinery, a fall on the rough surfaces or a flying splinter of

metal. Because the blood vessels are torn and twisted they tend to close down on themselves so that bleeding is usually slight. Torn wounds particularly liable to become infected with germs.

3. **A bruised (contused) wound:** This is a wound with very much bruising of surrounding parts. It is caused by a direct blow with something blunt, by sudden pressure or by steady crushing. There is usually very little bleeding.

Health may he slow on account of damage to the edges of the wound, and the immediate, surrounding tissues. There is often serious damage to the deeper parts beneath *e.g.*, there may be a fracture when it is on the head, chest or belly, should also suspect and examine for injury of the brain or internal organ.

A stab (Punctured wound): The opening in the skin is small, but the wound may be very deep. It is caused by the penetration of some pointed instrument; such as occurs knife stab or a fall on a spike. Bleeding from the skin is usually slight, but if a large blood vessel is damaged, deeper inside, loss of blood may be hard to control. Particular dangers of such wounds are that there may be serious damage to internal organs beneath and foreign materials and germs may have been introduced to the deeper tissues.

Dangers of wounds: Wounds cause two great dangers: (1) Bleeding and (2) Infection. Bleeding is the immediate danger and should be treated promptly. It also is a common cause of death in accidents. It is caused by rupture of the blood vessels due to severity of the injury. It may vary in from severe to slight.

Severe bleeding comes from a torn artery or a torn vein or both combined. Many large arteries and veins lie close together and are frequently injured together.

1. Blood from an artery in the systemic circulation is bright red. If the injured artery is near the skin, the blood is seen to spurt out in corresponding with the pulsation of the heart.
2. Blood from a vein is dark red. It flows in a brisk continuous stream.
3. Arterial and venous bleeding combined usually gushes out from the depth of the wound.
4. Slight bleeding comes usually from injured capillaries and may flow in a continuous stream or merely ooze from all part of the wound.

Management

Place the casualty in suitable position bearing in mind that escapes with less force when the patient sits and still less when the patient lies down.

1. Elevate the bleeding part, except in cases of fractured limb.
2. Do not disturb any blood clot already formed.
3. Remove any foreign body which is visible and can be easily picked out or wiped off with a piece of clean dressing. This should not open up the wound again which will cause more bleeding.
4. Apply a dressing, pad and bandage.

EXTERNAL AND INTERNAL BLEEDING

If the bleeding is from the surface of the body, it is called external bleeding. If it is within the chest, skull or abdomen etc. it is called internal bleeding. This cannot be seen immediately. But later blood may ooze out through the nose or ear, or coughed up from the lungs, or vomited from the stomach.

Signs and Symptoms of Bleeding

1. The causality feels faint and may even collapse.
2. Skin becomes pale, cold and clammy.
3. Pulse rapid but very weak.
4. Breathing becomes shallow, causality gasps for breath and sighs deeply.
5. There is profuse sweating.
6. The causality feels thirsty.

In Case of Severe External Bleeding

1. Bring the side of the wound and press firmly.
2. Place the causality in comfortable position and raise the injured part (if no bone Fracture is suspected).
3. Press on the pressure point firmly for 10–15 minutes.
4. Apply a clean pad larger than the wound and press it firmly with the palm until bleeding becomes less and less and finally stops.
5. If bleeding continues, do not take off the original dressing, but add more pads.
6. Finally, bandage firmly but not too tightly.
7. Treat for shock.
8. Get the causality to hospital as soon as possible.

In Case of Internal Bleeding

The aim is to prevent the condition from getting worse.

1. Lay the causality down with head low. Raise his legs by use of pillows etc.
2. Reassure him and keep him calm and relaxed. Do not allow him to move.
3. Keep up his body heat with warm clothing, blankets etc.
4. Do not give anything to eat or drink as he may have to be given an anesthesia later.
5. Do not apply hot water bottle or cold bags to chest or abdomen. This might make things worse.
6. Take him to a hospital as soon as possible; transport gently.

Pressure Points to Stop Bleeding

This is adopted when direct pressure becomes a failure. Pressure points is an area where an artery along its course can be pressed against an underlying bone, so as to prevent the flow of blood beyond the point. Generally, pulsation can be felt at such points.

Femoral pressure point which is felt in the groinfold and runs a little inwards upto the upper two thirds of the thigh. The thumb is placed at about the centre of the groin. Press the left thumb over the right and apply pressure directly backwards against the pelvic bone.

Bracheal pressure point: Apply pressure on the middle third of the arm by passing the finger under the area. It is compressed against humerus.

Infection

A wound is not initially infected even though it may be contaminated by the dirt and infected material which contain germs. Infection occurs after a lapse of time when germs have time to multiply and invade the tissue. Pus formation as part of the body's method to fight the infection.

Prevention of infection is, promptly attending to wounds:

1. Wash hands thoroughly with soap and water.
2. Clean the wound with plenty of good clean water. Wipe gently the surrounding skin and remove dirt sticking to wound. Use plenty of running water.
3. Do not use antiseptic in water to clean the wound.
4. Pick away the foreign material from the wound.
5. Dry the surrounding of wound gently with dry sterile gauze or freshly laundered kerchief
6. Cover the wound with dry sterile gauze if available or a clean soft handkerchief. Do not let the cotton come in contact with the wound.
7. Bandage.
8. Don't apply any antiseptic in large wounds. This may get absorbed and cause reactions. Infection is caused by germs getting into the body through the broken skin. The germs multiply in the wound and make it infected or septic. They may get into the blood stream and cause blood poisoning. The aim is to prevent number of germs that get into the wound.

Management

1. Stop bleeding
2. Handle the injured part as gently as possible.
3. Wash hands thoroughly or clean them with an antiseptic lotion if available.
4. Remove foreign objects if it can be removed easily.
5. Place clean dressing over the wound and bandages firmly.
6. Get a ready Medical Advice.

Dressings

A dressing is a protective covering applied to a wound to prevent infection:

- absorb discharge
- control bleeding

Avoid further injury. An effective dressing should be sterile and have a high degree of porocity to allow for oozing and sweating.

1. **Adhesive dressings:** These sterile dressings are of different kinds and consist of a pad of absorbent gauze of cellulose, held in place by a layer of adhesive material. The surrounding areas of the skin must be dry before the application.
2. **Non-adhesive dressings:** Readymade sterile dressing: The dressing consists of layers of gauze covered by a pad of cotton wool with an attached roller bandage to hold it in position.
3. **Gauze dressing:** Gauze in layers is commonly used as a dressing for large wounds, as it is very absorbent, soft and pliable. This may assist the clotting of blood. The dressing should be covered by one or more layers of cotton.

4. **Improvised dressing:** They can be of any soft absorbent, clean material such as a clean handkerchief, a piece of linen, a paper kerchief or cellulose tissue. They should be covered and retained in position by such materials as are available.

Great care must be taken in handling and appling dressings:

(*a*) Wash hand

(*b*) Avoid touching any part of the wound with fingers or any other part of the dressing which will be in contact with the wound.

(*c*) Do not talk or cough over the wound or the dressing.

(*d*) Dressing must be covered with adequate pads of cotton, extending well beyond them and retained in position by a bandage or strapping.

(*e*) If a dressing adheres to wound, do not try to remove it. Cover it with sterile dressing after cutting away whatever can be removed.

BURNS AND SCALDS

1. Dry heat *e.g.*, Fire, flame, a piece of hot metal, the Sun, contact with wire carrying high tension electric cur rent or by lighting or friction.
2. Chemical burns are caused by strong acids like Sulphuric acid and Nitric acid (or) strong alkalies like caustic soda, caustic potash, quick lime (or) strong Ammonia.
3. Nuclear burn is caused by the instantaneous flash of intense heat given off by a nuclear explosion. It is capable of causing superficial burns on the exposed area of persons several miles away.

Degree of burns: The degree of burns indicate the degree of damage to the tissues. There are three degrees of burn:

First degree burns produce pain and redress. They are not serious unless a large area of the body are involved.

Second degree burns are those associated with blister formation in which superficial layers of skin are destroyed, but the deeper layers escape injury. Patient with large area of second degree burns require hospitalization.

Third degree burns imply destruction of the full thickness of the skin and often of the underlying fat, muscles, and even bone.

The danger from the burns depends on the area of the burns rather than the degree. Any of the burn of over 30% irrespective of deep degree should be hospitalized as priority. The first two degrees are seen in scalds which are caused by moist heat due to boiling water, steam, oil, hotter etc.

The estimation or percentage of burns: Front-18 %, Back -18%, Head and Neck - 9%, Each arm-9%, Each Leg-18% and Perineum – 1%.

Results of Burns

Immediate	Later
1. Intense pain	3. There may be infection in the damaged area.
2. Shock	4. After healing, it will leave scars causing disfigurement and restricted.

Management of Minor Burns and Scalds

1. Clean the area gently with clean water.
2. Submerge the burnt area in cold water.
3. Cover with dry dressing.
4. Do not apply cotton wool direct to the burnt.
5. Do not apply any greasy substance.
6. Luke warm drinks.

Management of Serious Burns and Scalds

1. Keep the casualty quiet and assure him.
2. Wrap him up in clean cloth.
3. Do not remove adhering particles of charring clothing.
4. Cover burnt area with sterile or clean dressing and bandage. In case of burns covering a large part of the body, it is sufficient to cover the area with a clean sheet or towel.
5. Keep him warm but do not overheat him.
6. If the hands are involved, keep them above the level of the victim's heart.
7. Keep burnt feet or legs elevated.
8. If the victim's face is burnt, sit up or prop him up and keep him under continuous observation for breathing difficulty. If respiratory problems develop, an open airway must be maintained.
9. Do not immerse the extensive burnt area or apply ice-water over it because cold may intensify the shock-reaction. However, a cold pack may be applied to face or to the hands or feet.
10. Do not open blisters.
11. Shift him to the nearest hospital, if he is fit to be moved.

Management of Chemical Burns

1. Wash off the chemical with the large quantity by using a shower or hose if available as quickly as possible. This flooding with water will wash away most of the irritant.
2. Cut out contaminated clothing.
3. Do not touch burnt area with base fingers.
4. Treat as for burns.

Acid Burns

1. First aid for acid burns of the eye should be given as quickly as possible by thoroughly washing the face, eyelids and the eye for at least five minutes.
2. If the causality is lying down, turn his head to the side; hold the eyelids open and pour water from inner corner of the eye outward. Make sure the chemical does not wash into the other eye. Cover the eye dry clean, protection dressing and bandage (do not use cotton).
3. Caution the victim against rubbing his eye.
4. Get medical help immediately.

Alkali burns

Can be caused by drain cleaner or other cleaning solution. An eye that first appears to have only a slight surface injury, may develop deep inflammation and tissue destruction, and sight be lost.

5. Flood the eye for 15 minutes with water.
6. If the causality is lying down, turn his head to the side. Hold his lids open and pour water from the inner corner outward. Remove dry loose particles of dry chemicals floating on the "eye by lifting them off gently with a sterile gauze or a clean handkerchief.
7. Do not irrigate with soda solution.
8. Mobilize the eye by covering with dry pad or protective clothing. Seek immediate medical aid.

Although the local effects of a burn are most evident, the systemic effects pose a greater threat to life. Therefore, it is important to remember the ABCs of all trauma care during the early periods, Airway, Breathing Circulation.

Airway

Immediate therapy is directed towards establishing an airway, possibly through Orpharyngeal suctioning following by the administration of 100 % Oxygen. Such a high concentration of oxygen may not be available under emergency conditions; however, Oxygen by wash or nasal prongs is given initially.

When orpharyngeal edema is present, it may be necessary to incubate the patient. Next to handling respiratory, difficulties, most urgent need is paid and anxiety related to nerve injury and fluid replacement.

Fluid replacement	:	The following formulas are a guide to replace fluid:
On the first day	:	Wt. of pt. × percentage of burns × 2 mL.
Second day	:	Wt. of pt. × percentage of burns × 1.5 + 500 + Urine output.
E.g., First day	:	70 kg. patient with 50 % body surface area burn :
		70 × 50 × 2 = 700 mL/2rhrs.
Second Day	:	70 × 50 × 1.5 = Urine output.

Remember: Formulas are a guide. Patient response is primary determinant of actual fluid therapy and must be assessed at least hourly.

The Patient should be for:

1. The presence of hematuria (Bloody urine)
2. Urine output below 30 mL./hr. which suggests inadequate rate of fluid restriction.
3. Urine output above 100 mL./hr. which may presage pulmonary oedema or imminent water intoxication — suggested by following signs: Tremor, stitching nausea, diarrhoea, salivation and disorientation.
4. Blood pressure below 90/60 mm of Hg. If all extremities are burned, a strike dressing applied under the blood pressure cuff will protect the wound from contaminator.

Pain Relief

Intravenous morphine is usually prescribed to relieve pain in burn patient. Subcutaneous or intra-muscular routes are dangerous because of impaired circulation for individuals with burns of the head neck, morphine, which can depress respiration should be given in smaller dosages intravenously. In the presence of shock and hypovolaemia (narcotics) sedatives may cause cardiac arrest. It is wise to give very small doses at short intervals *e.g.*, (30 – 60 minutes) in early stages.

Temperature Maintenance

Burn patients are prone to chilling, cotton blankets, ceiling mounted heat lamps or heat shields or aluminium coated 'space' blankets are useful in maintaining patients' body temperature. An efficient approach to dressing, removal and wound care shorten the time during which patients are exposed to the ambient temperature and reduces shivering and metabolic stress.

Infection

Localised infection must be identified and eliminated. While caring for patients well extensive burn, in order to prevent infection, a mask and sterile gloves are worn.

FOREIGN BODIES IN THE EYE AND SKIN

Foreign body under the skin may be, thorn, glass, iron pieces, needles etc.

Treatment

1. Unless very easy to deal with, should not interfere.
2. Dress the wound
3. Immobilise the part with splints and get medical aid.

Foreign body in the eye: Wings of insects, dust, coal, metal particles and loose eye-lashes are common object which get lodged under the eyelids. They cause pain and later-redness, if they are not removed at once. Sometimes, iron particles and wood splinters get lodged in the cornea causing serious trouble. Penetrating foreign bodies, however, are a danger to the eye itself.

Management

1. Avoid rubbing the eyes in case his hands at the back.
2. Seat the casualty so that light falls on the eye. Pull the lower lid down. If the foreign body is floating and not embedded, remove it with a narrow, moist swab. 'Corner of a kerchief twisted to a fine point will also do.
3. If foreign body is not visible, it may be under the upper eyelid. Ask the causality to keep clean water in the hand and blink quickly in the water. If unsuccessful, pull the upper lid forward, push the lower lid upwards and let go of both the lids. The lashes of lower lid usually dislodge the foreign body. Try this two or three times.
4. If the foreign body is embedded in the eye, particularly the cornea, don't touch it. Apply pad bandage the eye, ask the causality not to rub the eye; and take him immediately to a hospital.
5. Penetrating foreign bodies are easily made out by bleeding pain etc. It is for the doctors to handle. The fist aider should first put a pad and bandage.
6. When corrosion acid alkali or juices, from plants are suspected, blinking eyelids underwater a number of times or flushing with large quantity of water is the last thing to do. Then apply a soft pad, tie a bandage and take the casualty to the hospital at once.

SHOCK

Shock is a state of circulatory failure due to there being insufficient blood which cannot fill in all the blood vessels in our body thus leading to collapse.

Causes

1. Haemorrhage
2. Diarrhoea
3. Vomiting
4. Burns
5. Injection reactions
6. Burst internal organs-appendix Intestines
7. Heart attack
8. Bacterial infections causing toxaemia
9. Crush injuries

Signs and symptoms

1. The victim feels faint or giddy or he may be in stupor or unconscious.
2. The victim may complain of blurring of vision.
3. The skin is cold and moist
4. The victim will be sweating
5. The victim may complain of causes or may even vomit
6. The pulse is threay (a threay pulse is one where the rate is very much increased but the volume is low or weak)
7. The respiration is shallow (the depth of respiration is decreased and the rate is increased.)
8. The victim is thirsty and may ask for water.
9. The face and lips look pale or the victim may even be cyanosed (bluish discoloration of skin, tongue, lip, ear lobes)
10. Urine output is lowered.

Management for shock

1. Reassure the victim
2. Put him in lying down position on his back and turn him to a side.
3. Guard airway breathing and circulation
4. If the victim is unconscious, put him in a three quarter prone (or recovery) position.
5. Elevate the lower limbs or tilt the body so that the head is at a lower level.
6. Loosen tight clothing at neck, chest, abdomen and waist.
7. Wrap the victim in bed sheets or thin rug to maintain body's warmth.
8. Do not give anything by mouth if the victim is unconscious or seriously injured or if he needs to undergo anesthetic or surgical procedure or if he is nauseated or vomiting. Otherwise you can give warm sip of coffee or tea or water.
9. Shift him immediately to the hospital on a stretcher.

Diarrhoea Vomiting and Dehydration

Children and infants may suffer from vomiting and diarrhoea. At that time, the child may even be febrile (having fever) and refuse to accept feeds. All these factors tend to cause dehydration. This disease is known as acute gastro enteritis. This disease frequently affects the adults also and we often read in the newspaper as to how many people died of this disease. Acute gastro enteritis is caused by bacterial, viral or parasitic infections of the intestines. Dehydration is due to loss of water and electrolytes or salts from the body.

Signs and Symptoms

1. The child looks dull, irritable, cries continuously, is not playful and is apathetic.
2. Urine output is reduced. Both the frequency and the amount of urine passed each time are reduced.
3. The tongue is dry (or parched)
4. The skin loses its elasticity
5. The fontanalle is depressed. (This sign can be elicited only in infants, whose fontanalle have not yet united)
6. The child loses its body weight
7. The child's eye balls are sunken

Diarrhoeal disorders, are the commonest causes of death in children in developing countries like India.

The aim of first aid is to prevent or correct dehydration. In presence of nausea or vomiting, give frequent small sips of oral fluids in a teaspoon. Control fever. Foods that are tolerated better are water, curds, buttermilk, boiled rice kanji, tender coconut water and mashed well ripe banana of green variety. It is better to avoid all other foods unless advised by our doctor.

Take the child to your doctor at the earliest and administer medicines as advised by him.

Unconsciousness

An interruption in the normal activity of the brain causes unconsciousness.

Degree of unconsciousness

(*i*) *Drowsiness:* The victim is easily roused but only for few moments. He often answers questions about his condition. Put again he passes back into sleep like state

(*ii*) *Stupor:* In this degree of unconsciousness, the victim does not answer questions easily or does so incoherently.'

(*iii*) *Coma:* The victim cannot be roused. He does not respond to calls. His pupils may be widely dilated and may not react to light. He is motionless and even application of pain (by pricking with a sharp needle) does not elicit any response.

Causes for unconsciousness

1. Head injuries
2. Cerebro-vascular accidents or stroke or apoplexy
3. Fit
4. Fever and fits in children

5. Fainting
6. Heart attack
7. Poisoning
8. Bleeding and shock
9. Fracture spine

First aid in unconsciousness

1. Guard airy breathing and circulation.
2. Do not give anything by mouth
3. Do not induce vomiting
4. Check whether the victim is breathing alright, his heart is beating alright and he does not have any fracture spine. Then put the victim in recovery position or three quarter prone position or unconscious position. This particular position is life saving because in this position the tongue will not fall back and obstruct respiration. Also in this position any vomits and throat secretions come out without choking the victim.

   To put the patient in recovery position:

   (*a*) Empty the pockets of the victim and remove his wrist watch.

   (*b*) Loosen his clothing at neck, chest, abdomen and wrist

   (*c*) Cross his further foot (with reference to the first aider) over his nearer ankle.

   (*d*) Line his arms alongside the body but tuck in nearer hand and forearm well into his side.

   (*e*) First aider kneels further back to leave a space into which to turn him.

   (*f*) First aider's hand cradles the nearer side of his face to protect it as it meets the floor, his other hand grasps the victim's further hip and rolls him on to the first aider's knees.

   (*g*) If the victim is on the ground, put a thin clean cloth or hand kerchief under his head. But never use a pillow which would raise the head.
5. Treat for and prevent shock.
6. Don't be ashamed to attempt to go on talking to the unconscious victim. Nobody knows when the victim will respond and will be grateful for the first aider's reassuring words.
7. Shift the casualty on a stretcher to the hospital immediately.
8. Check whether the victim is a diabetic or a blood pressure patient or an epileptic or a victim with previous history of attacks of strokes or heart attack.

HEART ATTACK

Coronary artery is a branch of aorta arising from base of aorta and goes back and supplies to heart itself. When this, blood supply to the heart through tile coronary artery is diminished or completely obstructed, it causes heart attack.

Contributing Factors

1. Sex : More in male
2. Age : More in elderly people
3. Smoking
4. Sedentary habits
5. Overweight
6. High blood pressure

7. Diabetes (or sugar complaint)
8. Genetic predisposition
9. High cholesterol level in blood.
10. Alcohol consumption
11. Long term use of oral contraceptive pills in women.

Symptoms

1. History of previous attacks
2. Chest pain
3. Nausea, vomiting
4. Breathlessness
5. Fainting
6. Tiredness
7. Sweating

Signs

1. Shock
2. Paleness of face
3. Extensive sweating
4. Cyanosis (bluish discoloration of skin and mucous membrane due to increased amount of un-oxygenated blood circulating through them)
5. Breathlessness Complication to be avoided. Death due to heart stopping

First Aid

1. Absolute rest in a position most comfortable to the casualty. Preferably lying with back propped up on pillows.
2. Reassurance
3. Avoid panic and confusion let loose
4. Guard airway breathing circulation
5. Loosen the clothing around neck and thorax
6. Shift the casualty immediately to the hospitals on a stretcher in an ambulance
7. or, if available, call the doctor who can supervise the entire operation on the spot.

Strokes

Strokes of apoplexy or paralysis is the disturbed functioning of the brain due to diminished blood supply or nature of blood vessels supplying to brain.

Contributing factors

1. Sex-more common in men
2. Age-More common in elderly people
3. High blood pressure

Signs and symptoms

1. Disturbed level of consciousness
2. Paralysis of part of the body or whole
3. Incontinence of urine

4. Fits
5. Difficult or no speech
6. Shock

First Aid

1. Guard airway breathing circulation
2. Give the care to the unconscious if the casualty is unconscious
3. Treat shock
4. Shift to the hospital on a stretcher

Choking

Choking is the obstruction to breathing due to a foreign body in the air passages. For example, food particle, chewing gum, chocolate, false tooth, toys, etc. Choking is more common in children,

Signs and Symptoms

The victim is in distress, coughing violently. The face and neck are congested. He keeps his hands on his throat (Heimlich's sign) even if he is unconscious. He will make violent efforts to breathe. The victim may fall forwards and become unconscious.

First Aid (Heimlich's manoeuvre)

In the upright position: Stand behind the victim. Keep your right clenched fist midway between the victim's navel and the lower end of the breast bone. Hold your clenched right fist with the other hand. Give a powerful and abrupt thrust with your right fist in an upward and backward direction. Repeat the procedure till the offending foreign body is out.

The same manoeuvre can he carried on a victim who has collapsed in a sitting position. The first aider in this instance, positions himself behind the chair in which the victim has collapsed.

If the victim has fallen on the floor: Heimlich's manoeuvre is most successful when carried out with victim standing up. However, if the victim is unconscious and fallen down and is too heavy to hold, we have to modify the manoeuvre.

Kneel aside the patient's side and keep the heel of the right hand on the apigustrium. Keep the heel of the left hand over the heel of the right hand. Deliver a series of abrupt powerful thrusts upwards and backwards till the offending foreign body is expelled.

In the absence of first aider, the victim can render self aid on himself. The victim must clench a fist and place it's thumb side, against the abdomen slightly above the navel. With the other hand, jerk it inwards several times.

Alternatively, the victim can render self aid by using the back of a chair or table edge. The victim must lean over the edge supported by his hands on either side. He must thrust inwards and upwards just above the navel till the offending foreign body is out.

POISONING

Poisoning is usually due to substances taken by Inhalation or by mouth or by Injection.

Types of Poisons

Solids	Sleeping tablets, Opium, Ganja, Datura, Ergot, Lead, Cyanides, Mercury, Metal Salts, Acids, Alkalies etc.
Liquids	Alcohols, Sprit, Insecticides, Kerosene, Acids etc.
Gases	Carbon monoxide, Ammonia, Sulphur dioxide, Methane, Chlorine etc.

Principles in Treating Poisoning

1. If the patient is conscious, determine the type of poison.
2. (*a*) If the patient is unconscious and breathes, put the patient in the recovery poision.

   (*b*) If the patient is not breathing, begin resuscitative measures.

Poisoning

The Patient is Conscious	The Patient is	Unconscious
↓	↓	↓
Determine the type of poison	And Breathes	Do not Breathe
↓	↓	↓
At this stage, the treatment begins	Put in Recovery Position	Begin Resuscitation
	Do not give anything by mouth	

Important

Do not induce vomiting in an unconscious patient—the danger is, there may be chances of aspiration into the lungs. Ascertain the type of poisoning and as per that proceed to resuscitate the victim.

Alcohol Poisoning

Alcohol in any form may cause symptoms of poisoning. It is usually due to Ethyl Alcohol, less frequently to Methyl Alcohol. Predisposing causes are:

1. Occupation:	Public housekeepers, barman and commercial travellers.
2. Hereditary:	A neuropathic disposition
3. Worry and Pain:	The habit may persist after alcohol has been ordered by a doctor during illness.

Acute Alcohol Poisoning

Alcohol is rapidly absorbed in the stomach and intestines and passes to the blood and cerebrospinal fluid. Post-mortem studies have shown hyperaemia (*i.e.*, congestion and reddening due to bleeding) of the gastric mucous membrane and the brain. A person is usually considered to be drunk, when, as a result of taking alcohol, he is not able to perform with his normal skill or facilitate the ordinary action of his life. The higher centers are first inhibited (*i.e.*, functions are diminished). The individual loses any sense of shyness and may become talkative or emotional. The onset of muscle fatigue is delayed. That is the person is little bit violent and aggressive. The finer acts of coordination are distorted and later there is unsteadiness of gait, inability to walk, and finally unconscious. In some cases, there is not exited stage, but marked depression from the onset.

The breath smells of alcohol, the face is usually flushed but may he pale, the skin is moist, the pupils are dilated and rapid and bounding pulse.

The speech may be blurred or over precise, and there may be difficulty in pronouncing certain words or in counting backwards. Coordination may be impaired as shown by the finger -nose test and inability to walk steadily along a line.

When the patient becomes unconscious, the pupils are constricted, the respiration is slow, the pulse is feeble, the body temperature is subnormal, the limbs are flacid, the deep reflexes dimined and the plantar response is flexor.

Drunkenness must be distinguished from excitement due to other causes, coma due to other causes such as hypoglycemia after a dose of Insulin.

Complications

In the majority of cases, the patient recovers after a night's rest but coma and pulmonary edema (which causes breathing difficulty) always causes great anxiety. Pneumonia may occur when the patient is exposed to cold, of which he is unawake.

Prognosis

This always good, but death may occur in coma from respiratory or cardiac failure.

Management

In the majority of the cases, no treatment is required excepting a night's rest. In severe cases, the patient should be given plenty of warm or cold fluids preferably water to dilute the blood concentration of alcohol. As the alcohol is absorbed very rapidly in the stomach and intestines, there is not much use in giving gastric ravage or administration of emetics. As the body temperature goes below normal, the patient should be covered with warm clothing, so that the body temperature is brought up as well as ensures good circulation. A close observation of the patient is essential — *i.e.*, monitoring the pulse, blood pressure and respiration. If the patient is comatous, inhalation of 10% CO_2 and 90 % O_2 helps to wash out the alcohol. This treatment restores the patient to consciousness in half-an-hour.

Drugs Poisoning

Normally Drug Poisoning is with the following drugs:

1. Barbiturates
2. Aspirin
3. Morphine
4. Sleeping does *e.g.*, Diazepam Largaetil

Barbiturate's

In barbiturates, the patient will be in coma; pupils are usually dilated and sluggish reaction to light. In very severe cases, the pupils are constricted and reactionless. The reflexes are exaggerated. The urine output is much less. The temperature will be high and there will be lung complications. Mostly, patient dies of lung complications and coma.

Management: Prophylactic careful monitoring of the dose.

Curative

Since the patient will be comatose, gastric ravage should be avoided. Artificial respiration should be applied and O_2 with 7 % CO_2 if there are symptoms of respiratory failure. The airway should be maintained. If facilities are available, Hyperbasic O_2 therapy should be adopted. Alkaline-Diuretics is the treatment (*i.e.* I.V. fluids with $NaHCO_3$ and administration of diuretics will flush the poisoning). Electrophytes should be administered through intravenous route and antibiotics through I.V. Line to combat lung complications. Severe cases requires Haemodialysis.

Aspirin Poisoning

Poisoning is usually the result of an overdose with suicidal intention. The lethal does is between 50 G to 75 G. Toxic signs appear after 9 hours.

Symptoms

The patient is usually sweating, and there may be exophthalmos. The pulse and respiration rates are raised but the temperature will be subnormal. The patient will vomit blood stained fluid. The pupils are small, not reacting to light and may be fixed. The deep reflexes are depressed. The patient may be alternatively violent and drowsy and, complain of irritation of scalp. In more severe cases, patient sinks into coma and there is extreme air hunger. The urine shows the evidence of the typical positive purple ferric chloride colour indicating the presence of salicylates. There may be a trace of protein.

Prognosis

Depends upon the amount taken, the patient's susceptibility and the treatment given.

Curative

If the patient is restless, he has to be sedated with S.C. Morphine coma. Stomach wash with warm sodabicarb solution 5 % should be given till the return is clear. Potassium salts are given through mouth or by I.V. route. As salicicotes causes bleeding, vitamin K should be administered parenterally.

Morphine Poisoning

Morphine poisoning may result from an accident, an overdose being prescribed or dispersed or owing to crash, on the part of the patient. Morphine may also be used as a suicidal poising.

Clinical findings: The patient is comatose, with small inactive pupils, depressed reflexes, a cold and clammy skin, feeble pulse and weak respirations. The same features are found in poisoning with opium, omnopon, pantopon, pethidine, physepton, heroin etc.

Curative

Even though the above drugs are in injectable forms, the stomach wash should be administered as these drugs are execrated into the stomach. "Stomach wash in these cases are with warm water containing $KMnO_4$." To stimulate the respiratory centre, O_2 & 7% CO_2 combination should be given. The specific antidote for morphine poisoning is nolarphine. The dosage ranges from 10 mg–40 mg depending upon the severity of poisoning. Route of Administration slow I.V. or S.C. Injection.

Food Poisoning

PTOMAINS Poisoning was the name formally applied to poisoning by tainted food especially tinned meat and fish. It was thought to be due to protein decomposition products produced by the action of microorganisms. It is now believed that the majority of cases are due to infection with microorganisms themselves and their toxins.

Clinical Findings

The on set of symptoms are sudden with shivering, nausea, vomiting, and abdominal pain. If the offending food passes into the intestine, diarrhoea follows. The motions are at first loose and foul smelling. Later they become watery and contain mucus and blood.

Symptoms

The tongue is furred, the temperature often raised and erythematous rash may appear. In severe cases, the patient collapses and cramps occur in the legs. In some instances, the causative organism can be recovered from the stools during the first few days and a specific agglutination reaction is obtained with the patient's serum in the second week of illness. The urine may show acetone bodies if there are prolonged vomiting.

Complication: The illness passes off in a day or so in slight infections. In more severe cases, the symptoms increase in intensity for several days. A tendency to gastroenteritis may persist subsequently.

Prognosis: Recovery is complete.

Treatment

The patient should be in bed and kept warm. In early stages, the stomach should be emptied by warm water or warm normal saline solution wash. A small dose of (15 mL) castor oil should then be given. If diarrhoea is severe, this should be avoided and Bismuth carbonate should be given. Mild antibiotic will be used as a precautionary method—as the case is due to Bacterial toxins.

Mouth to Mouth of Artificial Respiration

A person may stop breathing during the following conditions:

1. Heart has stopped beating/Heart Attack
2. Choking
3. Unconscious person in whom the tongue falls back and obstructs the airway.
4. Drowning
5. Carbon dioxide and Carbonmonoxide poisoning in big fires.
6. Hanging
7. Epilepsy
8. Acute poisoning, snake bite
9. Crush injury to the chest
10. Head injury
11. Strokes

Important

Please note that the above medical emergencies does not always cause respiratory arrest. The First Aider must give artificial respiration only when the respiration is stopped because of these causes, For example, a victim may be having chest pain, sweating due to a heart attack. But he may still be breathing alright. A person might have had fits or fainted and is unconscious but he may be breathing alright.

In these instances, one must not attempt artificial respiration.

Diagnosis

1. The victim is usually unconscious or will soon become unconscious.
2. His heart may continue to beat up to 12 minutes even after the breathing has stopped.
3. The upward and downward movements of the abdomen and chest along with the respiration are absent.
4. Listen to the breath sounds over the mouth, nose, wind pipe and chest wall. There will be no breath sounds audible.
5. Feel for the breath and the warmth of the breath over your cheeks held closely in front of the mouth and nose of the victim.
6. There may be bluish discolouration of the skin and mucous membrane of the victim in whom breathing has stopped. This is easily noticeable on the tongue, lips and in the nail beds.

Procedure

1. Give first 5 breaths rapidly to replenish the patients blood with oxygen.
   (*a*) Remove the cause or remove the casualty from the cause, *e.g.*, drowning, hanging, fire accidents, building collapse etc.
2. Relieve the choking, if any
3. Clear out all secretions: foreign bodies, dentures, food particles and vomits from the airway.
4. Place the casualty on his back and tilt the head back as far as possible.
5. Press the angle of jaw forwards and upwards.
6. Loosen all clothing at neck, chest and waist
7. Maintain body warmth
8. Take a deep breath with your mouth open wide.
9. Pinch the nostrils of the casualty
10. Cover the mouth of casualty with your mouth smugly.
11. Watching the chest blow into the casualty's lungs until the chest bellows up, withdraw your mouth.

    Note: The chest falls back.
12. Repeat the procedure 15–20 times a minute
    (*a*) If you interrupt the procedure, give 5 breaths rapidly and continue again.
13. If the casualty is a baby or a child the first aiders mouth must cover both the mouth and the nose of the casualty and the blow must be gentle yet sufficient to make the chest below up.
14. If the chest does not below up, the victim is choking. Relieve the choking.

15. If the mouth to mouth technique cannot be applied for any reason, use mouth to nose technique. But now the casualty's mouth should be kept closed by the first aider's thumb.
16. Confirm the presence of carotid pulsation, throughout the procedure.
17. If the heart has stopped beating, you must provide artificial circulation by external heart compression.
18. Continue artificial respiration till the casualty starts breathing on his own or a doctor comes on to the scene and takes over from the first aider.
19. *Warning:* It is blunder to resort to artificial respiration on a victim just because he is unconscious or fainted or having fits; yet breathing alright. However, it is a graver mistake if artificial respiration is not provided, when respiration is suspended.
20. Forget all other types of artificial respiration. They are too complicated to comprehend, remember and to practice. The mouth to mouth type of artificial respiration is the best and very effective and easy to understand. External heart compression can be given easily when needed along with this type of artificial respiration.

EXTERNAL HEART COMPRESSION

Cause for heart attack to stop. Causes are similar to those listed under the respiratory arrest, lightning, electric shock, injection reaction, heart attack, etc.

Diagnosis

1. The victim is always unconscious.
2. Respiration is always suspended in a person in whom heart has stopped beating.
3. Carotid pulsations at the root of the neck are absent.
4. Heart beats are not heard when you keep your ear to the left of the breast bone of the victim.
5. The pupils are dilated
6. Face is blue or pale
7. The body is cold to touch.

First Aid

1. Give 3 firms humpy over the lower half of the breast bone and check whether the heart has started beating. If not....
2. Lie the patient on a hard surface on his back,
3. Elevate the legs vertically and keep them propped up in this way.
4. Kneel alongside the patient. Put the heel of one hand on the lower half of the breast bone and the heel to the second hand on the top of the first. Do not use the lowest part of the breast. By leaving the last inch clear, the risks of damaging the liver, the stomach and other organs is minimised. Keep your fingers and palm off the chest.
5. Press the breast bone down by rocking your body forward and keeping your arms straight.
6. Relax the pressure by rocking your body back a little. The breast bone can be pressed back 1 to 1 1/2 inches in an adult.
7. Repeat the procedure 60 times a minute.
8. In small children increase the rate to 80 a minute and reduce the pressure by using one hand only.

9. In babies, increase the rate to100 a minute. Reduce the pressure by using only two fingers and raise the point on which you push to the middle of the breast bone.
10. Important, always provide mouth to mouth type of artificial respiration when giving external heart compression.
11. Properly administered external heart compression must produce distinctly felt pulse at the root of the neck and the femoral arteries in the thigh with each compression.
12. Continue external heart compression till the colour returns to normal and the pupils are contracting and the carotid pulse is restored.
13. If the pulse is not restored, continue external heart compression till the casualty reaches the hospital or a doctor takes over.
14. At least, two first aiders are needed to do all these things. Number ONE makes five heart compressions and watches the pupils. Number TWO gives two lung inflations and feels the carotid pulse at the root of the neck.
15. Unfortunately, if there is only one first aider, he must finish 15 heart compressions, rush to the head side, give two inflation to the lungs and go on repeating the procedure till the victim is revived.
16. **Warning:** Make sure that the heart has stopped functioning before you start external heart compression.

Cardio-Pulmonary Resuscitation

Don't delay. Place the victim on his back on a hard surface.

Step-A: Airway

If patient is unconscious, open the airway; thereafter make sure it stays open.

- Lift up neck
- Push forehead back
- Clean out mouth with fingers

Step-B: Breathing

If patient is not breathing, begin artificial respiration, mouth-to-mouth.

- Before beginning artificial respiration, check the pulse in the neck arteries. It should be felt again after the first minute and checked every five minutes thereafter.
- Give four quick breaths and continue at the rate of 12 inflations/minute.
- Chest should rise and fall. If it does not, check to make sure the victim's head is tilted back as far as possible.
- If necessary, use fingers to clean the airway.

Step-C: Circulation

If the pulse is absent, begin heart compression. If possible, use two rescuers. Don't delay; one rescuer can do the job.

- Locate pressure point (cover half of sternum)
- Depress sternum 4–5 cm, 60–80 times/minute

- If one rescuer-15 compression and two quick inflations
- If two rescuers-5 compression and one inflation.

Pupils of eyes should be checked during heart compression. Constriction of a pupil on exposure to light shows that the brain is getting adequate blood and oxygen.

		Contents of first aid box		
Sl. No.	**Articles**	**Quantity**	**Usage**	**Remarks**
1.	Sterilised roller bandage 7.5 cm and 5.0 cm each	2 nos of each	To bind on the cut portion	For external use only
2.	Sterilised cotton wool	4 pkts	For washing	For external use only
3.	Dressing cloth	4 pkts	For dressing the wound	For external use only
4.	Eye pad	Required quantity	For external eye covering	For external use only
5.	Dettol/Savlon	1 bottle	For disinfection of the wounded area	For external use only
6.	Tincture Benzine	1 bottle	For use in the wound	For external use only
7.	Tincture Iodine	1 bottle	For use in the wound	For external use only
8.	Eye ointment	1 tube	For internal use of eye	For internal use
9.	Antiseptic cream /Neosporin ointment/powder	1 tube 1 bottle	For protecting against infection and healing	For external use only
10.	Adhesive tap	1 Roll	For bandaging	For external use only
11.	Rose water	1 bottle	For use in the side eye	Internal use
12.	Burnol cream	1 tube	For use on burning wound	For external use only
13.	Scissor	one	For cutting the bandages/ dressing cloth etc.	For external use only
14.	Silver sulfadiazine ointment	1 tube	Application on the cut portion	For external use only
15.	Moov	1 tube	To apply on the parts, affected by hurt	For external use only
16.	Tablets for common cold fever, headache, acidity etc.	Sufficient	To swallow	Internal use

7

Environment

INTRODUCTION

In this chapter, environment is divided in two parts *i.e.*, environment hazards and pollution at work places and global environmental change that leads to total destruction of human and plant life in the globe. I have stretched the issue little further and provided wide scope for discussion on the present burning problems of global warming, climate change, emission of Green house gases and imminent danger that is likely to fall on the habitants of Earth.

WORK ENVIRONMENT

The work environment should be conducive for the workers to work without any hazard. Industrial development certainly contributes towards economical growth. But industrial growth brings along with a host of environmental problems. Improper sitting of industries can seriously affect environmental features such as air, water, land, vegetation, forest growth and animal population. Proper care should be needed at the time of sitting rather than going in for measures at a later stage. In economics, the preliminary requirements are land, labour, capital and organisation. So land comes first. In selecting the land/site for an industry, the entrepreneur must look to the following matters.

(*a*) Availability of raw material near by
(*b*) Availability of water source
(*c*) Close proximity to the nearest rail/road/water transport system.
(*d*) Considering the hazardous nature of Industry, the entrepreneur is going to commission, the site should be quite a way from the human settlement.
(*e*) The entrepreneur should see that the site is selected primarily on barren non-agricultural land and in case of its non-availability, care and caution be taken up for displacing the local habitants to the minimum and acquiring the agricultural land as little as possible.
(*f*) Similarly, the forest growth should be spared as much as possible.
(*g*) Access to the markets
(*h*) Availability of space and means for affluent disposal

(*i*) Interlinkage with other plants.

(*j*) Above all, the entrepreneur must consider government policies and plan accordingly.

On safety point of view, he should look to the following:

(*a*) Distance between site and built up area

(*b*) The site may be classified as rural or urban, taking into consideration of the site, being surrounded by agricultural fields and waste land or it's close proximity to the urban centre near by or the intermediate area.

For maintaining ideal environmental safety, free from pollution, the distance to be maintained from the area listed below.

1. 25 km away from ecologically sensitive area.
2. 0.5 km away from flood prone river system.
3. At least 0.5 km away from high way and railways.
4. Major human settlements, if those are 25 km from the projected growth boundary of the settlements. As per G.O. ms No-1 dated 6.3.84, no industry causing intensive water pollution shall be located within 1 km. radius from the banks of rivers, streams, lakes and dams.

This aspect has already been discussed in previous chapter under safety in Construction Industry. I reiterate here that the maintenance of environmental safety starts on the Drawing Board.

In nutshell, around 60% of environmental problems are solved at the stage of site selection and plan layout of the plant. The focus should be more on environment protection not profit, and the emphasis should be on ecology and not economy.

Factors leading to accidents due to environment hazards

Industry wise

(*a*) Non disposal of garbage/trash at the stage of site clearance.

(*b*) Establishing godowns/store houses of chemicals, acid, paints, close to the main building.

(*c*) Putting eating houses close to the godowns/main building.

(*d*) Choking the site with construction leaving very little pathway for egress and ingress of Fire Tenders into the construction premisses.

(*e*) Blocking ways, paths, working platforms, with boards, forms, wood nails and other building materials.

(*f*) Not properly providing/illuminating the workplace, paths, run ways etc.

(*g*) Keeping the work place floor, wet, dingy, slippery, oil spill etc.

(*h*) Dumpling of coal, in open spaces of the factory premises without cover.

(*i*) Mishandling gas cylinders/acid drums and paint cans and other inflammable items.

(*j*) Non cleaning of floor coverings, non spraying of water at intervals in respect of chemical and dust producing parts to minimise toxic affects.

Due to the above environmental lapses, accidents do occur.

REMEDY

The best remedy is good housekeeping, good planning with regard to situational change/rectification of plan, keeping ample space for free movement at work places.

Now, we must discuss on broad environmental policies undertaken by the management so far functioning of the machineries, it's type, the emission hazards, the radiation hazards, pollution etc. of the plant and effluent hazard due to improper disposal.

1. At the beginning, we may take up the product kind of the plant. The environmental hazards differ from product to product. As I have stated earlier, the raw material required for the product in a factory may pose a problem for environmental hazard. For instance, in a paper industry, the raw materials are bamboo, straw, wood pulp and the waste paper. The former, kind of raw materials are considered bad items as their processing creates pollution and ecological imbalance. But waste paper produces, comparatively less dust, fume, toxic gas, emits less, harmful elements. The affluent disposal in the waster paper is easy after due treatment than bamboo, wood, straw etc. So the management has to decide which type of raw materials are to be used for the product and what type of machineries are to be procured. Machinery revision is very important for which the management has to take a decision.
2. Nowadays, due to paucity of power and supply situation very much erratical in respect of government managed hydro electrical as well as Thermal Power Plants, the entrepreneurs invariably put their own thermal captive power plants for smooth uninterrupted running of the plant. On the event of such an arrangement takes place, the power plants run on coal power and on the long run, they largely pollute the environment. The green house gas emission causes environmental hazard not only to the locality but also plays havoc in the global scenario. As on today, around 57,000 million ton CO_2 gases are emitted annually by the thermal plants. Besides this, the polluting gases like Sulphur dioxide, nitrogen oxide , carbon monoxide, fly ash etc., are being a created from these plants and emitted to the atmosphere. The oxygen content in the atmosphere was 312 PPM in the year 1940 but now it is around 400 ppm. The coal based thermal plants now emit around 40 million annually into the atmosphere and around 50 tons of fly ash per hour by a power plant, meant to produce 400m.w. of electricity. Likewise, a power plant meant to generate 1 m.w. of electricity, requires one and half areas of land, to accomodate fly ash deposits in 20 years. So, for depositing this fly ash only, thousands acres of land are required. Now restrictions are being imposed on production of power, through thermal power plants and Australia is pioneering this movement in its country. It is selling coal to India and China and engaged in other alternative task to produce electricity. Some of the western countries, have put emphasis on solar power, wind power generation for this purpose.

   A government committee on energy co-ordination has proposed a tax on polluting industries as part of its efficiency initiative of the national action plan of climate change. A carbon tax may encourage industry to scaledown emission of heat trapping carbon dioxide and turn towards cleaner alternative energy options. So thermal plants, which have become part and parcel of todays industrial establishments, pose a threat to ecology imbalance, global warming and large-scale pollution of both air and water. It is likely that polluting industrial unit having to pay contemplated carbon tax, would transfer the additional cost to the consumer. That might turn consumer away, unwittingly leading them to greener cheaper options. The carbon tax paying units might have to eventually close down or set about reducing its emissions or it has to find a way to pay the tax or purchase emission allowances from other more efficient plants inorder to continue operation. In any way, carbon taxes, would prepare the ground for industry to become more efficient in their use of energy.

   Measuring and analysing the carbon footprint of polluting industry to be able to gauge the extent and volume of pollution generate is a step forward to punish culprit units. A standardised carbon footprint calculator for industry sector wise would help authorities do the necessary number crunching to formulate details of arriving at the taxation figure. It is also necessary to prepare a time print over a specified target period. Once a measurement and verification system is institutionalised, monitoring industrial emissions becomes simpler. The Nicholas Stern Committee report recommendations had said that carbon tax need be imposed on all coal, natural gas and oil production units. The taxed and collected fund may be utilised in developing alternative energy options like solar and wind power. The U.K.'s experience in this regard could be instructive. It has introduced a range of carbon taxes and levies including on air travel. Sweden enacted carbon tax in 1991 on power generating plants and aviation fuel used in domestic air travel and other scandinavian countries have since followed suit.

3. Alternative greener coal power plant has now come in to existence to save the large-scale pollution hazard. A Swedish Company has developed clean coal fired power plant. Instead of pumping CO_2 into the atmoshphere, the plant converts the emissions into liquid for underground storage below around 3500 meters in natural gas caves. In the process, the level CO_2, which escapes in to the atmoshphere is minimal. But nobody knows the impact of such storage and environmentalists are apprehensive of increase in temperature underground. But Swedish Electricity provider vatten fall is determined to go ahead with this green coal power plants and investing around 70 million Euros in the project in Spremberg Germany. All nuclear power stations are due to close by 2020 as these have posed greater threat to ecology imbalance due to radiation and their affluent disposal. If this project is successful, vatten fall will equip all these new installations with the technology. Countries like China and India could also start using it. By 2015, Vatten fall hopes, this cleaner form of electricity will be used commercially as part of the national grid.
4. Another pollutant factor is 'fly ash' common to atmost all industries. The fly ash, be collected using bake up technologies and could be used in Brick Making. The fly ash is dumped somewhere near the human settlements, which causes environmental hazards. Recent accident in NALCO ash ponds at Anugul, Orissa is an example on this score.
5. The exhaust chiminies of industrial units emit dangerous chemicals, gas and dust and largely pollute the air in the periphery. It can be controlled by installation of modern ESP in exhaust chiminies. The emission product waste and its presence in the air must be reduced to tolerable limit. Chemical, cement and stone crushing units must take steps to stop emiting harmful toxic gases, monoxide and dust, which causes life threatening diseases, enumerated above.
6. Water pollution in and around the peripheral area of the industrial units have posed a serious health hazard for the colony dwellers of the industry as well as the general public, settled near the units. Water is polluted in two ways. The effluent waste water is drained out in to the near by water sources like rivers, streams and nallahs. The underground water is polluted by indiscriminate exploitations of minerals like coal etc. The water treatment plants have eased the problem to some extent though, yet it's efficacy has not been lessened. The solution to some extent lies with innovation and application of modern technology in matters of recycling the mill water by treating it and it's reuse. Since there is large scale depletion of ground water due to uncontrolled excavations, bore well digging and pumping the huge stock for use by Industrial units, the time has come to put a check in the wanton destruction of underground water. State should frame suitable legislation to curb this uncontrolled exploitation. There has been agitations by the agriculturists of the locality, who suffer this calamity in relation to paucity of water for their irrigation purposes and demanded prohibition for the Industrial units to lift huge quantity of water from the underground store. Time will come,

when there will be water rationing and meters would be fixed in all units like power and the excess drawees suitably punished. To combat this hazard, number of water recharge centres be created in the locality.

The polluting industries, who flush out their affluent water to the nearby rivers/streams, create enormous health hazards for the people and other waterborn creatures. People who depend mainly on these water sources for their daily use, even for their drinking purpose, suffer from various diseases. The proportion of water and chemicals in the final raw material mix, is found always above the danger level.

The water bodies of Patancheru, the hub of biggest bulk drugs and pharmaceutical industries in Andhra Pradesh, have the world's highest level of pharma residues. A startling revelation, no doubt, but what is more horrific is this: researchers analysing vials of treated waste water have found 21 different active pharmaceutical ingredients used in generics mixed in them. However, this should not surprise us at all: in the last 10 years, there have been enough warnings about the deteriorating environmental condition of the area.

The 1990 Status of Environment in Andhra Pradesh, Citizen's Report, said that ground water pollution due to effluent discharge led to an increase in diseases, including cancer. In 1995, the district judge of Medak (Patancheru is located in this district) submitted a report to the Supreme Court, which said that over 100 industries were responsible for causing ground-water pollution in the region. Recently; the Central Pollution Control Board identified Patancheru as one of the 22 critically polluted areas in the country. However, what is clear now, thanks to the latest report, is this: everyone—the government, the companies and the pollution authorities did not bother to reverse the trend.

Meanwhile, it is the poor who have been paying the cost of this dumping: the district health officer's 2002 report found morbidity rates in the area have increased from 10.18 percent to 25.49 per cent over a period of 10 years. Today; many have become drug-resistant and a high incidence of cancer has been reported. Experts say that since the water of this area flows into the Godavari, the affected area could be much larger.

The Indian pharmaceutical industry is part of the 'sun-shine sector'. It is competitive because the manufacturing costs are low here. Needless to add, cost-cutting on pollution control is an important factor behind this advantage. While the developed countries outsource production to India to cut costs, the big Indian companies parcel these out to smaller firms, which are notorious for lack of pollution control. And lax laws and officers only perpetuate this problem. While the West has shunned Indian products that involve child labour, in the future, similar stringent laws could force the Indian pharmaceutical companies to clean up their backyard. Before that happens, the companies would do well to invest in pollution control and adhere to international norms. Just planting a tree or opening a school in the name of corporate social responsibility will no longer do.

The environment (Protection Act) 1986 has been enacted for protection and improvement of environment and for matters, connected there with. This include, water, air, land, human beings, other living creatures, plants, microorganism. C.R.Z. (coastal regulation zone) has been declared and prohibition imposed for establishment of certain industries. It envisages the industries to have some sort of spill prevention control and counter measures plan (SPCC). Environmental auditing has been made compulsory from 1993. Any industry found contravening the provision of above laws could be punished by imprisonment upto 7 years or fine up to one lakh. Additional fine upto 5 thousand for every day violation has been provided.

Hazardous Waste (management and handling) Rules, 1989 has been framed under environment (Protection Act 1986) to regulate generation and disposal of hazardous substances. Assessment has been made statutory for 29 categories of developmental projects under various sectors like industrial mining, irrigation, power, transport and others. The aforesaid Act has classified the industries as Red, Orange and Green, depending on the degree of pollution caused by them. If further specifies the various pollution control measures to be adopted by these industries. The important measures taken by the govt. to control air pollution are:

(*a*) Establishment and notification of standards for major sectors of polluting industries and enfrocement of pollution control measures.

(*b*) Establishment of ambient air quality standards and monitoring of air quality in different parts of the country.

(*c*) Identification of 24 critically polluted areas of the country and drawing of action plans for these areas.

(*d*) Schemes to set-up common effluent treatment plants for small scale industries.

(*e*) Establishment and notification of gross emission standards for on road vehicles, introduction of unleaded petrol and catalytic converter fitted vehicles.

(*f*) Installation of pollution control devices by 1259 Industrial Units.

(*g*) "Eco-mark" lebel for environment friendly products and development of software package "paryavaran," to help industries in environmental audit.

(*h*) The rivers of India are part of its cultural heritage. Over the years, the quality of water has deteoriated due to uncontrolled release of affluents by industries in to the rivers. A pioneering effort to control the pollution of Indian rivers was launched in the year 1985, by implementing Ganga Action Plan (GAP). The scheme implimented under the plan are interception, river front development and provisions of low cost sanitation etc. The pollution caused by Industries was regulated by enforcement of existing Acts and Regulations. A total of 253 schemes costing Rs.449.21 crore have been implimented. Besides national river action plan, launched in Haryana an U.P., National River Conservation Plan (NRCP) have been formulated to control pollution in other rivers of the country other than GAP. Another authority *i.e.*, national river conservation created to review implimentation of the programme. The NRCP covers 14 towns located along 22 inter state rivers in 14 states. The total cost of the scheme is around Rs. 2053.26 crores. Similarly, a national lake conservation plan has been approved in principle, which envisages the conservation of lakes by prevention of pollution made by polluting industries. A total 480 water quality monitoring stations have been set up of which, 50 are part of the Global Environment Monitoring System (GEMS).

(*i*) Rverting back to disposal of waste by the industries, particularly toxic waste, some of the western countries have evolved recycling plan and exporting these waste to under developed countries of south east Asia including India. They send around two million tons of waste every year to India via Rotter dam port and Netherland's share is around two lakhs ton. The European union forbids export of hazardous chemical waste to non-OECD countries. Para 49 of the new Europian Waste Shipment regulation lays down strict guidelines to treat the waste in an environment friendly manner and that there are no irregularities. The Basel Convention Treaty too, puts restriction on export of chemical waste. Some unscrupulous nations export their industry's waste under the veil of scrap. E-waste is also partly hazardous. It has circuit boards that are not in the green list. Hongkong bans second hand electronic goods and China bans e-waste.

In nut shell, this waste disposal problem could be effectively tackled to some extent by recycling, treating and reuse.

POLLUTION OF MARINE ENVIRONMENT AND PREVENTION

Effects of operational or accidental pollution of the marine environment.

The Pollutants that enter the marine environment from the ships can be broadly classified as follows:

(*a*) Oil
(*b*) Noxious liquid substances
(*c*) Harmful substances in packaged form
(*d*) Sewage
(*e*) Garbage.

These pollutants may enter the Marine environment operationally (*e.g.*, Tank washings and Ballast discharge by Tankers and Chemical carriers) or be accidentally discharged into sea. The Marpol (Convention for the prevention of Pollution from ships) details the procedures, treatment and amount of these Pollutants that can be discharged into the Marine environment legally.

The contracting States formulate their own laws to take action and impose deterrent penalties on any ship violating the convention requirements.

Ecological effects resulting from a spill may include physical and chemical changes in habitats, changes in growth, physiology and behavior of individual organisms and species, toxicity and increased mortality in individual organisms and species and destruction or modification of entire communities or organisms through the combined effects of toxicity and smothering.

Floating of oil may contaminate mammals and birds that swim or dive through the surface of the water. In the open sea, many birds and animals may be able to avoid contact with an oil slick, but in coastal areas, the organisms may be trapped between the shore and the approaching spill and become contaminated. If fishing activities take place in the immediate vicinity of a spill, both fishing gear and catch are likely to be contaminated as a result of contact with dispersed oil or water-in-oil emulsion, that may be floating on the surface, suspended in the water or settled on the sea bottom.

On fine sandy beaches, the oil is likely to remain on the surface where it can be removed: On beaches of cobble, gravel or even coarse sand, the oil may penetrate with the tidal water table down to hard substrata. In sheltered tidal flats, mangroves and salt-marshes, oil penetrates into the anoxic muds, facilitated by animal burrows, decaying root masses and interstitial water movement, where it may cause local biological effects for long periods. In such areas, biota are susceptible to physical smothering in the early stages of a spill and are likely to exhibit chemical tainting of their tissues for extended periods.

Noxious liquid substances from Chemical tankers are toxic to the marine organisms and much of the mortality that occurs are due to the toxic nature of the chemicals involved. Sewage water can carry diseases that may be discharged into the waters of the harbour and may cause epidemics. They are not permitted to be discharged unless treated. Sewage water may bring in harmful organism from other sea areas, which may affect the fragile bio system of the harbour.

Plastics are not biodegradable and can last for hundreds of years in the Marine environment. When discharged into the rivers, they enter the sea suction of the ships and block the sea chest.

Marine organism die by consuming or getting trapped in them. Lost fishing nets at sea account for the mortality of many organism. Plastics are, prohibited from being discharged into the sea.

A major oil spill occurring within or in close proximity to ports and Marines, could have far reaching physical and economic effects. In major ports, the potential for serious consequences is high, due to indirect costs associated with measures taken by the designated spill response authority.

Aquatic birds are the most visibly vulnerable to oil spills. Many species depend on the sea and coastal areas for food and nesting habitat. Some species reside permanently in one area, while others migrate seasonally over long distances. Species that spend most of their time on the water are particularly vulnerable to oil spill effects. In addition some species cannot fly during their moulting period and others do not fly at all. Sea birds may come, into contact with floating oil at sea by swimming into the slick, diving through it or surfacing into it from below.

BASIC ENVIRONMENTAL PROTECTION PROCEDURES

Anti pollution measures for oil

1. When any transfer of oil is planned on board ship then all scuppers should, be plugged.
2. Drip trays should be placed under hose connections or flanges. The drip trays to be drained or emptied periodically.
3. Dry material such as sand or saw dust should be always available to deal with any small spills.
4. Limited amounts of spill containment materials are usually stocked on board and these must be kept in readiness.
5. Spilled oil should not be washed overboard. They must be transferred to a storage tank and await proper disposal.
6. Oily rags should not be thrown overboard. They must be collected and incinerated or sent ashore for disposal.
7. All operational discharge should be according to Marpol Law and through a Oily discharge Monitoring system.
8. A positive means of communication should be continuously maintained throughout oil transfer operation between ship and shore, barge, sea loading station or lighting vessel. Communication should be checked and all signals used should be thoroughly understood by both parties before commencing operations.

Anti pollution measures for noxious liquid substances

In addition to the relevant anti pollution measures as stated above:

1. Operational discharge of Chemicals can only be carried out as per the provisions of Marpol.
2. Depending upon the toxicity of the Chemical involved, there are recommendations for their disposal or storage on board.

Anti pollution measures for harmful substances in packages

The Marpol refers to the carriage of above goods.

Harmful substances means any substance which if introduced into the sea is liable to:

(*a*) Create hazards to human health or
(*b*) Harm living resources and Marine Life or
(*c*) Damage amenities or
(*d*) Interfere with other legitimate use of the sea. Nearly 600 such Chemicals have been identified as Marine Pollutants in the IMDG code.

1. Certain harmful substances, which are Hazardous, may be prohibited for carriage by sea or their quantity on board may be limited depending upon the type of the vessel.
2. Jettisoning of such substances shall be prohibited unless it is done for the safety of ship or life.
3. Washing overboard of any leakage's shall be regulated according to the physical, chemical and biological properties of the Harmful substance, without impairing the safety of ship or life.

Garbage Disposal at Sea

Garbage includes all kinds of victual, domestic and operational waste, excluding fresh fish, generated during normal operations of the ship and liable to be disposed off continuously or periodically.

Discharge of Garbage is governed by Annexe V of Marpol.

Special areas are Mediterranean sea, Baltic Sea, Black sea, Red sea, Persian Gulf, North

Sea, Wider Caribbean and Antarctic.

A Ship of = 400 GRT or a ship carrying 15 persons shall:

(*a*) develop a Garbage Management Plan, outlining procedures for collecting, storing, processing and disposing of garbage.
(*b*) provide equipment as per the above plan
(*c*) designate a person to implement the above plan
(*d*) Maintain a Garbage record book

Garbage record books and receipts for shore discharge is to be preserved on board for 2 years after the date of last entry.

The following table is indicate discharge of garbage.

Garbage Type	Outside Special area	In Special Areas
Plastics-includes Synthetic ropes and Fishing nets and plastic Garbage	Disposal prohibited	Disposal prohibited
Floating dunnage Lining and Packing Materials	>25 miles offshore	Disposal prohibited
Paper, rags, glass bottles, metal, crockery and similar refuse	>12 miles	Disposal prohibited
All other Garbage Including paper, rags, glass etc., communicated or ground	>3 miles	Disposal prohibited
Food waste-communicated or ground	>3 miles	>12 miles

Other considerations in Port

(*a*) Funnel exhaust emissions should be maintained clean.
(*b*) Spill of Paint in the water from over side work is a pollutant.
(*c*) Hydraulic machinery Oil, deck must always be kept free of leaks.
(*d*) Inert gas scrubber overboard effluent on tankers should be checked clean.
(*e*) Clean up small quantities of oil immediately.

Today, 310 million people live in India's cities. By 2030, this number will be 575 million. By 2045, it will be 800 million.

In every minute, 30 Indians leave rural India for it's cities. Most Indian cities are already filled to the brim. In some areas, Mumbai has 101.066 people packed in to every square kilometer. Our metros generate 21.275 tons of waste everyday. 60% of the sewage goes untreated, to our rivers, oceans and lakes. So a sound policy on garbage disposal may be worked out and dumping industry waste, in the sea far from the recalcitrant nation's coast line be seriously dealt with.

We have very effective legislations in respect of environmental protection, ecological imbalance and global warming as detailed below.

(*a*) The Water (prevention, control of pollution) Act as amended in 1981.
(*b*) The Air (Prevention and Control of Pollution) Act as amended in 1998.
(*c*) The Environment (Protection) Act 1986.
(*d*) The Water (Prevention and Control of Pollution) Cess Act 1977 as a amended in 1991.
(*e*) The Public liability Insurance Act 1991.
(*f*) National Environment Tribunal Act 1995.
(*g*) National Environmental Appelate Authority ordinance, 1997.
(*h*) Wildlife (Protection) Act 1972
(*i*) Forest Conservation Act 1980.

PROTECTION OF ENVIRONMENT IN GLOBAL SCENARIO

1. India has played a significant role in the global environmental campaign. It has actively participated in various environmental conferences and programes launched since the UNEP (United Nations Environmental Protection) conference held in Stockholm in the year 1972.
2. India is a signatory to various conventions and agreements like the:
   (*a*) Ramsar Convention on wetlands,
   (*b*) Convention of International Trade of Endangered Species (CITES),
   (*c*) Vinna Convention for the Protection of Ozone Layer,
   (*d*) Monteal protocol on substances that deplete ozone layer,
   (*e*) Convention of Biodiversity and Climate Change,
   (*f*) Basel Convention on the Transboundary movement of Hazardous Substances. India hosted the global environmental facility assembly in the month of April 1998.

Since the Earth Summit in Rio in 1992, it has been a long march for the world to reach a consensous and to commit together, on the road to action in combating global warming. The 1st international approach to climate change had taken shape with the development of U.N. frame work convention on climate change (U.N. F. C.C.S.) adopted in 1992. The action was aimed at stabilisation of Green

House Gas (GHG) concentration in the atmosphere. The first conference of the parties to the convention was held in Berlin (March-April, 1995). The developed countries were pressurised for reduction of GHG emissions. Accordingly, a protocol of climate change convention was adopted ion Kyoto in the year 1997, now popularity known as KYOTO PROTOCOL. The objective of this Kyoto Protocol is aimed at bringing down the global GHG emissions by 5.2% during the year 2008–2012.

GREENHOUSE GASES

The Earth's atmosphere primarily consists of oxygen and nitrogen, but neither plays any significant role in what is called the greenhouse effect, as both are essentially transparent to terrestrial radiation. The gases currently known to cause the greenhouse effect include carbon dioxide (CO_2), methane (CH_4), nitrous oxide (N_2O), chlorofluorocarbons (CFCs), and two CFCs substitutes, hydrochlorofluorocarbon (HCFC-22), and perfluoromethane (CF_4). Of these, carbon dioxide, methane, nitrous oxide and ozone are the naturally occurring greenhouse gases. Certain human activities however add to the levels of most of these naturally occurring gases. Carbon dioxide is responsible for over half the enhancement of the greenhouse effect. The greenhouse effect is primarily a function of the concentration of water vapour, carbon dioxide, and other trace gases in the atmosphere that absorb the terrestrial radiation leaving the surface of the Earth, and act like a blanket over the earth's surface, keeping it warmer than it would otherwise be. Changes in the atmospheric concentration of these gases can alter the balance of energy transfers between the atmosphere, space, land, and the oceans.

CLIMATE CHANGE IMPACTS

The vulnerability of human societies and natural systems to climate extremes is demonstrated by the damage and death caused by events such as droughts, floods, heat waves, avalanches, windstorms etc. While there are uncertainties of such changes, it is however established that the geographical extent of the damage or loss will increase with the magnitude and rate of climate change. The effects of climate change are expected to be greatest in the developing countries in terms of loss of life and the economy. Many regions that are vulnerable to climate change are also under pressure from forces such as population growth, resource depletion, and poverty.

Climate and Biodiversity

Biodiversity refers to the variety of life on Earth, and its biological diversity. The number of species of plants, animals, microorganisms, the enormous diversity of genes in these species, the different ecosystems on the planet, such as deserts, rainforests and coral reefs are all part of biologically diverse earth. Biodiversity actually boosts ecosystem productivity where each species, no matter how small, all have an important role to play and that it is in this combination that enables the ecosystem to process the ability to prevent and recover from a variety of disasters.

Climate and biodiversity—the connection:

- Climate change is affecting species already threatened by multiple threats across the global habit fragmentation due to colonization, logging, agriculture and mining etc., are all contributing to further destruction of terrestrial habits.
- Individual species may not be able to adept-species most threatened by climate change have small ranges, low population densities, restricted habitat requirements and patchy distri-

bution.

- Pollination may be disrupted-higher temperatures and earlier. Snowmelt may trigger earlier flowering plants.
- Bird species with limited ranges may be lost-for example, the Kirtland Warbler nests in northern Michigan, whose habitat is the declining jack pines in the region. Melting of polar ice is causing problems for polar bears.
- Coral reef mortality may increase and erosion may be accelerated. In 1998 alone, 16% of the world's coral reef died from higher temperatures. Increase level of carbon dioxide adversely impacts the coral building process (calcification). Scientists estimate calcification could decline 17 to 35% below pre-industrial levels by 2100.
- Sea level may rise, engulfing low-lying areas-causing disappearance of many islands, and extinctions of endemic island species.
- Whale populations may decline-significant reductions in seal ice could adversely affect the abundance of krill, the primary source of food for whales in the Southern Hemisphere.
- Invasive species may be aided by climate change-exotic species can out-compete native wildlife for space, food, water and other resources, and may also prey on native wildlife.
- Droughts and wildfires may increase-an increased risk of wildfires due to warming and drying out of vegetation is likely.
- Sustained climate change may change the competitive balance among species and might lead to forests dieback.
- Climate change induced droughts may push humans into wildlife habitat.

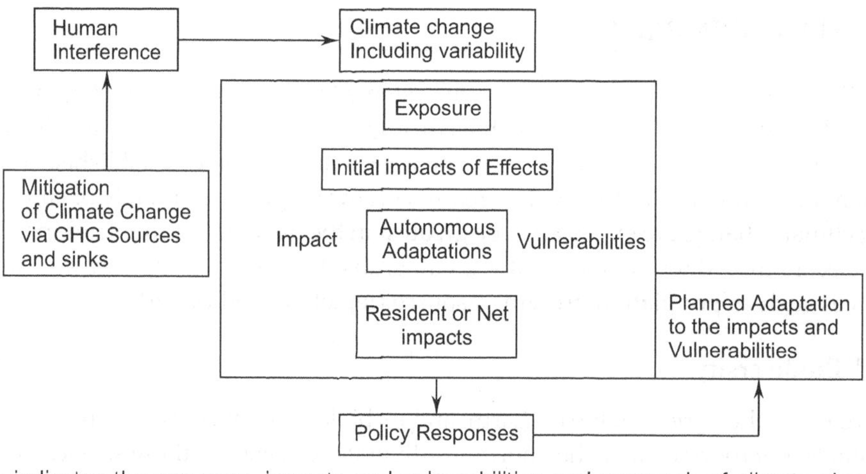

This indicates the exposure, impacts and vulnerabilities and approach of climate change.

Climate Change Negotiations

Global warming issue became a part of the international agenda in 1988. The climate issue came up at the UN General Assembly in December 1988, as part of a discussion on 'the common heritage of mankind' and was initiated by the small island nation Malta. The UN resolution of 1988 thus recognized climate changes as a common concern for mankind. The resolution set up a preparatory committee to work towards an international agreement. The concern for global warming particularly by the industrialized countries heated up since then and 'climate politics' matured with a series of international conferences and formal negotiations that followed. The momentum culminated in the signing of a Framework Convention on Climate Change (FCCC) and opened for signatures at the Rio Earth Summit in June, 1992. The FCCC aims at stabilization of greenhouse gas (GHG) concentrations, in the atmosphere at a level that would prevent dangerous anthropogenic interference with the climate system. Subsequently, the parties to the FCCC adopted the Kyoto Protocol in December, 1997. In the eyes of the developing nations the Protocol is fraught with loopholes as economic concerns, rather than ecological or social justice concerns are invariably the driving force. The main area of contention between the developed countries and the developing countries lies in the sectors pertaining to equity and sustainability. However, the operational details of the Kyoto Protocol have now been finalized after intensive deliberations at Marrakech, on November 10, 2001, which was participated by 171 countries. The way for widespread ratification by governments and the Protocol's entry into force is now made which is acceptable to governments across nations more or less [As of 17 September 2002], 94 countries have so far ratified the Protocol.

The protocol has been guided by Article 3.0 of the FCCC, and marks the first global attempt to place legally binding limits on greenhouse gas emissions from developed countries. In addition to carbon dioxide, the primary GHG, the protocol focuses on five other greenhouse gases viz. methane (CH_4), nitrous oxide (N_2O), Hydrofluorocarbons (HFCs), Perfluorocarbons (PFCs) and sulphur hexafluoride (SF_6). The Protocol calls for 5.2% reduction from their 1990 level of GHG emissions by the developed countries during the period 2008–2012. It also specifies the amount each country must contribute toward meeting the reduction goal. Nations with the highest CO_2 emissions like the United States, Japan and most European nations are expected to reduce emissions by a range of 6 to 8 percent. Key decisions taken at COP-7 in November 2001 in Merkakesh, Morocco included the following:

- Allowing emissions credits generated under the Clean Development Mechanism, Joint Implementation and Emissions Trading among (developed) countries to be treated equally;
- Creating the removal of units to represent credits generated in countries via sinks;
- Establishing guidelines for banking unused emissions credits;
- Allowing developing countries to unilaterally undertake a CDM project and market the resulting emissions credits;
- Giving Russia an increase in the ceiling for forest management credits; and
- Requiring developed countries to report on efforts to protect bio diversity in the context of sink activities.

GHG MITIGATION OPTIONS, SINKS AND BARRIERS

Mitigation Options

The challenges of climate change mitigation involves diverse issues-economic, political, social and environmental. As the National GHG inventory for India shows, the major increase in GHG emissions over the next 20 years would be related to energy consumption. As India has abundant coal deposits, it is beyond doubt that coal will be the dominant source of energy. Therefore, an energy efficiency measure in this sector remains our prime concern. Power generation in India is expected to reach a peak demand of 176 GW by 2012, and the total energy requirement will be 1085 billion units. Therefore, energy efficiency and increasing use of renewable energy or a move towards low carbon options are the two main measures that can greatly reduce GHG emissions. With the rising demand for energy, power generation specially coal fired power plants, will form a major share of capacity additions. Mitigation option in the power sector includes clean coal technologies and renewable energy sources. Options such as bagasse-based co-generation and combined cycle plants generate lesser emissions per kWh of electricity than the conventional power generation system. With a vast rural population and several remote areas, renewable energy could be the means to clean energy.

Similarly, small hydro, wind and biomass-based power, though more expensive than conventional coal-based power plants, provides significant abatement opportunities. Options for energy efficiency in all sectors, shall however be selected on the basis of three main criteria viz.

(*a*) consistency with national development priorities

(*b*) relatively high level of energy conscription in the base activity; and

(*c*) the relatively large GHG reduction potential offered by the abatement technology. Specific abatement strategies for the energy sector include fiscal incentives and taxes, voluntary emissions reductions, green rating, and capacity building etc. Another area of important from mitigation aspects in the power sector in the Indian context is the transmission and distribution losses, which is energy loss, and hence, emissions. There is considerable scope of reducing losses, which eventually translates into a large mitigation potential. Electricity demand in India has been increasing at an annual average of 8.8% during the past 35 years. Thermal Power accounts for most of this electricity generation. Its share in the total generation was 82% in 1997 and is expected to increase to 89% by 2015. Therefore, the use of using energy efficient appliances can also contribute a lot in indirect reduction of CO_2 by way of reducing the demand of power consumption. Barriers however still exist that hinder adoption of electricity conservation and demand management in India. Two major categories of barriers are: (*a*) Macro-level barrier—either policy induced or that which exists in the absence of appropriate policies and; (*b*) Micro-level barriers — that related to the individual characteristics of consumers and the economic environment they face.

In the forestry sector, IPPCC Second Assessment Report categorises three broad options viz.

- *Conservation management:* This strategy attempts to conserve the existing carbon storage capacity of forests by halting or slowing down forests deforestation and forests degradation
- *Storage management:* This strategy attempts to increase carbon strategy in woody vegetation and soil in existing degraded forest, as well as to create new carbon sinks in areas where forests do not exists or have been cleared. These may be achieved by promoting natural regeneration, reforestation on deforested lands, forestation of non-forest lands and agro-forestry on crop and pastureland.
- *Substitution management:* This strategy attempts involves the replacement of fossil fuels by renewable fuel, wood or other biomass products. In the agriculture sector, methane emissions from rice cultivation remains the major contributor of GHG emissions. Other sources being enteric fermentation, manure management, agricultural soils etc. Abatement strategy in this sector in India can be achieved through the following:
- increasing the digestibility of animal feed by supplementing it with molasses
- replacing open pit method of manure treatment with small scale digesters.
- using improved paddy varieties and draining fields frequently and;

- encouraging cultivation of rice varieties that emit less CH_4 per unit of output.
- Improving application efficiency of nitrogenous fertilizers etc.

The industrial sector, as the national inventory of GHG shows, major contribution came from energy intensive sectors like iron and steel, fertilizer, cement, aluminium, paper and pulp etc. Therefore, substantial mitigation option in the industrial process besides change of more efficient process methods include switching energy production options. A few option available for energy efficient options in power, industrial and domestic sector is given below:

Power sector	Industrial sector	Domestic sector
Combined cycle plants	Diesel cogeneration	Efficient refrigerators
Integrated Gasification	Iron and Steel	Efficient air conditioners
Combined Cycle (IGCC)	Basic oxygen furnace	
	Ultra high power electric	
	arc furnace	
Intercooled steam	Continuous casting	Efficient lighting
injected gas turbine	Direct reduction process	
(ISTIG)	Dry quenching rout	
Pressurized fluidized	Pulp and Paper	
bed combustion (PFBC)	continuous digesters	
	in papar industry	
Pulverized coal	Cement	
super-critical boilers	Dry kilns	
	Dry pre-heater kilns.	
Industrial cogeneration	Caustic soda	
Amorphous core	Membrane process soda ash	
transformers	dual process and akzoline process.	
	Waste heat recovery.	
	High efficiency burners (low excess air)	
	Heat pump, High efficiency motors.	
	Efficient lighting.	

Sinks as Mitigation Option

Carbon dioxide is removed from the atmosphere by a number of processes that operate on different time scales, and is subsequently transferred to reservoirs or sinks. Or more simply, we may define 'Sink' as any system that removes carbon dioxide from the atmosphere. The fastest process of removal is absorption into vegetation and surface layer of oceans. Sinks are of mainly three types viz. oceanic, terrestrial and inferred or missing. The Protocol has provisions that allows industrialised countries (Article 3.3) to take into account changes in emissions resulting from human induced land-use, land-use-change and forestry (LULUCF) activities limited to afforestation, as a sink to reduce carbon dioxide levels in the atmosphere. Further, Article 3.4 of the Protocol states that additional human induced activities in the agricultural soils and LULUCF categories may be added to the three mechanisms (Joint implementation, Clean Development mechanism and Emission trading) already counted under the Protocol, subject to certain conditions. Under Article 3.7 the Protocol mentions the phrase land-use change and forestry, in the context of whether this sector was a net source of greenhouse gas emissions in 1990.

In most of the tropical countries, as in India, forestry is dominated by government based institutions. These institutions therefore need support and new insight in order to effectively incorporate mitigation policies and measures in their resource management activities. India has been relentlessly implementing one of the largest reforestation programs in the tropics with over one million hectares planted annually. Nearly half of this reforestation is on degraded forests and village common land. It is estimated that the carbon uptake inforests, degraded forests, and plantations is estimated to offset the gross carbon emissions from the forests sector. Carbon dioxide emissions in India are projected to increase from no-net emissions in 1990 to 77 million tones by 2020. This situation can be turned around by promoting agro forestry, which, according to a study by the Indian Institute of Science, Bangalore, can abate 700,000 million tons of carbon on a dedicated 69 million hectares of land.

Barriers to Mitigation Measures

Greenhouse gas mitigation measures are compounded by several barriers inherent to the process of development. Sustainable development in a participatory framework can minimize these barriers, but the inequitable distribution of income and wealth forms a core feature of barriers to effective implementation of any type of intervention, and those related to climate change are no exception. For any policy decision on GHG reduction, any individual country can choose from a large set of possible policies, measures, and instruments to limit domestic GHG emissions. These can be categorized into market based instruments, regulatory instruments, and voluntary agreements of which some may fall into the category of market based instrument. Also for any country and more particularly for the developing countries, domestic structural reforms and policies on trade liberalization and liberalization of energy markets have a great impact on measures that acts as barriers to GHG reduction or enhance their sequestration by sinks. These policies coupled with macroeconomics, market oriented reforms, set the framework in which more specific climate policies would be implemented. During the 1990s, several countries including India and China implemented drastic market oriented reforms that have had important effects on energy use and energy efficiency, and therefore on GHG emissions.

Another major barrier in climate change mitigation is the transfer of environmentally sound technologies from developed to developing countries. Article 4.5 and other relevant provisions of the UNFCCC defines the nature and scope of technology transfer, which include environmentally sound and economically viable technologies and know how conducive to mitigating and adopting to climate change. An effective transfer of technology requires fulfilling three major criteria viz. the technology holder should be willing to transfer the technology, the technology must fit into the demand of the recipient country and the transfer must be made at reasonable cost to the recipient. The IPCC Special Report on technology Transfer (IPCC, 2000) identifies various important barriers that could impede environmental technology transfer, such as:

- lack of data, information, and knowledge, especially on emerging technologies;
- inadequate vision about the understanding of local needs and demands;

- high transaction costs and poor macroeconomic conditions;
- insufficient human and institutional capabilities;
- inappropriate technology adopted and
- poor legal institutions and framework.

CLEAN TECHNOLOGIES BY DEVELOPING COUNTRIES

The G8 lost a good opportunity to come clean at the summit in Hokkaido, Japan. While admitting the developed world's historical responsibility for the stock of emissions polluting the atmosphere, member countries could have spelt out a detailed action plan that; I would set guidelines for country-wise targets for emissions reduction and so prepare valuable inputs for the framing of an alternative agreement to the Kyoto Protocol that expires in 2012. The G8 could have also discussed ways in which it would help developing countries leapfrog to cleaner technologies to enable sustainable development.

The climate vision statement issued by leaders of major economies who were meeting at the G8 summit—including outreach countries like India and China-made a general endorsement on the need to halve global emissions by undertaking deep cuts by the year 2050. The plan mentions neither a base year, nor detailed targets for the interim period. But it does acknowledge the fact that climate change is one of the great global challenges of our time. It goes on to state that both developed and; developing countries should commit to combat climate change in accordance with the common but differentiated responsibilities and capabilities principle, but again, minus the specifics.

India has been firm that, as a developing country, there is no question of being equated with the G8 in computing emissions; cuts targets. However, as a leader in the developing world, along with China, India is in a position to redefine development by leapfrogging-to cleaner technologies and methods. This can only be achieved with the help of those who have the know-how and money, and it is to this end that the UN Framework Convention on Climate Change had suggested setting up a technology transfer fund, to which the G8 has pledged an annual contribution of $10 billion - of which the US would contribute 50 percent—for technology research and transfer. Making a clean start is an option that's available to the current crop of growing economics-provided the developed world takes at least part responsibility and commits more funds to this end.

The US has not ratified the Kyoto Protocol. Yet states like California and New York have gone progressively green, voluntarily, bringing down US emissions a shade. But those developed countries who have ratified the protocol have not, only been unable to meet their emissions cuts targets, some have increased their emissions. Since greenhouse gases building up in the atmosphere would affect poorer countries more intensely, and considering the G8's historic responsibility, there is a strong case for member countries to step up technology transfers to the outreach countries to enable greening of energy use and development.

Just as John F Kennedy set his sights on the moon, Al Gore is challenging the nation to produce every kilowatt of electricity through wind, sun and other Earth-friendly energy sources within 10 years, an audacious goal he hopes the next President will embrace. The Nobel Prize-winning former vice-president said, Barack Obama and John McCain are "way ahead" of most politicians in the fight against global climate change. Rising fuel costs, climate change and the national security, threats posed by US dependence on foreign oil are conspiring to create "a new political environment" that Gore said will sustain bold and expensive steps to wean the nation off fossil fuels: "I have never seen an opportunity for the country like the one that's emerging now," Gore said. He said he fully understands the magnitude of the challenge.

The Alliance for; Climate Protection, a bipartisan group that Gore chairs, estimates the cost of transforming the nation to socalled clean electricity sources at $1.5 trillion to $3 trillion over 30 years in public and private money. But he says it would cost about as much to build ozone-killing coal plants to satisfy current demand.

"This is an investment that will pay itself back many times over" Gore said. It's an expensive investment but not compared to the rising cost of continuing in fossil fuels." Called an alarmist by conservatives, Gore has made combating global warming his signature issue, a campaign that has been recognized worldwide–from an Academy award Nobel Prize.

"I hope to contribute to a new political environment in this country that will allow the next president to do what I think the next president is going to think is the right thing to do," Gore said. "But the people have to play a part." He likened his challenge to Kennedy's pledge in May 1961 to land a man on the moon by the end of the decade.

To meet his 10 years goal, Gore said nuclear energy output would continue at current levels while the nation dramatically increases its use of solar wind, geothermal and clean coal energy. Huge investments must also be made in technologies that reduce energy waste.

Managing Climate Change Mitigation Versus Growth

The road to global cooperation on climate change mitigation at the forthcoming Copenhagen Summit is currently gridlocked by an apparent direct clash of interests between the mitigation priorities of the developed countries and the growth priorities of the developing world. Nobel laureate Michael Spence has an imaginative strategy for getting around the gridlock that is both fair and efficient.

The International Panel on Climate Change, the acknowledged global authority on the subject, estimates that 50 years down the road, the acceptable safe level of CO_2 emissions; will be about 14.7 billion tonnes or 2.3 tonnes per capita per year. The average emission today is about double that limit at 4.8 tonnes per head. In the absence of a serious mitigation effort this will rise further to about 8.7 tonnes per head or four times the safe limit by 2060 as; many large high-growth developing countries approach the living standards and consumption patterns of the advanced countries.

Much of the CO_2 emissions come at present from the advanced countries. USA and Canada in particular emit about 20 tonnes per head. The other advanced countries emit between, 12 and 6 tonnes per head. Most developing countries, including India, emit well below the safe level of 2.3 tonnes per head and China just exceeds that benchmark. The developing countries accordingly maintain that the advanced countries have the responsibility to curb their emission levels.

However, the emissions balance will continue to shift rapidly if the fast growing developing countries maintain their pace of growth. In 50 years the major share of emissions will come from the developing countries, especially China and India. Hence, the advanced countries maintain that climate change mitigation is not feasible without critical action being taken by these countries. But any commitment to contain CO_2 emissions will compromise growth and poverty alleviation in these countries. What is the way out of this impasse? In the Spence solution, much of the mitigation action at present has to be taken by the advanced countries in the form of a Carbon Credit Trading System (CCTS). First, there has to be agreement on a global carbon emission mitigation time path leading to average emission levels of 2.3 to 3 tonnes per head in 50 years. This determines the available volume of carbon credits or emission rights in a given year. These should be distributed among the advanced countries according to some agreed formula, say in proportion to their population size. The countries should be allowed to exchange these carbon credits between themselves, based on their national priorities.

They should also be allowed to earn additional carbon credits through mitigation activities. These would include behaviour adjustments that reduce per capita emission levels, introduction of lower emission technologies and mitigation activities such as afforestation either in the country itself or elsewhere where the cost of mitigation is the lowest. These two features would establish a price for carbon credits that supports the permissible level of global CO_2 emissions and is based on the global marginal cost of mitigation. Of course this will require the establishment of cross-border mitigation arrangements that are collectively described as the Clean Development Mechanism (CDM) under the Kyoto Protocol.

The Spence solution does not require developing countries to take any action at this stage except to commit to the international agreement on the time path of global carbon emissions and commit that they would join the CCTS once they reach the living standards and emission levels of the advanced countries. In the next 25 to 30 years many of the fast growing developing countries would 'graduate' to advanced country status and at that stage they would start participating in the CCTS.

In preparation for their eventual entry into the CCTS on favourable terms, it would be in the interest of the developing countries to also engage in mitigation efforts at present to the extent these are compatible with high growth. Their efforts would be facilitated by cross-border mitigation investments and technologies provided by advanced countries which choose to increase their share of carbon credits through the CDM.

The Spence solution is attractive because it is both fair and cost efficient. But can it be actually implemented? The G20 countries account for 90 per cent of global output and two-thirds of the global population. It is this group plus a few other large emerging economies such as Mexico, Egypt, etc. that will have to deal with all the economic and technological challenges of mitigation. That helps because the group is small enough to allow for effective negotiations yet large enough to determine global trends if they do reach agreement. However, the Spence solution also assumes that the technologies and behaviour adjustments required to move along the agreed global emission path will be available. Hence, the feasibility of his solution ultimately depends on whether or not the incentives built into the proposed CCTS will be sufficient to generate the required technologies and behaviour adjustments.

Green buildings seem the right measure to tackle environment hazards posed by constructions in the form of pollution, contamination etc.

The United Nations theme on World Environment Day reflects our individual responsibility to protect Planet Earth.

The countdown has begun. It could be doomsday very soon. It is estimated that within the next 10 to 15 years, drastic steps will need to be taken to avoid crossing catastrophic effects on Earth.

As good citizens and organisations it is now time to wake up to the unavoidable reality called global warming. The need to act responsibly and make up for lost time is being increasingly felt by the residents of the earth. And this is where new and scientifically advanced innovations are being introduced to counter the damages of carbon footprints. Creating Green Buildings is one of the very effective measures.

Green buildings are more 'eco-friendly' man-made creations that make optimum use of resources like energy, water and materials. Such buildings contribute by controlling the harm done by buildings and high rises like releasing waste into water, air pollution, toxic contamination etc. Impacts on human health has also been initiated through better siting, design, construction, operation, maintenance and removal.

Green buildings are also sometimes known as sustainable buildings or environmental buildings. Yet another concept which hovers close to the concept of green buildings is natural buildings which are the product of natural materials available locally.

Sustainable development is the driving force of a green building. The trend started in Hyderabad and since then it has only grown. Today, a variety of green building projects-such as residential complexes, exhibition centers, hospitals, educational institutions, laboratories, IT parks, airports, government buildings and corporate offices; across the country are coming up.

The Green Effect

Some of the effects of agreen building are: Less operating costs due to increased productivity and lesser use of energy and water Better public and occupant health due to improved' indoor air quality.

Reduced environmental impacts by certain means like lessening storm water runoff and the heat island effect.

According to industry experts, builders of such creations do not only aspire for the "ecodesign mandatories" laid out for a green architecture but also seek aesthetic harmony between the building and the surroundings.

According to some, the need to go green will be felt soon even more in the coming days among the aspects of green buildings; one is, harvesting rain water. At present, people might think there is abundance of rainwater but the way things are going, in another five years, the pinch will be severely felt, said Jitendra Khaitan, CEO, Pioneer Property Management.

He said the cost of green buildings was a little higher but it was worth it. "Installing air conditioners will cost about three times the normal rate but it consumes a lot less power," he said.

According to many others in the industry, green buildings were definitely the order of the day.

"Investment potential is very high as there are better returns. They are eco-friendly, user-friendly and do not incur any extra expense as far as maintenance is concerned. Two sets of fans take care of the ventilation to maintain the carbon dioxide level in the air," said a top executive of Forum Projects Pvt. Ltd. He said in office buildings, employees would be highly benefited.

"Our creation, Technoplis became the first green building in the world to be acknowledged by the UN. More than 300 green projects are currently being developed all across India," said Rahul Saraf, MD, Forum Projects.

"On prolonged exposure to the computer, people show signs like reddening of eyes, irritation of head, etc., that will be taken care of and it is only in these green buildings that a 20 cfm of fresh air per person is ensured," he said.

Harshvardhan Neotia, MD, Ambuja Realty, too believes in the concept. "All our projects are green inclined, though we cannot always comply to the platinum and gold standards due to various reasons, but I definitely think attempt should be made to conserve energy and proper waste management," he said.

The concept has evolved widely in the country, southern and western regions leading, and the numbers could go to almost a 1000 by year 2012.

The flagship buildings platinum rated are CM Godrej Hyderabad, Wipro Gurgaon, ITC Gurgaon, BG house Mumbai and Central Spectral services Noida, said Jagdeep Singh, global sales head, Desiccant Rotors International (DRI). Technopolis in Kolkata is already a gold certified building. There are some more projects like Infinity Benchmark, Hub, csc, Videocon IT Park which would be looking at similar certifications.

Hundreds of industry leaders, opinion makers, and climate change experts are gathering this week for the Delhi Sustainable Development Summit, dedicated to climate change. The summit, being held for the second year running, could not come at a more opportune time. 2009 is a defining year for the international climate change process. The world is poised to conclude one of the most complex international agreements ever to be negotiated, an agreement which has the potential to decide the fate of the human race.

India and the rest of the world cannot afford the worst predictions of the UN's Intergovernmental Panel on Climate Change to come true, for example, the flooding of coastal cities and extreme food shortages. Failure to first slow and then reverse current greenhouse gas emission trends within the next 10 years or so would lead to a rise in global average temperatures that would completely undermine all efforts to combat poverty and achieve sustainable development. But there are good reasons to believe that the international community is progressing in the right direction.

In 2007 in Bali, the 192 parties to the UN Framework Convention on Climate Change committed themselves to launching negotiations on strengthened action on climate change, to culminate in an ambitious and effective international deal in Copenhagen at the end of 2009. Since then, there have been several highly significant events which make every one optimistic that a solid agreement can be reached this year.

For example, at the end of last year, the European Union agreed to a climate and energy package with which it will be able to reach its target of a 20 percent emission reduction by 2020 and minus 30 per cent if other industrialised countries follow suit.

A new US administration has assumed office, with President Barack Obama agreeing on 80 per cent reduction of greenhouse gas emissions by 2050, and to reduce his country's emissions to 1990 levels by 2020. President Obama is not only firmly committed to taking action against climate change domestically; but has announced that he will reengage his country in the international climate change negotiations.

Many pundits warn that the current global economic woes could throw efforts to combat climate change off-track. But even the financial and economic crisis is being used by countries such as China and the US, as an opportunity to change direction and to shift towards the greening of their economies. For the US, this includes unleashing $150 billion over 10 years to create five million new 'green' jobs.

China recently announced a $586 billion economic stimulus package, some 25 per cent of which is to help bolster conservation, environmental protection, and renewable energy efforts. All this has injected energy into the negotiating process.

At Bali, industrialised countries reconfirmed their commitment to lead in reducing greenhouse gas emissions. In the context of the Copenhagen agreement, developing countries must receive measurable, reportable and verifiable funding for measurable, reportable and verifiable action on climate change. In 2008, the first year of the two-year negotiating process towards Copenhagen, India put forward important proposals. For example, a reinsurance fund to deal with the unavoidable losses arising from climate change and an international technology transfer mechanism.

Domestically; India has blazed a trail by drawing up an ambitious National Action Plan. In the field of mitigation, key elements of the plan are to boost solar energy; promote research and development into renewable energies and enhance energy efficiency. And the plan foresees effective adaptation measures, such as helping farmers by boosting the development of drought and pest-resistant crop varieties. The question is now how India can dovetail its existing actions with international cooperation in such a way that national development and the fight against climate change become two sides of the same coin.

India already benefits from international cooperation on climate change as a leading participant in a key mechanism under the Kyoto Protocol. The country globally ranks first in terms of the number of Clean Development Mechanism (CDM) projects. Under the CDM, projects—in developing countries can earn saleable certified emission reduction credits. Countries with an emission reduction commitment under the protocol can use these credits to cover a part of that commitment.

The current carbon market structures are a promising first step in the right direction. But in their present form, they are insufficient to meet the challenge of drastically-reducing green house gas emissions and need to be expanded. India could therefore now put forward proposals on how carbon market mechanisms could be considerably strengthened and scaled up.

The negotiating process of 2009 and the UN Climate Change Conference in Copenhagen, constitute a narrow window of opportunity that if recognised and made the most of, could set the world on a path towards a truly green economy. India is ideally positioned to grab the opportunities offered by the negotiating process and to help lead the rest of the world to an ambitious outcome in Denmark at the end of this year.

8

SECURITY MANAGEMENT

INTRODUCTION

Security of a State or Country means only against war, invasion, revolution, internal rebellion, sabotage, etc. Variety and needs of security also vary from country to country from large and powerful states to small and weak ones and these are known to all of us. I am not going to discuss here the state or for that matter, the global security. I am now concerned with the security of Industries, social and personal security like banks, offices, hospitals, factories, residential homes and personnel security of individuals. For public sector industries the CIS largely looks after their security aspects.

In all types of security three factors or agencies play an active part.

1. The beneficiaries ranging from individuals to the management and employees of large industrial and commercial establishments.
2. The consultants, who plan and the security staff, who give security coverage.
3. The causes, both human and non-human, which creates security hazards and problems.

The beneficiaries have to be aware of the security risks, they are exposed to and the ends for which, they need security coverage. If the beneficiaries are security minded, the second agency *i.e.*, the expert and security personnel are not only helped but encouraged to a great deal, if they are sure that the clients and beneficiaries have the right attitude and will follow the advice and precautions, administered to them.

Functionally, the state police is expected to manage the overall security system of a country. The police draws legitimacy for the functioning and laws provides ample power for the purpose. As per Barley's definition of 'police' is "people authorized by a group to regulate interpersonal relations within the group through the application of physical force". This definition is more appropriately applicable to the policing as a function in general and need not be restricted to the police as an organization. However, when it was felt that state police force is inadequate to safeguard the life and property of citizens in a country, the concept of "private security" system was born and developed in course of time as a full fledged guardian of security of personnel as well as institutions world over.

Before justifying the entry and existence of private security system, let us put a glance on the nature of duties and work load, the state police is saddled with.

Statutorily, the police is required to maintain Law and order, which is its primary duty. Prevention and detection of crime, surveillance over suspected criminals, patrolling of public places, regulating entry to public places, regulate vehicular traffic, protect public property and life of citizens, apprehends, prosecute the offenders, control riot and affray, fire fighting etc. Afflux of time has necessitated some more additional functions like providing security escort to V.I.P.s, designated individuals and potential meetings etc. Due to increase in Country's population, large scale unemployment, rapid urbanization, mushroom growth of Industries, slums in the periphery of urban centers, invention of technology in the field of communication, transport, production of consumer goods and services etc., the crime rate has increased 100 fold. Over and above, the terror thrown open by Naxalites, Maoists, the communal forces, religious zealots, obscurantist forces, cessationists, divisive forces, drug peddling, Economic and political Mafiosi etc., the work load of police has crossed beyond the tolerable limits. Increase of the force, time and again with the state's limited resources has failed to cope with the pace. The emergence of the golden era in trade and commerce, sustainable development, large scale investment in economics and financial sectors, blazeoning of commercial activities and cut throat competitions, gave birth to business espionage, piracy of trade secrets, intellectual property right, patents and products, danger of subversion and sabotage and hundreds of sundry other economic and cyber crimes have shaken the police force to its heels. The police force today is not considered as the be all and end all of the state.

All these factors gave rise to emergence of private security system which started with a concept of "watch and ward". In course of time, it has developed into a supplementary and complimentary force of state police. The private security Industry is now being called upon to play a crucial role in promoting place, progress stability and economic growth. Governments world over, have widely out sourced the private security organizations. In U.K., the citizens charter of 1991, visualized the off loading of some of the duties of police, naming escorting of the prisoners, guarding and managing prisons, management and immigration centers, regulation of traffic etc. to private security agencies. USA and some South American states have followed suit and the demand for private security personnel has increased tremendously in India and abroad. The Govt. of India is contemplating to introduce a legislation to regulate the quality and performance of the private security agencies.

A security man now, needs more in brain than muscles. Intelligence comes first and strength at last. Data is the key measure in the present economic scenario. It has to be protected against destruction, leakage, piracy or misuse to the detriment of productivity. This need arises from the phenomenal growth of high-tech devices line computers, Electric mail, Electronic money transfer and such like magic of the chip. The cyber crime has increased ten fold and the magic of individual criminals will invariably try to warm into such systems. To insulate industrial, commercial and financial institutions against the danger of infiltration or intrusion of intelligence operation an efficient and competent in vigilance, intelligence and investigation has to be the integral part of the system.

Today, the threat perception from the Naxalites/Maoists have gained momentum. The state police, the CRPF, in spite of their war against terrorism and mainly conceding their large scale casualties have not been able to create confidence in the minds of general public. Large numbers of police informers, special police officers have laid their lives. To me, the security industry is to come forward and face this challenge. At this stage, they may be engaged in collection of intelligence and work as combatants in the rear.

Finally, the institution of private security is as old as the concept of personal profession. The genesis of organization of public police lies in the entrustment of this basic service to an earmark segment of society, so as to enable the members of the society to freely follow their pursuits of life. This however did not divest the individual from securing himself and his possessions against any threat. The private security, therefore persisted despite the evolution of public policing. The emerging crime and security scenario is clearly out pacing the police resources to cope with it. The emergence and growth of private sector in security is clearly drawing a line between what a public police can and should do and what is to be left to private security. How far these two tiers of the security system would go along, the police attitude towards private security ranging from "Denial" to "Active partnership", would largely be dictated by the crime situation and security in a country. Mean while, cogitation and articulation of this important facet of law enforcement and security system should be continued to pave the way for a meaningful and healthy relationship between law enforcement and private security.

THE PRINCIPLES OF SECURITY

The following are some of the security principles laws:

(*a*) Informed,
(*b*) Directed,
(*c*) Independent,
(*d*) Co-operative,
(*e*) Monitored,
(*f*) Consistent,
(*g*) Unpredictable,
(*h*) Concentrated,
(*i*) Appreciated,
(*j*) Acceptable.

(*a*) **Informed:** This is a natural pre-requisite of any operation. Before it can commence and be intelligently directed, it must be based on sound and up-to-date information. With modern computer printout and other mechanized business systems, there is no excuse these days for management not to have accurate and timely information indicating precisely where any areas of loss may be occurring.

(*b*) **Directed:** Even the best information is useless unless it is somebody's clearly defined task to study this information and put into effect the necessary security measures to correct the loss situation. It is very necessary that security direction should be at Senior Management level or at least with direct access to a Senior Manager. It does not mean, of course, that this person is personally involved in the execution of security policy but is only the functional head who must ensure that the necessary direction is given to the individuals who prepare Assignment Instructions and train the staff that must execute the instructions.

(*c*) **Independent:** It is now generally accepted that the operations and the staffing of the security organization should be independent of the normal line hierarchy of the organization it is servicing. If this is not so, the security operation will become biased or influenced by line management and it will not be true security. This is in no way meant to abrogate the line manager's responsibility for the security of his part of the business. Security is an independent function and it must have its own independent and parallel reporting system and its day-by-day operation and method must be independent of line interference. This is where the tact and diplomacy of the guarding manager comes into effect.

(*d*) **Co-operation:** Having said that the security staff must be independent, this does not mean that they work in a vacuum. They are only there to assist management and they need the willing co-operation of many people. Not only the management and workers in the company concerned but also people outside such as the Police, Fire, Ambulance services, perhaps advisers, such as Architects and even the Company's suppliers and customers. Co-operation must be sought in the operation of security. Very important, also is the co-ordination of security with neighbours of one's own premises. It is recommended that security staff liaise with sites adjacent to their own.

(*e*) **Monitored:** The security system must itself be subject to check and supervision and this checking system should be in-built. It is not enough to merely instigate what may be thought to be a good security system. There must be a reporting system, which in turn, is monitored and supervised at the guarding company headquarters.

(*f*) **Consistent:** The operation of security must be consistent, consistent in time, in space and towards people. By this it is meant that security must exist at all times, not just when the Managing Director is upset, or when the Company is not too busy. It must also apply equally in parts of the building and its perimeter, car parks etc. It must apply equally to the Managing Director and the most junior factory-hand and of course, to the office staff. It is vital that the guards on site are trained and briefed to understand this policy of Operations.

(*g*) **Unpredictable:** This principle is not in conflict with the one above although it may seem so, at first sight. While everybody will know that security is to be applied continuously, they must not be able to predict exactly how and when, or how often, a certain measure will take place. As soon as the various operations of security are predictable, they cease to be secure. Patrols must be varied, searches done without prior warning and spot checks on vehicles at random.

(*h*) **Concentrated:** While it is fairly obvious that the concentration of security would be most effective, it is again obvious that this can usually only be achieved at the expense of some other considerations. If certain materials in the factory are of a very much higher value than others, it would pay to concentrate these in a small area and give it special security precautions and safeguards. This concentration may, however, cause some inconvenience in the practical operation of the plant or the flow of materials and the principle can only be applied as far as is practical and economical and with the clients approval.

(*i*) **Appreciated:** The reaction of the client and the staff on the site, that is being protected will allow you to gauge whether or not your efforts are being appreciated. Friendliness by staff to security will indicate whether the efforts of the guards is being appreciated. However it must be borne in mind that over familiarity with the client's staff will undermine the security operation.

(*j*) **Acceptable:** The client will through the operations, management, liaison, visits make the company aware of his acceptance or will make the necessary amendments to practices on a day-to-day basis.

RECRUITMENT OF GUARDING STAFF

Having established the quality of the management to cultivate and promote the standard of security that has been set as your 'bench mark', you should now aim to recruit men/women who can achieve the standard set for the security function, which should cover the following.

(*a*) The protection of people from accident and injury.

(*b*) The protection of loyal labour from subversion, intimidation, fear and allow them peace of mind.

(*c*) The protection of people from terrorism and kidnapping.

(*d*) The protection of assets from accidental or malicious damage or sabotage.
(*e*) The protection of assets from external and internal theft.
(*f*) The protection of technical and commercial information from espionage.
(*g*) The protection of essential services such as gas, electricity water, transport and communications from deliberate damage.

Recruitment should be directed at the calibre of person who can work to and promote the standard of security you aim to achieve.

Marketing a Security Service

The Marketing of a Security Service is of paramount importance. Having established the standard of service you intend to provide, you should address your marketing and selling efforts to those companies who require an up-market service and are prepared to pay for the service at the right level.

Negotiators for the right charge rate, should always bear in mind that in order to recruit the right operative to provide the standard you have set down to achieve, you must be prepared to pay the right wage in order to attract the right person. It comes to reason therefore that the charge rate to the client will always reflect the quality of service provided.

Management Controls

Management control of costs dictates whether or not your company is being run profitably. In the guarding industry, it is therefore vital that hours worked against hours contracted are strictly controlled. If the company is contracted to 100 hours a week then provided, the guard is paid no more than 100 hours, there will be a profit margin. If however due to lack of control a guards are paid 110 hours, you will find that there is no profit within that particular contract.

WHAT IS SECURITY?

Definition

"Security" is not confined to watch and ward or "chowkidari" alone, it has a very large dimension. That the science of Industrial Security is now a distinct discipline by itself. In the days of improved technology, and competition in the international marketing arena, the importance of fortifying security system in every industrial establishments has become a bane. Due to globalization of economy, various security hazards have surfaced in the production centres of goods and services. Followings are some of the hazards, so far security is concerned.

(*a*) Total loss prevention concept. A proper mix of technology and man power is sine quinon of cost effective security.
(*b*) Theft perception (General)
(*c*) Sabotage
(*d*) Espionage
(*e*) Threat perception including kidnapping of heads/key personnel
(*f*) HRD Security
(*g*) Trade Secret Theft

(*h*) Terrorism
(*i*) Computer Security
(*j*) Security against fire
(*k*) Product related security

To contain all the above security hazards, training, experience and education can be the triangle of power that provides quality leaders in security and safety for the 21st century.

In brief, it is meant for the prevention of losses of all kinds. From whatever cause, security has come to mean, all the measures, physical or procedural which are concerned with preventing losses, caused by man made or natural hazards, maintaining order and preventing crime or other wrongful personal conduct, detrimental to the organization.

What does Security Cover?

Security covers virtually every aspect of our lives and a client, who employs a guarding company expects the definition of security as given to you, to be interpreted literally and that level of service provided. It is therefore necessary to realise that, when servicing a security contract, you just not providing man or women in the numbers agreed but are providing trained security expertise, which meets all requirements laid down by the client. In order to achieve an efficient and accepted level of performance, it is necessary to build up an open relationship with the clients, ensuring that both of you understand each others' problems and inturn are able to compromise and co-operate to achieve the laid down proficiency goals.

Security Master Planning

As per P. Benjamin, "security planning in a comprehensive and formal manner is yet to begin in most of the Industrial environment in our country. Long term planning of the security has no reckoning. Thus proper security planning must be comprehensive throughout the facility and it must take into account the need for a commitment to security consciousness at all levels in the organization. The concept of a comprehensive productive master planning of security is the need of the hour".

The plan must be comprehensive and formal. The need is for a commitment to security consciousness at all levels of organization.

The broad understanding of the organization.

The broad understanding of the organization, the goal and mission.

Before planning, a detail survey must be undertaken.

The initial conceptual design, leading to final design. It is said that security master planning approach is similar to a through health care diagnosis. All symptoms major and minor are examined and traced to the malady. The four major phases:

(*a*) Asset definition
(*b*) Threat assessment
(*c*) Vulnerability analysis
(*d*) Selection of suitable security measures.

Detail Elaboration

(*a*) Assets like buildings, machineries, equipments, goods and informations and also intangibles, like the skills and knowledge, employees' morale, the image, the public good will and confidence and the whole flock of personnel in the organization. From this inventory, the planner could know the extent to which the organization would suffer, if such assets are lost and damaged. The planner must identity the critical assets. For example take the case of a power generation company. If continuous supply of lubricants is stopped either accidentally or intentionally, it may lead to very costly power shutdown and repairs. So interacting with middle management and frontline supervisory level is a must. Here, ergonomic theory could also be applied.

(*b*) *Threat assessment:* To study the potential threats from criminal activities to national disaster, potential harmful events must be identified. For example: In a carpet industry, the key harmful event is fire hazard. It can be both criminal act and natural calamity. So. It must be planned accordingly to install fire prevention measures. For this, the following dates be analyzed.

(*i*) Previous security records of the firm.
(*ii*) Records of history of reported losses maintained by the office.
(*iii*) Crime data on local, national and international.
(*iv*) Local police records, pertaining to crimes in and around the factory premises.
(*v*) Views and opinions of experienced law enforcing agencies, external environmental agencies and meteorological Deptt.
(*vi*) Data on history of Accidents in the firm.

(*c*) Vulnerability analysis is to identify potential points of attack. There may be different types of attackers. So, measures be defined clearly to as certain these.

(*d*) Selection of suitable security measures, that encompass the following most probable threats.

(*i*) Necessary policies and procedures.
(*ii*) Physical security hardware symptoms.
(*iii*) Security personnel and their tasks and responsibilities.

Policies and Procedures

These provide foundation for a security plan. Management's direction is stipulated in the form of policies and the process of implementation as defined in the procedures.

The policy evolves out of various pressures, the organization is subjected to, like political pressure, Govt. regulation, community pressure & economic consideration etc. Clear policy directives, should cover all functions and include the following.

(*i*) Protection of critical classified information, proper filing, storage and secure destruction after done with.
(*ii*) Security of employees and visitors.
(*iii*) Basic security procedures for control of persons, material and keys.
(*iv*) In the event of intrusion, detection, response procedures.
(*v*) General policies, directly related to security and prevention.
(*vi*) Procedures for good dispatch and reception areas, store yards and maintenance of ware houses.
(*vii*) Emergency procedures and resumption of activities.
(*viii*) Security awareness, information and training of employees.

Security of Hardware and Systems

Electronic security measures are to be planned. The slogan is Deny, Deter, Delay and Detect.

(Four D's)

Deny = Deny unauthorized entry/intrusion.

Deter = Take deterrent measures in smoke, toxic gases release.

Delay = Make delay in spurt of the hazard till relief reaches.

Detect = Investigate the causes of the hazard.

Remedy = Make use of Electronic security implements like alarm, C.C.T.V., Fencing hardened enclosures, vehicle barriers, bullet proof personal protective equipments and conveyances, safes and vaults etc. A successful system requires efficient planning. A committee, comprising of senior experts drawn from relevant functional group, which may include systems group, engineer, finance and security in charge.

Designing a Security System (Step by step)

1. The owner, Chief Executive should engage a qualified consultant/firm to design the security system of his/her factory or establishment.
2. The owner's interest is on money and his short term involvement in the security industry must be discouraged.
3. The consultant should obtain a copy of the site plan of his factory/establishment from the owner at the first place.
4. Then the consultant should start preparing the project quality plan (PQP).
5. The consultant should know from the owner in clear terms, the reasons for preparation of such a security system.
6. The consultant ascertains the reason of owners' worries on large scale theft of goods/finished products. Also, issues on occupational health and safety hazards, the safety of staff and visitors.
7. The consultant should sit down to draw up a security policy of the firm/establishment, after ascertaining the reasons of owners'/CEO's concern on above issues.
8. The consultant should then sit for identifying various risks and the useful informations he should get from his client on the following.
   (*a*) Who is his enemy?
   (*b*) How much harm the enemy may cause?
   (*c*) How likely the enemy is to cause the harm?
   (*d*) Whether the owner/CE can accept the risk of harm.
   (*e*) Whether he can reduce the risk of harm.
9. The owner/Chief Executive agrees that he can reduce the risk by an appropriate security system.
10. The consultant starts gathering the brief.
11. He must ascertain the approximate budget cost of the system, the owner/CE requires to spend on the following items.
    (*a*) Whether long term/short term system.
    (*b*) Carry out site tests to determine the most effective options for security equipments.

(*c*) Design an effective "electronic perimeter" for the site.
(*d*) Review the operation and layout of the company and recommend modifications.
(*e*) Document to carryout installation without interference to the operations.
(*f*) Develop an appropriate "Commission and test plan" for the system.
(*g*) Develop an "asset management plan" for the system.

12. The consultant has to make safe survey and defects as detailed below if any, be taken note of.
    (*a*) The company's buildings are arranged differently as shown in the plan.
    (*b*) There is no different passages for visitors, who are required to enter through factory and store areas.
    (*c*) The switch boards are exposed (installed outside) and prone to unauthorized use.
13. Site Test for installation of security appliances, like fixing CCTV, (Internal/External) Dictators (Internal and External), mock unauthorized Access to switch Board, response times in mock alarm situations.

Design installation and implementation

1. The PQP, the consultant would then prepare when design commences. The plan covers matters such as the scope of works, inputs, contact people, time and cost details and work method statement. All the resources of PEP should be involved in the technical design. "Equipment" may be a mix of technologies, but it needs to be manageable. "Procedures" for the design include requirements for operating and maintenance manuals, training and work method statements.
2. If the manufacturer of Equipment expresses reluctance for test, those should be out right rejected.
3. Within the framework of the security policy, it is good practice to get inputs into the design as it progresses and before it is finalized by a "Review Session" from people not directly involved in the project.
4. The next job of the consultant is to discuss closely with the owner/CEO in the development of the design and must obtain his consent to input from a review committee. This input includes:
    (*a*) Road diversion if any. Fixation of automatic gate on main entrance with access detector, providing early warning of unauthorized access. (If not already there)
    (*b*) Make use of control room in the city (city office if any) provision of a automatic gate at the entrance to the site to allow access to the site (preferably in the night) by authorized persons. Also provide a warning that an unauthorized vehicle has approached the entrance. Provision for dispensing with the staff, working overtime on security.
    (*c*) Provision for requesting local police to carryout patrols.
    (*d*) Provision for security guards for the duration of the special event.

Equipment components as per design

(*a*) Automatic boom gate
(*b*) Electronic perimeter
(*c*) Movement detectors in each room
(*d*) Sensors and audible alarms near vases
(*e*) Sensors attached on vases

(*f*) CCTV system covering each room with monitors in each room (medium quality).
(*g*) High quality CCTV system with a camera monitoring each vase and the entrance/exit doors at all times.
(*h*) Signage for CCTV system.
(*i*) CCTV and Detectors (including boom gate) connected to security control centre .

Selection of contractor

The consultant would assist the owner/CE for selection of contractor. He should follow the following criteria.

(*a*) A minimum experience of two years in this work.
(*b*) Demonstrated experience in the installation of security system.
(*c*) Acceptable quality assurance system.
(*d*) Sufficient financial capacity of the contractor to carryout the work.
(*e*) Ingenuity of job planning.
(*f*) Satisfactory report from at least three referees.
(*g*) Acceptable procedures for securing information.
(*h*) Manufacturer of appropriate items of equipment.

Installation

The consultant should look to the following aspects at the time of installation.

(*a*) The installation work may be conducted during night preferably, causing no interference to the company's usual operations.
(*b*) A photograph (time date) is taken of each room before and after the work is carried out.

The implementation

The consultant should see the following:

(*a*) Commissioning and Testing.
(*b*) Handover from the contractor of the system and all documentation.
(*c*) Maintenance and defects correlation by the contractor for a specified period.
(*d*) On going interest in the system by the owner.
(*e*) Keep a provision for an asset management plan for the replacement of equipment.

Tests

(*a*) Operation of cameras and detectors.
(*b*) Mock intrusion with perimeter intact determine response by security.
(*c*) Mock attempt to break into main switch board determine response by security.
(*d*) Allow access to inside determine response if painting is touched.
(*e*) Mock intrusion of boom gate determine response by security.

Outcomes

Some of the outcomes are listed below:

(*a*) The system is installed to the budget.
(*b*) The special event occurred without incident.

(c) Police patrol discouraged persons from loitering near the company's site.
(d) The boom gates functions properly.
(e) There has not been any successful penetration to the system.
(f) Better community relations due to publicity and liaison with local police.

Finally, the advantages of design system are:

(a) There is more to a security system than purchasing / installing equipment.
(b) The designer needs to have relevant skills.
(c) No two systems can be exactly the same.
(d) There are cost benefits.
(e) There is always need to upgrade security without the need to use actual security equipment.

ECONOMICS OF SECURITY

With the liberalization of economy, the Industrial and business houses in India are gradually realizing the value of cost effectiveness of their operations. Gaining a competitive edge over the rivals, has become a sinequanon of the success of a business operation, be it public, private or Government sector. The basic element involved in being competitive is to keep the costs to the lowest. This is possible only when costs by way of wasteful expenditure, malpractices, pilferage, accidents, theft, sabotage, negligence etc., are drastically reduced, if not totally eliminated. To the extent these avoidable losses are reduced, the competitiveness as well as the profitability of the organization will improve correspondingly. For, one run saved is one run scored. Security cannot prevent all losses. However, it is one of the most important components of the loss prevention effort. Security today has assumed much importance as a total management function having direct bearing on loss prevention, productivity and profitability of an enterprise. When we relate security with productivity and profitability, a question arises in mind; can security performance be measured? Measuring a security performance is necessary not only to determine its cost effectiveness but also to assess its utility as a measure of loss prevention. It identifies various measures to facilitate constructive action in minimizing losses. The criteria of measurement of security performance would be both tangible or measurable as well as intangible or non-measurable. For instance, it is possible to measure in monetary terms to what extent prevention of pilferage, thefts and recovery of stolen goods through intensive security measures, has resulted in savings for an enterprise in a particular year.

Take the case of five star and seven star hotels, where international celebrities stay. They basically prefer hotels, even bearing thc logo and flag of less stars, where the security arrangement is exemplary, fool proof and satisfactory. They never bother for good food or beverages either. We will assume that the security professionals attached to them make comprehensive survey on the security arrangements of a number of five star hotels and finally fit select the one for stay of the VIPs to be fit for stay from security angle. In anyway, it may be remembered that there is a cost to the effort involved in reducing the losses and the cost should obviously be lower than the cost of the loss prevented or reduced. The operations must be cost effective. A modern and effective security setup today is the one, which is designed to utilize an optimum man machine mix of performing efficient, effective and reliable operations at the minimum cost. The following factors are responsible for cost reduction, after up gradation of security system.

(a) Attraction: Infuse pulling power of the establishment mixing glamour. The workers prefer to accept less wages.
(*b*) Job security
(*c*) Security in work places
(*d*) Protection
(*e*) Moral climate
(*f*) Good management
(*g*) Role of security personnel
(*h*) Organizational power
(*i*) Corporate culture
(*j*) Insurance premiums
(*k*) Energy consumption: The lower the electricity consumption, the lower the cost.
(*i*) Time and attendance: The time keeper (security), if replaces another function, there will be saving in cost.
(*m*) CCTV and good lighting
(*n*) Good Asset management
(*o*) Provision of security compartments. Direct cost benefits like reducing the staff by upgrading security system in introduction of modern technology.
(*p*) Reducing injuries by introducing good security system. The cost of treatment, loss of man days etc.
(*q*) The up gradation pays back, whatever is spent in it.

If the indirect benefits are taken into consideration, the capital outlay would be recovered in about 1/2 years. In reducing the cost and making it cost effective, the following steps are very much necessary.

1. Identify the problems — Human, Technical and financial.
2. Design the security up grade to suit the needs of the venue.
3. Realize that the security upgradation, will result in economic benefits.
4. Identify the benefits to the owner, security staff and manager.
5. Identify other non security benefits, which can be integrated to the system.
6. Sell the system upgrade to the owner and security staff.
7. Implement the upgrade and provide feedback to the owner and the security staff.

Allocation of adequate resources is a must for security measures. Much money of the company could be saved in eliminating or reducing the internal and external thefts, prevention of unauthorized movement of personnel and Goods and increase of worker productivity, by creating a peaceful and incident free working environment.

Types of Security Hazards

1. Theft perception (General)
   (*a*) Goods,
   (*b*) Classified documents,
   (*c*) Trade secrets,
   (*d*) Cyber thefts

(*e*) Bank ATM.

(*f*) Sophisticated machineries

2. Terrorism
3. Sabotage
4. Kidnapping of heads/key personnel/VIPs for ransom, extortion
5. Industrial espionage
6. Violence at the work places
7. Hijacking
8. Campus killings
9. Communal riot.

THEFT PERCEPTION

(*a*) Theft in an Industry/establishment may be of various kinds like stealing finished products from the store yard, raw material, sophisticated machineries, furniture and fixtures, building materials, money from accounts and finance section. For prevention, there should be night vigil at the point of exit of workers, after the shift hours. Through search may be made by the security men at the gate. Similarly, night vigil may be arranged to guard the both inner and outer side of the premises.

(*b*) There are danger for theft of classified documents, confidential records etc., from the office by the members of staff as well as unauthorized intruders. They sell these documents to the peer competitors for gain. To prevent this type of theft, it is incumbent on the part of the line managers to adopt fool proof preservation procedures of records, documents and confidential papers and maintain double lock systems. Both manager and the clerk in charge of records maintain strict security measures when handling such papers. Presence of guards at the time of opening and closing of safes and vaults and maintenance of log books must be made mandatory.

(*c*) Theft of Trade secrets is a very common in business and industrial sectors. Cut throat competitions gave rise to such threats. These types of crimes are generally perpetrated by the employees of the Industrial establishments for gain. They secretly colluse with the rival competitors in the business and pass on the trade secrets, being heavily bribed. In such situations, it is necessary to maintain the exhaustive profile of the employees, who are entrusted with this type of sensitive documents, formula sheets, plans etc. The 24 hour vigil equipment like CCTV, should be there to monitor access to these papers/records/documents. Pre verification of antecedents of the employees in charge of handling of these secrets is to be done before their employment. So, also, strict vigil may be maintained by the security organization on their movement, their association, telephonic communication with outside agencies. To prevent theft of cash, valuables, A.T.M. cards, credit cards, investment papers, insurance documents, shares and debentures etc., physical security measures by application of locks, grills, steel bars, reinforced glass and such type of hardening deterrents be provided.

(*d*) Sometimes, sophisticated machineries are stolen by the employees and outside intruders. These items are very light, small and therefore could be smuggled out easily. To prevent this crime, the security men must be provided with metal detectors, and censors, to check the suspected employees, who maintain shaddy history sheets and other visitors/intruders at the time of their exit.

The above application of physical security measures (target hardening) is just one component of package of measures, which should be addressed, when consideration is being given to the prevention (and of course detection) of criminal activities.

This multi-pronged attack on criminal activity has been termed "Situational Crime Prevention" (Poyner 1983) and an important element of this approach is "Crime Prevention Throught Environmental Design (CPTED)".

Situational Crime Prevention

Situational Crime Prevention refers to a range of pre-emptive measures which:

1. Are directed at specific forms of crime
2. Involve changing, managing or manipulating the environment in which these crimes take place
3. Are systematic, and as permanent as possible
4. Reduce the opportunities for commission of these crimes, and
5. Are seen to do so by the potential offender

It is sometimes known as "Opportunity Reduction" or "Primary Prevention". The methods used to fall into three main groups according to the level at which they are designed to operate, *i.e.*, those aimed:

(*a*) At the individual

(*b*) At the community, and

(*c*) At the physical environment.

Crime prevention measures aimed at the individual (*i.e.*, the potential victim) include target hardening (*e.g.*, physical security measures) as well as publicity campaigns such as free property marking schemes, and so on.

Measures aimed at the community include those initiatives, which encourage neighbours to look out for each other's property, and measures aimed at reducing crime through changing the physical environment are based on architectural solutions.

In 1980 a British Home Office research unit paper was published which identified eight key aspects of situational crime control measures. The first three of these reduce the opportunity for crime by making it harder to commit and the remaining five vary the costs and benefits of offending (Clarke and Mayhew 1980). The eight key aspects are:

Target Hardening

This refers to the increase in physical security measures by the application of locks, grills, steel bars, reinforced glass and so on.

Target Removal

Obviously removing the potential target from the environment will have an impact on crime levels. An example of this is the payment of employees by direct bank transfer rather than by cash to remove the opportunity for a payroll robbery.

Removing the Means to Crime

This relates to measures that are taken to remove the tool or equipment needed to commit the offence. Examples include the security screening at airports, which has reduced "hyjacking", gun control measures and preventing children from buying spray paint cans.

Reducing the Pay-off

In property crimes, this strategy refers to making the goods less valuable to the offender by taking such measures as property marking or security coding initiatives.

Formal Surveillance

This category includes formal surveillance by the police, private security firms, the use of CCTV as well as Community Surveillance Schemes.

Natural Surveillance

"Natural" surveillance takes place as people go about their everyday tasks and it can be enhanced by planning and architectural solutions, which increase the opportunity for this to occur.

Surveillance by Employees

Natural surveillance may have limited effectiveness and formal surveillance can incur significant costs so another approach is to consider enlisting the help of certain types of employees who by nature of their work are well placed to observe criminal activity. These groups include shop assistants, doormen, janitors and early morning delivery men.

Environmental Management

This refers to measures, which involve manipulation of the social rather than the physical environment. It includes such examples as the organization of football matches where there is liaison between the two clubs and the police to avoid confrontation between rival supporters and it also includes housing policies, which attempt to avoid high concentrations of children on certain types of housing developments.

More recently (Home Office 1993) this list has been expanded to include.

Deflecting Offenders

Diverting offenders away from particular localities or offences (*e.g.*, locating clubs away from residential areas or reducing congestion in markets to lessen the opportunity for street crime).

Entry/Exit Screening

Identifying potential offenders at entry and exit points (*e.g.*, baggage screening and merchandise tags).

Removing Inducements

Making expressive crime less worthwhile (*e.g.*, repairing vandalism and removing graffiti quickly to prevent repeat incidents).

Rule Setting

Explicitly setting out the legal or acceptable social behaviour (*e.g.*, regulating employee's private use of telephones; preparing and circulating codes of practice).

Defensible Space

Many of the concepts underpinning environmental crime control, derive from the work of Oscar Newman in 1972, entitled "Defensible Space". Newman identified a hierarchy of defensible space showing how it was possible to move from public space, through semi-public and then semi-private space to finally arrive at private space which was under the sole control of a specific individual, or small groups of individuals. So, for example, the street is public space, the entry to a block of flats would be semi-public, the lobbies at each lift landing would be semi-private with the interior of the flats themselves being private space.

By analyzing crime in New York public housing, Newman identified three principles which helped the commission of crimes and which made them difficult to prevent.

The first of these 'was that the impersonal character of areas promoted "anonymity" which allowed criminals to feel secure in the knowledge that they would not be identified. As intruders are indistinguishable from authorized personnel, there is clearly no reason for residents (or employees) to question their presence in a particular area. Therefore, as the anonymity of an area increases with size, it is helpful to reduce the number of people passing through a particular entrance. This increased access control in itself enhances the crime prevention effort.

Also, as the degree of sharing of semi-public space increases, so "ownership" of it decreases and the less defensible it becomes as confidence to challenge an intruder declines (Coleman 1985). Too much public space is seen as the responsibility of "other" people (Wilson 1980).

The second principle was that of surveillance. Clearly, criminals like to operate unseen and residents (and employees) are safer and more in control of their environment if they can clearly see who is approaching. By providing surveillance opportunities, the group is more likely to be vigilant and more willing to challenge intruders.

Finally, the availability of alternative escape routes increases the confidence of the offender as he is more likely to believe that even if detected, he would be able to escape. Newman (1972) used the term "defensible space" to include a wide range of mechanisms, such as:

(*i*) Real and symbolic barriers
(*ii*) Strongly defined areas of influence
(*iii*) Improved surveillance opportunities.

All of which combine to bring an environment under the control of its occupants. The following elements of defensible space were intended to interact rather than operate independently, although there were times when this was possible:

Territoriality

The physical environment may create perceived zones of territorial influence, for example by:

1. Site design - to indicate that an area is for the private use of a particular group of people, and
2. The mechanisms, which create boundaries, which in turn indicate a gradual shift from public to private zones.

However, it may not be "territoriality" that reduces crime but rather it may be "accessibility", *i.e.*, that by reducing access opportunities (a feature of enhanced territoriality) levels of criminal activity can also be reduced (Poyner 1983).

Real Versus Symbolic Barriers

Real barriers include such mechanisms as high walls, gates, fences and doors. Symbolic barriers include open gateways, shrubs and even a change in the texture of surface material. These both serve to give notice that one is passing from a public space to a private space where one's presence may require justification. The success of symbolic barriers depends on the existence of four conditions:

1. The intruder needs to be able to read and understand the symbol.
2. The inhabitants need to be able to maintain control through surveillance.
3. The defined space must make the intruder declare his intentions (*i.e.*, there must be a low tolerance to ambiguous use), and
4. The inhabitants need the capacity to challenge the intruder and follow this up with further action if required.

Thus a combination of symbolic barriers can denote that one is clearly passing through several zones of transition into a private area, without the need to actually use real barriers.

Natural Surveillance

The physical design of the environment can provide surveillance opportunities for the occupants. The ability to observe the public areas and the feeling that when in them, one is continually under observation, enhances the security of the environment.

TERRORISM

Terrorism is as old as William Shakespeare, who referred to "terror" in his plays-Kinglear and Richard III. From the age of interstate wars through the decades of revolutionary evidence and insurgencies, the world has moved into a new phase of intensive violence in the form of terrorism. The discussion on terrorism will cover the following:

(*a*) Definition and historical background.
(*b*) Types of terrorism
(*c*) Cause of terrorism
(*d*) Factors of promoting terrorism
(*e*) Terrorist, philosophies and methods
(*f*) Nuclear, chemical and biological terrorism
(*g*) Combating terrorism

(*a*) **Definition:**

(*i*) "A person, who indulges in the wanton killing of persons or disputes the services or means of communication essential to the community or damages property, which adversely affect the harmony between different religions, racial or regional groups or castes or communities, or coercing, or attempting to over awe the Govt. established by law or endanger the sovereignty or integrity of a sovereign nation.

(*ii*) Libya's version of terrorism is "To station American forces is terrorism." To monopolies the wealth of countries is terrorism. To dominate out let of seas and ocean is terrorism. To provide aging regimes with sophisticated weapons and oppress the people is terrorism. To use wheat and gold as political toys, when the world is starving is terrorism."

(*iii*) Historians first traced terrorism with one Arab religion teacher, Hassan Ibn Sabah, who founded the "Society of assabins". The original Arabic word being Hashshasin, which indicated the addition of terrorists to hashis. The members of this group (religious + political) believed that killing bad people on the command of their leader was a sacred duty, which ensured their personal salvation. This was in the middle of eleventh century.

(*iv*) Western historical traced political terrorism from the "reign of terror" inaugurated by Robespierre in 1793, during French revolution, which cost the lives of 40,000 people and led to the imprisonment of around 3 lakhs.

(*v*) The Tsarist administration in Russia killed five thousand revolutionaries and imprisoned thirty thousand people, which was also termed as terrorism unleashed by a cruel monarch.

(*vi*) During chiang Kaishekh's regime in china, civil war ensued between right terror group and left terror group led by chiang and Maotse Deng. Wide scale messacres were the accepted tools. After Mao came to occupy power, in 1949, he used terror in a bigger scale and established people's republic. He himself felt victim of this terror by his opponents in the name of cultural revolution. The arrest and trial of gang of four was the out come of this terror, where innumerable lives were lost in Tijiamin square, Bejing.

(*vii*) The genocide of Jews in Germany, where around 6 millions were killed by Adolf Hitler.

(*viii*) In 1968, there were US and European terrorism unleashed. Student unrest swept like grass fire. Loss of lives of US soldiers in Vietnam, aroused antiwar movement and president Lyndon Johnson renounced his second term. Students of Paris rose to near by destroy Charles de Gaulle's fifth republic. Uncontrolled 'Red Army' swept Tokyo. A sudden wave of violence swept in Spain's northern basque provinces forcing General Franco to proclaim emergency. Waves of street fighting broke out in Turkey, resulting imposition of martial law in 1970. Northern Ireland moved to the thresh hold civil war, resulting induction of Royal Army into Ireland in 1969. George Habbash, the Palestinian Guerrila, exported terrorism from middle east to Europe, sending commands to hijack an E1 A1 plane in Rome.

(*ix*) In the middle east Scenario, after creation of Israel as a result of settlement at the end of 2nd world war, the displaced Palestinians formed terrorist groups and fought against. Israel. From Japan to Ireland, terror networks established. This was visible in 1972, when terrorists of the Japanese Red Army indulged in indiscriminate shooting of innocent passengers at lod (Israel) airport, killing 26 and wounding 80. This was followed by assacination of eleven Israel sports men and six others in munich Olympic. In Latin America, two insurgencies separated by 20 years took place in Cuba in 1959, and Nicaragua in 1979. After Castro's victory, cuba became a communist state and for economic reasons, the country allied itself with the soviet block. This area produced terrorists like Che Guevara and Carlos Marighella. The biggest event of terrorism happened in world history is detonation of two nuclear bombs in Hiroshima and Nagasaki, Japan.

(x) Coming to the Indian subcontinent, Afghanisthan and Chechnya, the terrorist organizations like Lasker-e-Toiba, Alkaida, Taliban, ULFA and hundreds of small and big religious cessationists organizations are operating in Kashmir, Pakisthan, Chechnya, Philippines, Indonesia, Assam, Nagaland, Mizoram and Naxalities, maoists are operating in Orissa, Bihar, Chhatisgarh, Andhra West Bengal and Jharkhand.

(b) **Types of terrorism:**

State as terrorist: Practicing terrorism against the nationals of one's state by other. This is what is known as "Terrorism as a state policy". When a State promotes or practices terrorism in some other nation, it is known as "Terrorism as a foreign policy".

The other form of state as a Terrorists is Terrorism, as a state policy. This type of terrorism is practiced by Governments in several dictatorial states like chile, Nicaragua, Argentina, Pakisthan (when popular democratic Govt. was not there) and Philippines etc. The states under stalin, Franco, Tao, Hitler, Mussolini, Ayub Khan, 'Musar off' have been termed as Terrorists. Terrorism has been practiced against Tibetans in China, Kurds in Iraq and Iran.

Terror, as an instrument of foreign policy is practiced by states like US, the then USSR, Pakistan, China in third world countries *i.e.,* Nicaragua, etc.

Revolutionary organizations are

(i) Anarchist Terrorism,

(ii) Nihilism,

(iii) Nationalist ethical separatist aspiration terrorism,

(iv) Terrorism as a war by proxy,

(v) Terrorism by religious fundaments. This type of terrorism is spreading like wild fire in these days.

(c) **Causes of terrorism:**

(i) *Individual level:* It is psychological. Due to aggression, frustration surfaces, which hundred in attainment of goal. The discrepancy between demand and fulfillment, drives the individual to aggression which may be expressed on physical violence. Another cause is biological. Lombroso, the founder of modern criminology opines that there is direct connection between terrorism and vitamine defficiency.

(ii) *National level:* Imaginary or real deprivation is a factor in inciting a group of people to be terrorists. For example : The Maoists and Naxalites are indulging in terrorism in the hope of achieving faster economic development.

(iii) *Cultural environment:* The existence of tradition of violence (as in Latin America Afghanisthan and Indian tribal areas) conceived as a catalyzing factor of terrorism.

(iv) Over reaction in the form of severe repression or under reaction, implying incompetence on behalf of reacting govts, may both attract and encourage terrorism.

(v) *Minority group and terrorism:* The minor group in a nation justify terrorism, arguing that they cannot win conventional war against powerful majority groups, who oppress them on religious, cultural, social and economic fronts and are not willing to grant concessions or concede independence to them. For example, the kurds in Iran and tribals in North eastern states in India.

(vi) At times, the threatened regime recruit volunteers secretly doing for the state what the state prefer to avoid doing itself due to international pressure sanction and aggrandizement. For example, Pakisthan sending volunteers to Kashmir in India (Kargil war).

(*vii*) Corrupt and inefficient Admn. and poor law and order situation conflagrates terrorism. Besides events in one state bound to spill over into other countries and repercussions felt at some other place. For example, the impact of recent communal violence between highurs and Han Chinese in China evoked reaction in the middle east.

(*d*) **Factors promoting terrorism:**

(*i*) *Due to strong ideological motivation:* The political theorists have largely provided these theories like Rousseau's social contract theory, Herbert Marcus Theory and Marxists Leninist theory. The youth are attracted to these theories, particularly from the unprivileged sections and harbors romantic visions of terrorism for a cause.

(*ii*) Terrorism finds a fertile soil in a democracy to grow, than in totalitarian and dictatorship form of Govt. This is primarily Mao's, conviction. It has thrived in China, where as annihilated in USSR, Germany and Italy.

(*iii*) *Technological Advancement:* Terrorism thrives not only due to invention of sophisticated arms and ammunitions but also the sophistication of news media, communication network and availability of yet attractive targets. Confucious said, "kill one frighten ten thousand". If he has television he could have said "kill one and frighten one million".

(*e*) **Terrorist philosophy and methods:**

(*i*) Theories propounded by Carlos Merighella, "Guerrilla Warfare" by Mao, "Communist Manifesto" by Marx and Lenin, Guevara, Grivas of Cyprus and host of others are the sources of inspiration to give birth to terrorism. As for methods, general grievas of Cyprus has suggested, blowing up numerous public institutions and places and organized violence all around the 1st phase of offensive be started by blind sabotage and terrorism to make an impressive beginning. The students, should be utilized to distribute subversive literature and teacher to spread the ideology.

(*ii*) Those who do not cooperate or are visibly against the revolution, are to be kidnapped tortured and some times liquidated. For example, the Naxalites/Maoists are doing now in Orissa, Andhra, W.B. Chhatisgarh and Bihar.

(*iii*) According to them, violence should be aimed to bring the terrorists and their cause to lime light.

(*iv*) It is only the masses alone but the cadres of their own organization, against whom terrorism provides an effective tool. Measured and timely doses of coercive terrorism prove helpful to counter any hesitation to face hardships, fear of office reprisals, demonstration due to defeats, tendency towards desertions, erosion of faith in the hierarchy and general disenchantment.

(*v*) *Competitive terrorism:* This may arise between two or more competitive terrorist organizations and groups. The one whose intelligence liquidation and sabotage system and squads operate more efficiently and ruthlessly is able to put the others out of business.

(*vi*) *Subversive terrorism:* In this, the terrorists aim at hitting the authorities where it hurts the most by subverting their officials and forces for their own good. Mao succeeded in this venture.

(*vii*) *Agitational Propaganda:* The final stage of the demonstrative terrorism is in the form of agitational propaganda. By successfully organizing rallies, demonstrations, picketing, local raids, isolated attacks and wide spread civil strife and sabotage over protracted period, they aim at ridiculing the system. After reducing the establishment to farce by incapacitating its teeth, they hope to ultimately overwhelm it.

(*viii*) Other common methods are kidnapping, assassinations, brutal bombings, hijackings and so on.

(*f*) **Nuclear chemical and biological terrorism:**

This happens like the criminal attempts to steal a missible sub marine and fire its nuclear war heads in a city. The water supply of a city is poisoned with highly lethal radio active material by a terrorist group.

A local commander seizes nuclear weapons and threatens to use them against central Govt.

(g) **Combating terrorism:**

(*i*) The first strategy to suppress it, is to take a long time measure and should be directed to root out the causes and finally destroy the terrorist organization. For example, the Naxalites and Maoists work for upliftment of poor and downtrodden and in the process, they kill the corrupt bureaucrats, politicians, the repressing police who, according to them are largely obstructing the progress and development of the poor and tribals in the country. The Late Prime Minister Rajib Gandhi had publicity said that out of one rupee spent on so called development of the area and raise economic conditions of the unprivileged lot, only 15 paisa is actually utilized for the purpose and the balance 85 paisa is misappropriated by officials in the execution process. This concept still holds good and the agenda of Naxalites, Maoists is to bring the poor landless and the unprivileged section of the society into the main stream of national development process. The vested interests *i.e.*, the corrupt politicians, the bureaucracy and the elite sections, who combinedly loot the govt. exchequer, thwart their just movements and curb their activities by using the state force. As a result, the general public (poor and unprivileged section) are over whelmingly supporting them and directly or indirectly adding these terrorist organizations, and joining in their cadre. This is the reason, why Govt, fails to suppress them. It is now spreading like wild fire and may be, the whole country would come under its grip. Socio-economic inequalities in distribution of resources and wealth should be effectively curbed.

Therefore, the best thing to combat it, is to gain confidence and support of the people of the area, by implimenting development projects in the area and take steps to root out corruption.

(*ii*) To combat with political/religious/cultural terrorist organizations, international cooperation, bilateral cooperation, regional cooperation are necessary. This may express itself formally or informally in executive as well as legislative and judicial domains. It may include a regional intelligence pool, police coordination in apprehending terrorists, cooperation in control of frontiers and common training etc.

(*iii*) *Effective law enforcement:* The basic component of law enforcement is the 'legal sanction'. Since terrorism is crime, sanction is, its correlate. States can and often have armed themselves with additional powers to check terrorism. This is possible, if the terrorism is confined to the state boundary. If it becomes inter state/international, difficulty arises. No international legal repressive machinery exists to apprehend and try the terrorists, who generally go unpunished. Absence of international cooperation through judicial means, encourages recourse to terrorism. For example, the case of Dawd Ibrahim, who after committing terrorism, in india has escapted to pakisthan for the last one decade and enjoying a lavish life style there.

(*iv*) *Intelligence:* Inter State/international intelligence system should be upgraded.

(*v*) *Awakened public opinion:* Public opinion should be awakened to help the security forces and the authorities in meeting terrorism, public should be educated on this front.

(*vi*) *The role of mass media:* (*a*) Editors and journalists must apply self censorship. They should play down the event as a method of self censorship.

(*b*) The acts of terrorism must be condemned with ferocity to dissociate general public from the terrorists.

(*c*) The media should amplify Govt, efforts to fight terrorism.

(*d*) Media could be used as a public information tool to disseminate the rules of behavour to be observed in cases of terrorist assaults or the security measures to prevent terrorist acts.

(*e*) Regulating propagandas on an international Govt. level, agreements may be made between the countries involved in the task.

Police and Army

Terrorism is basically a law and order problem and it is mainly the job of police. However, due to its international in character and the severity of the happenings, military may have to be called in. Govt, of India, however, refuses to engage army and airforce, may be for political reasons, in suppressing the terrorism.

Avoiding over reaction: The state authorities should learn to control their reaction. Over reaction will mean playing into the hands of terrorists. If a state avoids killing them it is stronger state. Harsh penalties against terrorists are seen as a relation beyond the law and separation of law from justice assures that, order will be in jeopardy. Police should abide by the law, even if the terrorists are violating it. This can be tackled in the beginning and for its efficiency, there should be an excellent intelligence machinery.

No negotiation with the terrorists: Negotiations with the terrorists not only encourages them to continue with violence but also may amount to granting them a special, recognized political status. This is the view of Mr. Noemi Gal, an Israelli writer.

Raid and Counter Terror

Sometimes it ends successfully when raid and counter terror is launched against terrorists. Israel successfully launched raid and rescued all passengers from hijacked plane in 1977. But USA failed in the Iran front in 1985.

Counter measures against nuclear, chemical and biological terrorism: The counter measure dealing with explosive devices and the hijacked planes has not been very successful. In this scenario, it is very difficult to imagine that only effective measures could be taken against the future generation of terrorists, who may well be equipped with the horrifying atomic, chemical and biological weapons, to make their threats more dangerous and demands less practicable to meet. Therefore, reliance for survival against such new forms of terrorism may have to be placed mostly in the institutional arrangements. It may still possible to initiate some measures at international, national and individual level. Countries, fearing chemical terrorism can contrive some advance warning systems like the "Nerve Agent immobilized Enzyme Alarm and Detector" adopted by the USA, which can give adequate preparatory time to the people.

SABOTAGE

Every entrepreneur strives to continue process of measuring his products, services and practices against his toughest competitors or those companies known as leaders. Some blacksheeps in the organization, deliberately, take some adverse measures, there by the quality of the products and its quantity does not reach optimum level. As a result, the entrepreneur fails to compete with his peers in the open market. Therefore, it is incumbent on the part of the entrepreneur is to keep a strict vigil on saboteurs/potential saboteurs in his herd and see that, the rotten eggs don't pollute the whole bunch. They are no doubt a dissatisfied and disgruntled group for various reasons like stoppage of increments and service benefits for adverse character record, issue of warning, favoring other of their colleagues, there by raising jealousy and heart burning, personal animosity with their colleagues at work places and for which, they want to make the targeted employee, a scap goat by damaging the machineries in his section.

Sabotage may be of various kinds.

(*a*) Secretly striving to spoil the quality of goods.

(*b*) Theft of trade secrets and sell to the competitors out side.

(*c*) Damaging machineries and causing frequent accidents on the factory.

(*d*) Secretly inciting other workers to go slow and reduce production.

(*e*) Forming groups and placing various demands before management and indulge in frequent strikes.

(*f*) Misappropriation of company's fund in many ways inserting false bills/vouchers expenditures/ tender documents.

(*g*) Always remaining busy in working against the interest of the plant.

Remedies

1. The management, through the trained security staff, identify such dissatisfied and disgruntled employees and keep a watch on their activities.
2. The management must install sophisticated equipments like CCTV etc., in designated places to watch them from the control room, from the office of the G.M. etc.
3. Tighten the plant's safety system.
4. Try to stop theft by the help of security personnel.
5. Detection and investigation system should be made fool proof.
6. The H.R. Department of the plant should start various employee/labour welfare programmes to contain dissatisfaction among employees, relating to service matters and sanction of various service benefits.
7. Counter measures against the saboteers may be taken as far as possible.
8. The employees/leaders, who are found sabotaging against the interest of the company be identified and gradually weeded out.
9. Management should try to gain over the supporters/sympathisers of the saboteers, by extending various benefits and segregate their association with the recalcitrant lot.
10. Security system in Record section, Accounts and Finance Section, Safes and vaults, where trade secrets/classified documents are kept, are tightened.
11. Provisions of rewarding the employees for their good work, dedication and trust worthiness be started. Over and above, the management should deal with employees by application of principles of ERGONOMICS.

KIDNAPPING OF HEADS/KEY PERSONNEL

The task of executive protection has become increasingly challenging and demanding day by day. The operatives make use of the latest technology and have developed worldwide ramifications through the nexus of terrorists, smugglers and gunrunners. Their modus operandi is not only sophisticated but has acquired deftness, dexterity and speed. Therefore, this challenge now calls for specialization with a high degree of professionalism.

Thousands of senior executives of the large number of corporations throughout the world, who are of any significance in the commercial world, have become vulnerable to the threat of executive protection. Similar is the plight of thousands of other high profile personalities, who have achieved excellence in their respective fields like art and culture, music, sports, theater, cinema, journalism or politics. Various terrorist and criminal groups know it very well that due to their significant position in the society, kidnapping of these executives and high profile personalities can fetch them handsome ransom money, wide media coverage and also, an added advantages of negotiation power to get their gang members released who might be in jail. Terrorists indulge in these activities to create panic and to undermine the confidence of the people in the establishment.

The Threat

The threats, which a normal high profile executive faces today, are the following:

(*i*) Threat of physical assault or humiliation at work place by a violent mob of workers;

(*ii*) Threat of mental torture to self and family through anonymous calls and letters;

(*iii*) Threat of mugging while moving on the streets;

(*iv*) Threat of kidnapping and abduction for ransom or for black-mailing; and

(*v*) Threat of assassination.

An important point to note, when we discuss about threat potential is that, all the possibility of the threats mentioned above, do also apply to the family members of the executives. Another interesting point to note is that as we raise the protection level of an executive, the probability of the family members becoming "Soft targets" rises. The reason being that the moment a targeted executive becomes a difficult target due to increased level of protection, the attention is automatically shifted to the family members, who are likely to be less protected.

Reasons for Increasing Threat

A number of reasons may be attributed to increasing threat perception of the executives, during the past few years. Some of the important reasons are as below:

(*a*) Increasing Expectation;

(*b*) Diminishing Loyalties;

(*c*) General Lawlessness;

(*d*) Unemployment;

(*e*) Terrorism;

(*f*) Irresponsible trade unionism or political movement/disaffected employees.

(*g*) Unethical competitors;

(*h*) Disproportionate reasons; and

(*i*) Ineffective law enforcement.

Threatened Executive

People ask me how to identify a threatened executive? It is a difficult question to answer. However, we may profile a threatened executive in the following words: "Any executive who is required to take such decisions on behalf of the company which may substantially benefit or harm an individual or a group of people, whether workers or outsiders, is a potentially threatened executive today." However, this is a general statement and this definition may vary according to changing situation. Therefore, in executive protection, maximum weightage should be given by any protective agency to this identification of the threatened Executive, the level of threat perception and the persons/groups from whom the threat is anticipated.

Vulnerability

There are a number of vulnerable points in executive protection and a number of situations when even highly protected executive might become vulnerable or a soft target.

The following few points are important to note:

(*a*) Casualness at home and office;
(*b*) Drive between home and office;
(*c*) Check-points and bottlenecks;
(*d*) Traveling executives;
(*e*) School going children;
(*f*) House wives when alone at home or while shopping; and
(*g*) Children in playground.

Critical Areas

The critical threat areas are near home and office and while travelling between home and office. Some examples:

(*a*) A German businessman killed 300 yards from his home.
(*b*) A French businessman assassinated at his door step.
(*c*) A Senior executive of TELCO at Jamshedpur killed in front of his house;
(*d*) Another senior executive of TELCO kidnapped from his work place.
(*e*) An U.S. military officer killed 300 yards from his home.
(*f*) A Colombian businessman was kidnapped 200 yards from his office
(*g*) A Romanian Diplomat kidnapped between his home and office.
(*h*) An Indian diplomat kidnapped outside his residence.
(*i*) An Indian businessman was abducted while getting down from his car in his own portico.
(*j*) Mrs. Indira Gandhi was assassinated within her residence.

Precautions and Alertness

An executive cannot be protected by the security forces, if he himself is not aware of the threats and does not co-operate fully with the protective agencies.

Observations

Observation is an important aspect in executive protection. It is the responsibility of the security expert to ensure that the executives are trained to observe. Some of the important areas of observations are:

(*a*) Watch for disaffected employees;
(*b*) Watch for abnormal movement near office/Home;
(*c*) Train family members to discuss abnormal incidents;
(*d*) Debrief children periodically;
(*e*) Establish identity of visitors; and
(*f*) Do not hesitate to take protective agencies into confidence.

Countermeasures

There are numerous countermeasures in the area of executive protection. However, some of the general points which every security officer must keep in mind are

(*a*) Periodical review of threat perception;
(*b*) Evaluation of threat level;
(*c*) Notify concerned executives;
(*d*) Strengthen intelligence network;
(*e*) Train executives to observe;
(*f*) Train PSO's to protect and not to fight; and
(*g*) Discourage arming of PSO's.
(P.S.O. = Personal security officer.)

Do's and Don't

No exercise of executive protection can be successful, if the executives themselves do not adhere to the norms. Therefore, the executives must be trained and advised to follow the minimum norms. Some of the DO's and DON'T are mentioned below:

Do's

Executives are advised to follow the instructions as far as possible:

1. Evaluate and recognise security threats to self and family periodically as a matter of routine.
2. Trust your Security Manager/Consultant and follow his advice religiously.
3. Take your family and children into confidence about the possible security threats.
4. Train children to observe the surroundings and debrief them regularly. Discuss a few recent case studies, with them in a manner that they become more cautious and better educated and not "scared".
5. Brief children, female members to be alert while playing/shopping.
6. Maintain records of visitors and telephone calls received at home and office. This record will be vital document for the investigators in case of any happening.
7. Train personal staff, family members and servants to converse briefly on telephone and divulge no specific information about your travel plans or movement.
8. Keep a watch on activities of disaffected employees/contractors/ex-employees with grouse etc. They can be potential trouble makers.
9. Train your driver to identify change in environment and to react quickly. Tell him what to look for.

10. Keep the doors locked while driving. Never give any lift on the way. Do not stop to help victims of an accident as it might be an ambush instead inform the nearest police station.
11. Train driver and self in routine vehicle inspection before starting the vehicle break pipes, tie rod ends, oil, water, tyres, locks, undercarriage, luggage boot, battery etc.
12. Keep on changing route and timing of driving to the work place. Watch for any vehicle following, possible ambushes on the way, usual signs like traffic-jams on highways, deserted roads etc., and be careful at bottlenecks and checkpoints.
13. Maintain a low profile while traveling.
14. Search thoroughly and look for suspicious objects while checking into a hotel room; ensure that the lock and key are working properly.
15. Keep in touch with family and office on daily basis while on tour.
16. Remain calm and cool in case of emergency.

Don't

Executives are advised to avoid the following:

1. Never take security casually. Maximum lives have been lost due to this kind of approach only.
2. Don't hesitate in taking your security expert into confidence. Never ignore advice of your security expert.
3. Don't ignore threats through an anonymous letter or call or caution signals given, by family members and friends. Every "Feeder" is important for "Executive Protection".
4. Don't meet any unknown visitor at office or entertain any stranger at home, directly and without screening the stranger guests.
5. Do not encourage casual and informal talks on telephone by servants or family members unless the caller is very well known.
6. Avoid divulging information about travel plans, movements, etc., on telephone.
7. Do not take into confidence even your driver or personal staff about timings and routes in advance.
8. Never be ignorant of the surroundings whether driving self or riding a chauffeur driven car.
9. Avoid things which might fully involve you while driving, like listening to favourite music on high volume, reading newspapers, magazine or serious discussion with co-travelers.
10. Avoid publicizing the programme of travel. Your programme should only be known to the people who 'need to know it'.
11. Do not pick up friendship with strangers while driving/traveling/flying or in a new city.
12. Never visit localities which are not very well known while on tour; avoid side walks and dark lanes.
13. Do not carry heavy cash while traveling
14. Avoid keeping valuables in hotel room.
15. Do not panic in case of any emergency.
16. Do not try to resist or react violently while in hostage situation.

Industrial Espionage

Espionage is typically an organized effort by one country's govt, to obtain vital national security secrets of another country. Espionage has focused on military secrets. But as cold war has drawn to a close, this classic form of espionage has evolved. Economic superiority is increasingly as important as military superiority. Despite an unhappy connotation of its being a some what objectionable activity, espionage is an essential and increasable human need. Information gathering being its end result, espionage has a role to play in most spheres of human life and not the least in the world of industry and commerce. Industrial espionage too is a hard fact of life, whose role and importance have grown continuously from the start of industrial revolution in mid eighteenth century. In recent times, the end of cold war led to a major shift of attention to the part of premier intelligence agencies to transnational economic/Industrial espionage, which has more recently, received a further spurt and increased width as a result of the economic liberalization and globalization currently sweeping the world, especially the developing countries. This phenomenon can now be over looked by any industrial undertaking only at grave peril to itself. Hence, the need for a close looks at the industrial espionage and measures to counter it.

Imperial power has always been based on a convergence of military power, used in the defence of trade. This convergence was at the heart of the gunboat diplomacy during colonialism. A similar convergence is now taking shape around the defence of trading interests in a period globalization and located free trade. The British Empire was built through the destruction of manufacturing capacities in the colonies, and the prevention of emergence of such capacities. During the technological superiorly of England, there were instances of cutting off the thumbs of weavers in Bengal and Orissa, the forced cultivation of Indigo in Bihar, the slave trade in Africa to supply free labour for cotton cultivation in US etc. England passed several laws to prevent technology transfer from 1765 to 1789. The skilled people were not allowed to leave England to ensure that England remained the world's industrial power. Samuel slater of US was probably the first person, violating English laws, secretly carried the knowledge of mechanical spinning and weaving to America. Surprisingly, US after building its economic power, passed the economic Espionage Act 1996, prohibiting espionage of technology. The punishment is 15 years of imprisonment and a fine of 10 million dollars. The Act defines intellectual properly infringement as a crime and justified use of intelligence agencies to deal with issues of science and technology exchange.

Every country has the right to protect its national security. Globalization now is being used to force other countries to give up their national interest, national security and national sovereignty. The US is the driving force in implementing its agenda accordingly to above norms. It has taken India to W.T.O. and challenged its patent Act. It uses super 301 and special 301 clauses of its trade Act, to force countries to undermine their national security in order to create opportunities for US corporations. Economic espionage in terms of intellectual property, or trade secrets infringement, spreads over the public domain too. The move to privatize the public domain and the intellectual commons through intellectual property rights is itself a "theft" of knowledge. The Economic Espionage Act even condemns an inventer/scientist committing theft of his own theory, formula, new technology, if he or she works part time outside or joins another industry/firm leaving his present assignment. The reasons, he used the lab and equipment of his Employer and therefore, his formula, so created becomes the intellectual property of the firm/company and cannot in any manner be transferred. It is being deemed as espionage. US corporations have largely privatise indigenous innovations and claimed it as their intellectual property. Examples include patents on neem and haldi (turmeric) etc.

Preventive Measures

1. Theft of trade secrets of a firm has been discussed in earlier chapters. Strict vigil may be kept on suspected employees who work in research and development wings. Their activities may be spied by the industry's security forces. The modern security gadgets may be used to keep track of their outside contacts.
2. Sometimes, vulnerable employees are bribed by the competitive Industrial establishments and procure secret formulas. The security force should curb this hazard by adoption of counter espionage formula.
3. An industrial and commercial intelligence wing be created by the industrial and corporation heads and make them active to prevent this hazard.
4. Electronic espionage is conducted for profits to steal and sell proprietary designs, pending merger and acquisitions secrets, the result of research and development, experimentation and confidential information about company litigation etc.
5. The best preventive measure is the art of managing people in an industrial establishment. For achieving optimum result in this score, the human aspects like sense of identity, sense of importance and sense of development may be attended to.

VIOLENCE AT WORK PLACES

Definition

The attempted or actual exercise by a person, other than a worker, of any physical force so as to cause injury to a worker and includes any threatening statement or behaviour which gives a worker reasonable cause to believe that the worker is at risk of injury. The companies, generally more or less experience a work place fatality as a result of violence. Deaths and other serious incident taking place included physical assault, robbery, weapons possession and vandalism.

Work place violence prevention

These are classified into:

(*a*) Programme and procedures
(*b*) Training
(*c*) Policies
(*d*) Reporting/Recording.

Programme and procedures

Applicant back ground and screening: All the companies routinely perform background checks of employees focusing particularly on employment history.

(*a*) The employer shall eliminate risk.
(*b*) If not possible, minimize risk.
(*c*) Establish procedure for reporting, investigating and documenting incidents.

Factors of violence at work place

(*a*) Work location
(*b*) Staffing level
(*c*) Work place layout
(*d*) Lighting/Security Provisions
(*e*) Hours of operation
(*f*) Work processes *i.e.*, money transactions etc.

Work place conduct

Improper activity or behavior includes:

(*a*) The attempted of actual exercise by a worker towards another worker (co-worker) of any physical force
(*b*) Any threatening statement or behaviour
(*c*) Horse play or practical joke
(*d*) Teasing, abusing, distracting
(*e*) Frustration and conflict
(*f*) Jealously
(*g*) Quarrel some.
Improper activity or behaviour at a work place is prohibited and must be reported and investigated.
(*h*) Violence attributes to worker
  (*i*) Competency based training/experience
  (*ii*) Health
  (*iii*) Personality/attitude.

While investigating the following aspects may be closely looked into:

(*i*) Number/frequency
(*ii*) Type/severity
(*iii*) Time/location

Work place violence prevention measures

(*a*) A plan should be prepared taking into consideration the above facts
(*b*) Senior management participation
(*c*) Employees Assistance
(*d*) Crisis management/Trauma team: The members should be drawn from departments like human resources, security, employee assistance, legal, medical, labour relations, labour law, communications, employ support/benefits, operations and psychological services.
(*e*) Threat Assessment team
(*f*) Critical incident review
The main purpose of formation this team is to analyze, learn from and improve procedures.
(*g*) Out placement services for terminated employees
(*h*) Relations with law enforcement agencies
(*i*) Termination guide lines
(*j*) Grievance process for employees
(*k*) Union involvement
(*l*) Dissemination of work place violence information like formats, included tip sheets in pay envelopes, company news letters, bulletins, Electronic mail and briefing sessions with field employees.

(*m*) Engagement of work place violence experts
(*n*) Assessment of organizational climate
(*o*) Assessment of site/facility security: The companies should regularly assess site facility/security through reviews of security policies and procedures, personnel and equipment needs, prevention efforts etc.
(*p*) Assessment of Readiness
The companies should assess their abilities to handle work place violence.
(*q*) Assessment of employee/facility risk level
(*r*) Protection of threatened employees

Training on

(*a*) Detecting "Red flags"
Providing mandatory training to detect behaviours which indicate the potential for violence.
(*b*) Domestic violence response
(*c*) Conflict resolution
(*d*) Cultural diversity
(*e*) Personal safety
(*f*) Dealing with different customers
(*g*) Effective communication
(*h*) stress management
(*i*) Diffusing violent situations
(*j*) Liabilities issues
(*k*) Zero tolerance of threats/violence
(*l*) Code of conduct/standard of corporate behaviour
(*m*) Code of sanctions for non-compliance with company policy
(*n*) Safeguarding access to facilities by non employees
(*o*) Weapon prohibitions
(*p*) Drug and alcohol prohibition
(*q*) Sexual harassment prohibitions.

Reporting/Recording

(*a*) *Notification process:* Employees be trained to notify the appropriate department or individual when they observed situations with a potential for violence.
(*b*) *Incident reporting:* The companies should sympathies reporting procedures, when threats or violent activities occurred at work sites.
(*c*) *Computerized database:* Companies are to maintain computerized database to report incident.
(*d*) *24 hour reporting hot line:* The companies should have 24 hour hot lines for employees to report emergencies.
(*e*) *Communication equipment:* The companies should have effective communication equipment for personnel, who responded to emergencies. These may include pagers/cellular telephones, and e.mails.

HIJACKING

Hijacking is a form of terrorism which has been discussed earlier. It is used as a tool by the hijackers to settle scores, for ransom for settlement of political affairs and release of prisoners/offenders from the custody. Earlier, hijacking was being done relating to airplanes and ships. Now, instances of hijacking public buses, trains, are rampart. For example: recently, the Naxalites, Maoists tried to hijack a passenger train and kept it as a hostage nearly 8 to 10 hours near Midnapur, in West Bengal. Their demand was to release some of their comrades from judicial custody (Jail). Hijacking of ships are now a common sight in Somali coast as discussed earlier for ransom. With the increase of terrorist organizations in the political front world over, the incidents of hijacking of passenger aeroplanes have increased ten fold. This has been discussed elaborately in the subsequent chapter, under "Aviation Security".

Campus Killing

Campus killing is a Western phenomenon. It is as old as 1929; which is called St. Valentine day massacre in Chicago. There was subsequent massacre in the year 1970 and the killer was called as serial killer. Because of the open and naive atmosphere of the college campus, students are often the target of these people who prey on the unsuspecting and inexperienced. Most of these people, strike suddenly and without apparent motive. It is difficult to protect a community from such people.

Causes

(*a*) Mostly psychological imbalance.
(*b*) Frustration
(*c*) Diseased
(*d*) Sudden Provocation
(*e*) Failure in love
(*f*) Out to take revenge
(*g*) Alcohol and drug addiction
(*h*) Women haters (who kill only feminists)

America and other states in the western hemisphere are the venues of such campus killings and serial killings. In those countries, weapons are procured and possessed without license and therefore, it is freely available and there is absolutely no restriction on its carriage from one place to another. Some such killings are found to be the product of play and fun, when the minors/urchins handle guns and enjoy the fusillade.

Coming to India, the weapons like Gun, Pistol etc., are controlled by Arms Act and restrictions have been imposed for selling and buying. Possession, carriage, transfer and use of arms and weapons are regulated by state control. Nowadays, the state Army, police force who are authorized to posses arms, found largely misusing their privileges and killing their colleagues, superiors due to frustration, disappointment and to take revenge. The causes attributes to such killings are the same as we generally attribute in case of western killings noted above.

Incidents of serial killings are reported in many parts of India, where the killer used stones and iron rods for killing innocent people sleeping on foot paths during night. The "Stone man" episode that was reported in the press media, years back, revealed a gruesome picture of serial killing. The victims are slum dwellers and quite soft targets of the killer. Investigation revealed that, the killer had sexual inter course with the women, he mercilessly killed. The attack was sudden, without any provocation and there was no motive behind it. The man was possessed of deranged mind, evil and emotive predicaments.

The most recent instances of serial killings perpetrated are by the Naxalites Maoists and their henchmen. In such massacres, their plan and procedure are supposed to be fool proof. They have a network of intelligence agencies, to keep tidbit informations on police strategy, movement, stock of arms and ammunitions at their disposal etc. They detonate the gelatin in the nick of the time, the force, crosses the bridge, gun down the escaped lot and make the operation a success. So far prevention is concerned, police, public, partnership plans are being contemplated and executed. The public have been advised to be vigilant in relation to one or two serial killers and this menace to some extent has been curtailed. But killings of torturers and tormentors, the class enemies and the forces etc., this plan has not been very much successful. The members of public, who extend their support to the police force are branded as "Police informers" and immediately liquidated by the maoists. They expand their cadre not only through their ideology but on the point of gun too. The force has failed to provide protection to innumerable people living in remote jungles and mountains. When the state force has failed, it is beyond imagination how the private security personnel could able to face this hazard.

COMMUNAL RIOT

This is one of the most sensitive security hazards that happens on sudden spurt of group violence. This may be classified in so many ways as detailed below.

(*a*) Religious
(*b*) Caste considerations
(*c*) Cultural differences
(*d*) Colloquial feelings
(*e*) Language differences
(*f*) Regional differences
(*g*) Exploitation of minor communities by major
(*h*) Differences between various economic political and occupational groups
(*i*) People with common interest form a group and fight with the rival group
(*j*) Differences between various trade unions, service associations, professional guilds etc.

Handling the Riot

It is always very difficult to control the fire of mob frenzy, fanned by religions dogma. It spreads like wild fire and the casualties are largely enormous. The state police force and even Army some times fail to control this fury. The havoc it creates which in a very short space of time, immobile the armed police to nab the culprits. Communication system becomes paralyzed. The number of existing police and para military forces are found insufficient. The stock of arms too are soon depleted and the replenishment becomes impossible. Any strategy, carefully planned and adopted becomes in operative. The rioters target the opposite community and the potential attackers are instantly created making the number quite unweild to handle. In the inaccessible areas, the casualty is generally cent percent.

Under these difficult situations, the role of private security is totally zero. The affluent and elite section of people of both these groups who hire the security personnel for their safety, find themselves being left in the lurch as, the latter cannot face the combatants or willfully desert their masters. Due to damage of communications system, rescue operation becomes very difficult. Hence, the role played by private security forces are found to be minimal and microscopically limited.

9

INDUSTRIAL SECURITY (INDUSTRYWISE)

GENERAL SECURITY SYSTEM IN FACTORIES

This has been discussed in detail earlier. However some more useful tips are detailed below.

(*a*) Control of access and exit in gates.

(*b*) Crowd control in relation to visitors regulating the entry in a disciplined and courteous manner.

(*c*) Entry of vehicles inside the factory premises.

(*d*) Observance of parking regulation of vehicles.

(*e*) Patrolling.

(*f*) Handling of security equipment like detectors etc., on the gates.

(*g*) Use of force wherever necessary.

(*h*) Search, seizure and apprehending the intruders, suspected persons, forcing entry into the premises and reporting the matter to the management and local police.

(*i*) Use of fire prevention equipment.

(*j*) First Aid.

(*k*) Legal use of fire arms, where it is found absolutely necessary.

(*l*) Report writing.

(*m*) Regular inspection of factory premises, preferably weak points in compound fencing walls.

(*n*) Crime prevention of all types in and outside the factory area.

(*o*) Total loss prevention.

(*p*) Handling unruly persons and control all types of violence inside the workshop floors, office and other places of congregation.

(*q*) Maintenance of good public relations.

(*r*) Inculcation of safety habits, hygiene, health and environment.

(*s*) Develop perfect communication system.

(*t*) Equipment use and care.

(*u*) Receive regular training and updating the security agenda of the factory.
(*v*) Watch out for cigarettes/Bidis etc., inside the premises.
(*w*) Regular checking of building surroundings for fire/leaks/damage/hazards.
(*x*) Check ladders and equipments, which can assist the criminals, are secure.
(*y*) Checking all padlocks after office hours.
(*z*) Checking all lights including security one's.
(aa) Checking all machinery processes, effective by the negligent workers.
(ab) Checking stores/items left outside or in unusual places and see that all items are secured against weather damaged wastage.
(ac) Checking locks/seals/sheeting/immobilization of all vehicles/trailers owned by the factory.
(ad) Check all outside plant immobilized and secure, specially the ash ponds, raw material depots, metal waste dumping yards, garbage yards, affluent drainage systems etc., and see if warning tapes have been fixed all round these sites.
(ae) Black thread various areas/stick small objects in door/window cracks.
(af) Beware of any passive security warning devices which can be activated.
(ag) See, if patrolling areas with internal static guards be sure to know password and avoid surprising them to avoid being injured.
(ah) Use of dogs for security where ever necessary (Both patrolling and investigation of crimes in the factory premises).

This subject has been discussed on more detail in subsequent chapters under duties and responsibilities of security personnel (Security officer, supervisors/Guards).

AVIATION SECURITY

This type of security bas been discussed in detail under "Aviation safety" in earlier chapter.

However, some important tips are now provided, which are considered as most important. The basic aviation security agendas are:

(*a*) Wardens are to be posted to carryout duties like air craft guarding. Cabin checks and handling of precious cargo and diplomatic mail bags. They should be properly trained.
(*b*) They are expected to handle some dangerous jobs like escorting criminals, under trial prisoners, criminal deportees, gold, bullion in and precious stones without arm escort.
(*c*) The security people are left to guard and patrol the airport premises.
(*d*) The security personnel should be suitably trained, specializing in aviation security.
(*e*) Introduction of procedures manual and check lists. An exhaustive procedure manual has to be developed. This manual comes complete with checklists for each area of operations.
(*f*) Customer service campaign should be launched.
(*g*) Safe passage X-ray simulator. The personnel should be trained in X-ray detection with- in shortest possible time.
(*h*) Crisis management procedures are to be framed to aid airlines facing critical situations.
(*i*) Quality control and work improvements teams are to be formed.
(*j*) Security audit organisation is to conduct regular audit and readiness checks at random to ensure staff maintain a high level of alertness at times.
(*k*) Wide scale computerisation is a step towards perfection.

COMPUTER SECURITY

Industrial development and growth is the hallmark of economic independence and of its status in the comity of Nations. In India, the successive five-year Development plans have been responsible for a tremendous rate of growth in the Industrial development in general and establishment of key industries in the public sector in particular. Industrial security has, therefore, assumed a special significance in this growing industrial environment as it has a direct bearing on production, productivity and growth.

At present, a variety of agencies and organisations are involved in looking after the security functions in the Indian industry. There are private security agencies, the police and there is a specialised Para military force called Central Industrial Security Force designed to look after the security of the major industrial undertakings of the Government of India. However by and large, all these security organisations have been functioning more in the manner of the conventional watch and ward staff. The limitations of a totally manpower dependent security through conventional methods of gate pass checking and manual searching are being exceedingly realised by the modern managements of the industrial establishments. This has resulted in a demand for more reliable and cost effective methods of securing the industry from various types of internal and external threats some of which were not even envisaged when the conventional methods of security were evolved.

The Central Industrial Security Force of the Ministry of Home Affairs is, perhaps, the leading Force in making a beginning with ushering in an era of electronic gadgetry for the security functions. Concrete plans have been drawn up by this Force and furnished to a number of heavy industrial units for installations of security gadgetry like intrusion alarms, access control close circuit TV camera networks and communication network for prompt response. A host of devices have already come up in the area of security gadgetry which are mostly electronic based. It is encouraging to note that interest in electronic security equipment is growing rapidly.

The aim of this paper is to highlight the role of electronic computers, general or special purpose, for enhancing the capability of the security men to perform better so as to build up a high level of security environment in the industrial sector, which is most vulnerable among the national resources.

An industrial unit basically consists of personnel, materials, locations and information. While there are numerous systems and devices to ensure security in these key areas of industry, the unique capabilities of a computer can be usefully harnessed in ensuring higher grades security over and above various other important technique and technologies.

Computer Based Personnel Security

Computer applications in the field of personal management need hardly be over emphasized because it is already in vogue in various industrial establishments. The computer applications can be extended to cover security aspects in regard to personnel in meaningful way. Computer files can be created assigning various security gradings to the personnel keeping in view, their experience, expertise, reliability and previous record of performance and conduct. These files can be created after thoughtful consideration by the management and can be updated periodically for changing the security gradings as per requirements. This can help the management in allocation of duties particularly in times of emergency of difficult situations. Presently, the management has to bank upon the impressions of various managers and supervisors through hurried consultations, which may not be the ideal method of doing it. Another application could be to develop computer files for personnel who are specifically debarred from being assigned certain duties or from entering certain areas/locations.

A relevant area of computer application is the regulation and control of issue of identity cards or gate passes. Many industrial organisations already have a system of issuing such cards with different colour codes to different categories of employees permitting them to enter only restricted areas or during specific shifts only for duty. Management of such assignment becomes difficult particularly in large establishments. Backtracking of the issuance of such cards over a long period in the past for investigation purposes becomes almost impossible, as information is required to be retrieved from old manual records, which are not easily available. Misuse of identity cards/gate passes is another vulnerable area, where help of computer is relevant. It is very difficult in manual systems to keep track of all the cards/passes whose period of validity has expired and which have not been renewed or deposited back. It is equally difficult to ascertain, at a given point of time the number of cards/passes which have neither come up for renewal nor have been deposited and from which department and in respect of which employee. Such cards are liable for misuse. In the computer system, such lists can be generated with considerable ease and holders of such cards can be traced for further action by the concerned branches of the industry. Computer system can also help in prevention of issuance of an identity card or a gate pass of a specific type to a debarred employee. The technique of bar codes can also be adopted in regard to the identity cards, which will hasten the process of checking the identity cards at the gate, to ensure that no invalid or forged identity cards or gate passes are used for gaining entry.

The security organisations already have or should have system by which they rotate the security personnel from one duty point to another, from many points of view. However, the rotational deployment under manual system is not very effective and security personnel can manage to be posted at selected duty points again and again in quicker sucession as desired. Computers can very much help here in conforming to a given pattern of deployment in a fool-proof manner as also prevent the posting of security personnel at places from which they are debarred. An analogy existed, though in a different context. In a University in Australia, where computers were used to make seating arrangements in the dining hall for about 2000 foreign students in such a way that students of various nationalities sit with each other in rotation so as to promote better interaction and interrelationship.

Computerised Access Control

There are a number of technologies used for access control and to check illegal intrusions. Computerised access control has also come to play a significant role and has proved to be more effective and safe, by the very nature of computer functioning. Use of magnetic cards for opening the doors and gates of sensitive locations is now computer based. This helps in changing the codes of each access point, not only with greater ease but a systematic manner. Many hotels abroad have adopted this system. As soon as a guest checks out he does not even have to deposit back his magnetic card as this is of no use to anyone, the code of the room vacated having already been changed by the computer as soon as the checking out report was keyed in at the reception counter. A Micro processor controlled taut wire intrusion detection system has been developed and is in use with self adjusting sensors. The close circuit TV cameras based access control and anti intrusion system are fairly well known now. The latest development is that such a CCTV network is also new computer based for regulation and control. The status of operation of each camera is monitored by computer and any defects in the electronics system in the operation of cameras triggers off a signal for remedial action. Computers are capable of supporting such a monitoring function for 500 cameras at a time. Similarly, all other electronic devices can also be hooked on to the computers for monitoring the status of their working to avoid and rectify any failures in their working.

Fingerprint is well known and accepted to be unique identity parameter of a human being allover the world. The matching of fingerprints has also been established as an exact science and accepted by law courts throughout the world. However, this matching is done by fingerprint experts visually. The latest developments in computer technology has brought about the "Computer Vision" which enable the machines to scan a fingerprint, identify the ridge characteristics and match with the stored fingerprints to establish the identity of a person. This computer technology has been used for access control through fingerprints in very sensitive and vulnerable industrial locations. The person seeking entry to a room has to insert a specified finger in a slot of the computer scanner, which will read the fingerprint, match it with the prints of the various persons already in the computer file and automatically release the door lock to give entry to the authorized person. In case of a mismatch, an alarm is generated for necessary field action. This technology has also been extended to a similar scanning of the faces of the persons seeking entry after matching with the digitised photographs of those persons on the computer files. Conversely, the photograph scanning systems can be installed at airports or at the entry gate of large industrial establishments to track down undesirable person e.g., terrorists, saboteurs, anti national elements, etc., trying to gain entry to such places. One such system called 'OCTOPUS' is reported to be installed at one of the international airports in the USA. Similar developments are taking place in regard to voice analysis of human beings. Voice, which is said to be unique to every person, can be scanned and voice graphs stored in computer files. Voice of persons seeking entry can be similarly scanned, graphed and matched with file voice prints to ascertain his authorisation or otherwise.

Monitoring and supervision of the performance of security men can also be aided and enhanced with the help of computers. The patrolmen on mobile duties can be required to press the switches installed at vulnerable places which will be recorded in the computer along with the time. The computer can produce a list at the end of the day or night or in given periodicity, a graph of the patrolling done by the security men in various areas for the benefit of the supervisor which will give him the frequencies with which patrolling has been done and the sequence in which the locations have been visited along with the time of the visit. This can be applied to the visits of the supervisors also, so that the senior security officers can monitor the efficacy of the patrolling system.

Computer Aided Information Security

The techniques of micro films and micro fiche preparation of vulnerable records for information security are well known. However, location of specific information in a plethora of micro films or micro fiche cassetter, rolls and capsules has become difficult with the increasing numbers. We have now computer based micro film of micro fiche readers. An index of all that goes into micro films is maintained on computers. A manager has only to give out the file number or the topic or subject in which he is interested and the computer will indicate the cassette number and the frame number of the cassette in which the desired information is stored. This can then be manually keyed into the computer based micro film reader which will detect and project on the screen the exact information required. Computers can be used to directly control the micro film reader also in which case even this much of manual intervention is not necessary. More sophisticated computer based systems for encryption and scrambling of vulnerable information and subsequent description are also available which obviate the use of micro films or micro fiche. What is more, data storage on computer itself is more secure than the conventional manual records which are vulnerable to many more threats than the computer based records.

Material Security Through Computers

Materials and equipment are as important in an industrial environment as personnel, locations and information. Techniques of computer based inventory control and materials management are already well known and in operation. With the advent of bar codes reading and matching with the help of computers, equipment and materials can be secured better by spot reading of bar codes through electronic devices, connected to computer files to see if these are located at places where these should be and whether equipment going out of the premises, is the same which is authorised to go out. Periodic checking with the help of bar code readers in conjunction with the computer based file helps in ascertaining the missing equipment much quicker than in the existing system, in which loss of equipment and materials is reported after long periods when it comes to the notice of Computer based inventory management. The stock taking techniques can also help the security organisation in determining the pattern of consumption of raw materials, quantum of finished products, movement of materials from one place to another vis-a-vis requirements to assess possible leakages and pilferage.

Security of the Computer Centre

While the computers play a significant role in enhancing the security environment in an industrial organisation, the computer centre are themselves not immune from security threat, internal as well as external. Security of the computer and the Computer Centre should also, therefore, attract attention of the security organisations. An industry having a large computer environment, would be desirable to have a computer professional among the security personnel to look after the security of the Computer Centre depending upon the magnitude of the threat potential. In this context, it may be appreciated that the computer environment itself is a very valuable resource in monetary terms. Computer itself only accounts for 10% of the cost incurred on the total computer environment. Hence, securing the whole environment and not only the machine should be the target of the Security Organisation.

Apart from the physical protection of the computer centre, a very close check on the personnel and the system procedures, documents, computer files, and even the garbage going out of the computer centre consisting of invalid outputs, used punch cards and tapes, carbon papers of multi part stationary etc., deserve attention. There are a number of safeguards provided in the computer hardware as well as software and in the system procedures, it needs to be ensured that these are meticulously observed. Segregation of duties and functions of computer personnel is essential so that no single computer person should have access to the entire gamut of computer environment. For example, the programmers should not have access to data files and their access to the computers should be restricted only to programme testing and experimentation. The computer operators and supervisors should not be allowed access to the computer programme source listings, flowcharts, etc.

System documentation and operating procedures should be safeguarded from unauthorised copying of leakage and should be kept in the safe custody of responsible or reliable EDP employees. Operators in the night need special attention. Dismissed or ex-employees of the data processing departments are particularly full of threat potential to the security of the computer environment and need to be specially safeguarded against.

Conclusion

The objective of an industry is not security but production. The security is one of the many functions which fulfill that objective. Lack of security definitely makes a dent into the fulfillment of the corporate objectives. The threat potential to the industry from organised crime and professional criminals has definitely increased over a period of time. This cannot fail to generate the desired awareness among the industrial managements to raise the level of the security function to match the indigenous ways and means being adopted by the deviant elements in the society. It is reported that one electronic aided crime has cost the British Commerce and Industry, about 35 billion dollars in 1986. The industry in the field of security gadgetry is reported to be growing at a rate of 10 to 20%. It will not, therefore, be a day too early if Indian industry also grasps the threat potential and applies itself to the corrective and protective measures.

BANKING SECURITY

One of the most over worked words in current use is "Professionalism". It means many things used in various contexts. Does it indicate specialisation? Or in general does it confine to academic and other vocations. What do we understand when we read "Saurav Ganguly retires from professional cricket?" The above is just indicative of the difficulty in defining the word "Pro-fessionalism". For my part, I would say that apart from including a skill of the high order in the chosen calling, it is the contribution of that rare ability to relate the knowledge to the environment, its changes and demands.

When I was an youngster in college I used to visit the state Bank of India Branch where my relative was the Agent. I used to look with awe at the Dharwan who was immaculately dressed with all his liveries and standing his full height of over 6 feet. He would recognise me, give a broad smile and allow me to enter the premises. At the same time, I noticed him scrutinising all the new faces and questioning their identity before allowing them to enter. He was in fact exercising access control as we call it today.

You all remember the story of Alibaba and forty thieves. On learning the code word "Open sesame" he gained entry to the robbers, cave containing stolen wealth. The cave door recognised the code word but could not differentiate individuals using it. Today, Computer criminals can be based in Hamburg, for instance, steal money from a U. K. Bank Branch based in Spain via telephone net work in South America. How can this be possible? It has become possible due to technological advancement in computing and communications. The criminal could have access to computer network for transfer of money. Even when the criminal can be identified, prosecution becomes difficult due to international implications of the act. Such acts entail special aspects of security management in Banking Industry.

Before we look into the present utilisation of modern technology in banking industry let us first look into the existing security scenario especially in our country.

Security Scenario

As stated before, the rapid growth of Indian Banks has, in its wake, brought about a number of problems for the banks to tackle. One such problem that conspicuously cropped up, particularly during the 1990s, has been the increasing incidence of burglary, dacoities, holdups, terrorism and insurgency resulting even in loss of life of customers and bank staff besides cash and valuables.

The Reserve Bank of India's licensing policy (a shift from class banking to mass banking) was intended to spread the network of banks to the rural and unbanked areas. This has resulted in banks opening a large number of branches in the rural and semi urban areas without adequate planning and trained staff. A number of these centres lack even the basic infrastructure, such as transport, post office, police station etc. Coupled with this rapid expansion of branches, is the increase in the number and variety of customers that banks have to handle on account of the social banking policies introduced by the Government. For instance, Banks in rural centres accept gold ornaments as security for certain types of lending and a large number of such branches hold as security gold ornaments of substantial value carrying unimaginable risks.

The Banks found it difficult to cope with the increase in the volume and variety of work: control became inadequate and inefficient and this resulted in the increase in frauds and misappropriation. Enquiries show that the rapid growth in the number of frauds and misappropriation was caused by unscrupulous members of the staff and/or with their connivance, by resourceful outsiders.

The recent stock market scam is an indicator of the magnitude of loss that can result from collusion. The loss through collusion is difficult to detect. These crimes are not fully reported like external bank crime statistics as, to do so, would cause an ever greater loss of public confidence in banks and bankers image is a large factor in the reluctance to report all these crimes. However, I am not covering the white collar crimes in banks. And in order to reduce, if not, avoid such losses, banks, are progressively undertaking mechanisation and computerisation. In additions, the following aspects have increased the risks in banks:

- Increase in population
- Unemployment
- Effect of movies showing hold-ups etc.
- Currency not identifiable in case of a dacoity. Even the decoy money used, is often not traceable
- Procedural problems in obtaining licenses for weapons etc.
- Casual attitude towards security in the bank hierarchy.

All these factors have brought banks, more than any other organisation in our country to public attention in a pronounced manner and rendered them attractive targets for attacks.

The security arrangements in the branches cannot be deemed anywhere adequate much less ideal.

Incidents of bank robberies, dacoities and hold ups and other forms of crimes have become widespread and are no longer confined to any particular geographic area of the country. However, majority of dacoities, due to terrorism are happening in the entire northern belt of our country. A state wise distribution of dacoities in the last few years clearly shows that a few states in the north have had majority of dacoities linked to terrorism.

Planning Bank Security

The urgent need for proper planning, and designing of security measures in banks has assumed great significance. The availability of ready money and valuable have emboldened the criminals to resort to even daylight dacoities. Banks are no longer perceived as an impregnable bastions but have come to be regarded as easy targets for attack.

Bank Security and Risk Management involves a careful Management of existing and anticipated threats, arising from external or internal sources or a collusion of both.

The cardinal principles of bank security are DETER, DETECT, DENY and DELAY.

The principal aim of bank security is to create deterrents and if this is adequate, crime can be minimised if not completely prevented. The initial stage of planning security can be said as early warning stage which can be achieved by proper Access Control. If the criminal is able to overcome the initial defense like (access control), there must be a second line of defense which may be called as the detection and prevention line. The final stage would be to deny the target which in the case of the banks is the safe vault/strong room.

Security Designing

The security measures that are required to protect the assets vary from branch to branch and hence need individual profiles. Four major factors viz., location, operational and Economic, Psychological and structure and layout have been listed for risk assessment. Each factor has under it a group of sub factors. Having measured the security risks of each branch and converting the points into percentage, it can be categorized as under:

High risk	– 70% to 100% points.
Normal risk	– 50% to 69% points
Low risk	– 1 % to 49% Points

There is another way of looking at this problem. The method is to match the threat, target and environment with the anticipated degree of danger / loss and decide on the threat perspective. The marketability or liquidity factor depend upon the attractiveness of the target to a potential offender. While cash is fully convertible and cannot be easily traced, jewels pose not only difficulty in conversion but also may lead investigators to the robber. Therefore, normally jewels are not that attractive to a robber or a burglar. This also must be taken into consideration while designing a security system.

The layout of a branch itself is very important. Apart from the physical barriers, psychological barriers are also to be created to achieve the required deterrents. Staff vigilance plays an important part. The layout and grouping of the departments should be with a progressively increasing security control, as one proceeds from one zone to the other. The principle is shown diagrammatically below:

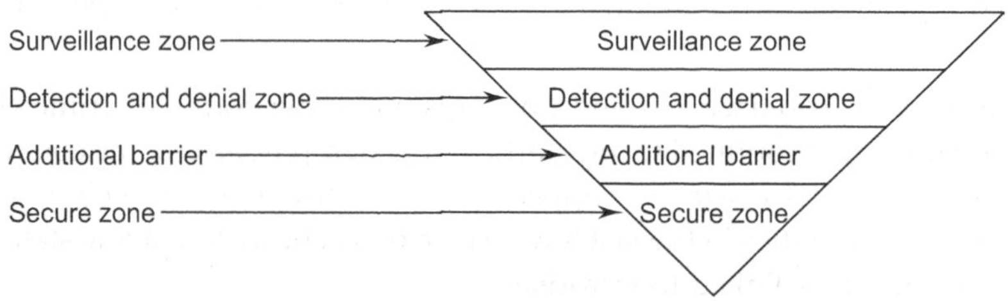

The main entrance and the customer lounge area should form the "Surveillance Zone", where everyone present is under the observation of the security personnel and bank staff. The next zone "Detection Zone" should be the area (Banking Hall) which consists only the staff and customers who are permitted inside and any other outsider is easily spotted. The next zone is the "Secure Zone" which consists of the cash department wherein even members of the other departments cannot get access to this area without permission. An additional barrier in the form of wooden partition with wire mesh can be created between the detection and the secure zone if required. The target (vault) itself could be further protected with an additional grill.

Traditional and Modern Methods of Countering Threats

A wind of change is blowing throughout the world. Day by day more and more countries are orienting their economy towards the world market. In this customer centered economy, the industries are competing on meeting the changing needs of the customer.

Banking is a service oriented organisation and is no exception. Banks are now bringing about changes to serve the customer at his convenience in terms of time and place. Customer in this sense includes not only the depositors but also the borrowers.

On Star TV Net work, you may have seen an advertisement of an International Bank. It says "There are still some places where you cannot reach your Bank. But yet global access with more than thousand ATM's twenty four hours a day is available". Unlike yester years, a bank account holder at a branch can withdraw money anytime at anyone of the ATM's. In a similar fashion transfer of large amounts of money can be made across international borders through Electronic Fund Transfer (EFT) system. There is hardly any delay in these transactions. Funds are positioned at the point of need and at the time of need.

Technological advancements have made these possible. In the main two areas of advanced technology are involved. They are:

- Computer technology
- Telecommunication technology.

A host computer at a central location maintains all the data. This computer is accessed by other distributed computers after identifying the operator. The host computer then depending on the data source, authorises or denies the payment. The operation is illustrated in a simplified manner below:

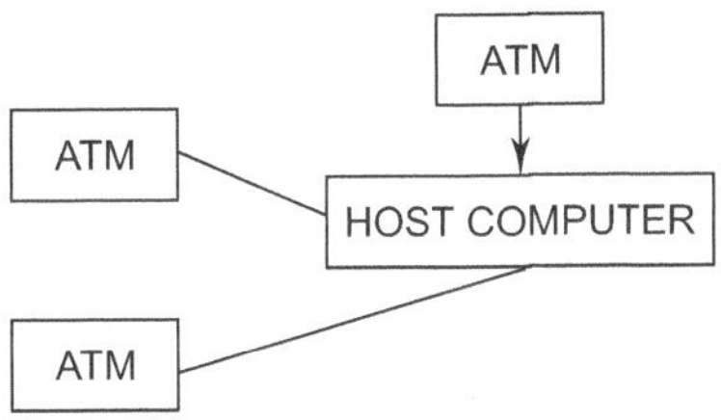

TRANSACTIONS THROUGH ATM

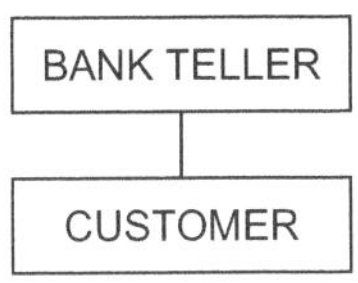

Traditional Transaction at Bank

In traditional transaction the customer had to go to the bank for withdrawal of cash. On the other hand, with ATM he can be at any location and yet can draw from his account. Participants will notice that in both cases it is necessary to:

- have access to the account
- establish authenticity.

Yet there is a significant difference. In traditional transaction, there is the physical presence of the individual at the branch. In the ATM transaction, not only there is a geographical distance involved, but also the anonymity of the individual except through a card claiming a particular identity. With minor changes for our understanding, EFT also operates in a similar fashion.

Conceptuality the threats and countermeasures remain the same. The threats that Banks face are:

- Fraud
- Theft

- Robbery/Dacoity
- Fire
- Terrorism
  (Bomb Threats etc.)

Since Banks are threatened in both traditional and modern methods, the security Manager has to draw a profile of the Bank and design its protection accordingly.

First let us discuss the traditional way of countering threats. The security measures can be generally divided into three types:

Physical

Procedural and

Motivational.

Physical

The physical measures can be categorised as defensive and offensive. Some of the measures taken by the Banks are:

- Security Guards
- Structural Safety and electronic fencing
- Alarm Systems
- Surveillance by Staff
- CCTV
- Training of the bank staff
- State Level Committees organised by the Reserve Bank of India to review the various security measures periodically
- Police patrolling
- Weapon system
- Time delay locks

Procedural

These could be categorised as operational safety and emergency measures like:

- Timing of Opening and closing of branches
- Strong room operations
- Cash Management by branches
- Liaison with Various authorities
- Custody of keys and documents
- Fire Safety
- Cash in transit (self extinct currency, vehicle incapacitation, bait money and tear gas/dye packs)
- Action during holdups.

Motivational

This involves motivating the staff members thereby creating a sense of security consciousness amongst them. Training is one of the most effective means of doing this. It can be undertaken to make certain that personnel understand the reasons for security procedures and policies.

A recent study on Training in banks, carried out by a working group nominated by the Reserve Bank of India revealed that there are serious short comings in so far as security training is concerned. The working group has suggested regular security training programmes to be conducted for various levels. With proper training, in addition to improvement in the awareness, knowledge, skills and attitude of bank personnel, the vulnerability of banks as perceived by the miscreants at present would also decrease. Security consciousness on the part of the banks is bound to act as a deterrent leading to minimisation, if not elimination, of dacoities, robberies, burglaries etc.

Modem transaction can be visualised as

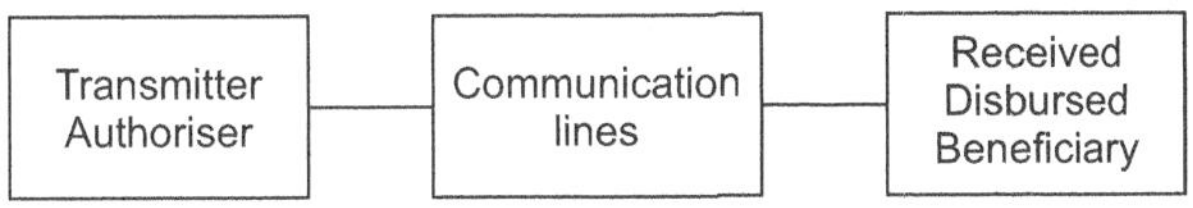

In such a situation the vulnerable points are:

Point of authorisation	–	unauthorised access and knowledge of passwords.
		Authorising authenticity at a low level in hierarchy.
	–	Hacker
	–	Logic manipulator
Communication lines	–	Interception (EMI)
Action Addressee	–	No confirmation of signatures

Now let us have a look at a few of the modem security aids available.

Personal Access Control System

Access control technology has undergone a sea change from the Hollerth coded card circuit or capacitance cards to Electronic smart card, RF proximity card and even cardless Biometric system. There are very high security cards offering high resistance to counterfeiting and duplication along with a self destruction facility. Owing to increased cost, the systems were only used in sensitive installations so far. Now that these systems are becoming cost effective, banks should be able to introduce them to strengthen their security measures. The checks introduced to prevent misuse of credit cards are:

(a) *Card Verification Value (CVV).* There is a code embedded in the card. Any attempt to tamper or overwrite the card will disturb this code. On presentation, the machine will reject.

(b) *Floor Limit.* This limits the amount up to which a person can use the card for a one-time transaction.

(c) *Point of Sale (POS).* The trader checks with the issuing authority (bank) on the validity of the card. The validation normally will be "Approved", "Declined" and "Pickup". Pickup indicates that it is a "Hot Card".

Intrusion Detection System

IDS detects a perpetrator's action before he seeks entry to a protected area. Basically it consists of three major elements — sensor (or detector), alarm assessment and alarm annunciation or reporting system. Most commonly used interior intrusion detectors are micro wave, infrared, ultrasonic detectors, vibration sensors, glass break sensors, photoelectric beam sensors etc. These detectors and sensors could be used in banks effectively providing depth to the security system.

Far and Nar

One of the fundamental considerations of an effective intrusion detection system is its false alarm rate (FAR) and nuisance alarm rate (NAR). FAR and NAR is expected to improve greatly using techniques of digital signal processing. For example the micro wave intrusion detector will be intelligent enough to differentiate most of the times a stray dog or bird movements or even swaying of grass. New RF leaky cable and fiber optic intrusion detectors are cost effective and provide good performance.

Explosive Detector

Today one half of a kilo of high performance explosive is enough to blow up a bank. The deadly plastic can be shaped into almost any configuration, including thin sheets that might look like a lining of a suitcase. Moreover, the preferred bombs are made of hydrogen, oxygen, nitrogen and carbon (H, O, N, C). The same elements that make up most clothing and plastic. It is understood that in U.K. a small gadget has been effectively tested to identify such plastic materials. This identifies the material through the vapour emanating from it.

Metal Detectors

Future metal detectors may be available which may combine both the principle of detection (*i.e.*, detection of a particular metal at a particular frequency) and also inform about the type and shape of metal.

CCTV Surveillance

All the three subsystems of CCTV — Camera, Transmission Media and Display Unit are undergoing rapid modifications. So far in India we have utilised this system only in selected bank branches.

ATM Surveillance

A new ATM surveillance system which captures a set of digital images of each user, via a pinhole camera, which can be concealed or mounted outside the machine as a visible deterrent has been introduced. The images, together with details of time, account and branch are down loaded to a tape streamer for archiving. Each unit monitor three ATM's and hold 8000 transactions on memory. If an act of vandalism occurs an anti tamper sensor is triggered.

Remote Monitoring of Alarms

As brought out by me earlier, the rural and semi-urban branches in our country find it difficult to mobilize an effective reaction force in time of any threat.

Silent Communicator

This is a small communication unit interfaced with a telephone line on one hand and to various sensors and detectors on the other hand. In case of a potential threat the sensors or detectors activates an alarm, a silent communicator is automatically activated which dials up to ten numbers and communicate a pre-recorded threat message. The lacunae in this system is that, it fails once the telephone communication is dead.

Integrated Monitoring System

There are various kinds of integrated alarms and sensors available which monitors the signals received from within the building. All these systems function on the principle of converting the signals received from the sensors into telephonic communication signals and carry these through telephones to a signal processing unit which identifies and classifies the nature of threat and alarms, the reaction forces accordingly.

The security arrangements in the bank have to withstand growing pressure from terrorists/ saboteurs etc., who will have access to upgraded technology. However a security man with little technical knowledge will be able to operate equipment efficiently and meet the threat response effectively.

From the reports available India should be able to produce the requirements of electronic based security system indigenously. We are self sufficient and able to offer computer-based system to others also in future. Infrared intrusion Detectors, CLIPS for making tamper proof identity cards, video image retrieval system, armoured vehicles for transportation, partation of cash and valuables, alarm reporting system, central controller with data based for access control system have already been developed by us.

We should be in a position to develop integrated security systems which will incorporate all the functions using indigenous resources.

Conclusion

We should remember that whatever benefit technology confers on law abiding citizens, is also available to a criminally mined person. He utilises this opportunity and consequently detection, as well as prosecution is becoming difficult. Therefore first and foremost law relating to such crimes need to be revised. Secondly, since different locales can be involved, international agreement on prosecuting such criminals must be uniform and mutually supportive. Coming to measures within the industry, the following can be considered:

- Have your own Network
- Have encryption facilities
- Internet
- Principle of complicity to be introduced in all transactions
- Transmission of Instructions or authorisation to be done only after positive identification of the receiver or action addressee
- Fiber optic linkages wherever possible
- Introduction of modem at both ends
- Messages in codes — encoded and decoded by two different personnel
- Introduction of Tiger Teams.

Security measures and devices will have to be constantly improved, and updated, in the light of the experience gained and new strategies adopted to prevent and detect crimes involving ever new "modus operandi".

V.I.P. SECURITY

The rigid VIP security measures in the capital city have been inviting severe criticism for quite some time now from all walks of life. Editorial and columns critisising the present VIP security arrangements are appearing repeatedly. Citizens in Delhi are becoming increasingly intolerant with the VIP security measures they see around and their intolerance came to the forefront in the wake of security arrangements for former Prime Minister Naramsimha Rao's trial in a city court. In this issue, we analyse a column, which appeared in the Hindustan Times. In his regular column, Musings, the Editor of the Hindustan Times, Mr. V N Narayan writes,

An intolerant society and an insecure state! How long can we get? Both are symptoms of same malady-pervasive dehumanisation. It is a society where people kill and maim in the name of God and religion and where a state protects criminals and oppresses citizens. The more laws we have, the less justice we get; the less secure we feel... To be sure 95 percent of the people who roam around with A to Z category security in tow, face no threat from anyone. And their lives are not so valuable that the nation should spend so much to preserve them. Security is useless even to the person who feels the need for it...

Cho Ramaswamy had no on guard him despite the fact he had been the bitterest and foremost critic of the dreaded L TTE. His logic was simple. With security you die every day, such a life is not worth living... The omnipresence of security is the most glaring proof of the terrorist, the killer or criminal, because the best result each of them seek is to sow fear, confusion and panic among the people to make them feel uncomfortable with life and with the world they inhibit...

Never before have cities and roads been made so conspicuously security-ridden as in the present century. In the process, security has become a self-promoting, self-perpetuating, multi-billion dollar industry which produces nothing of substance and very little of security. P V Naramsimha Rao was on his way to Vigyan Bhawan in his bulletproof limousine accompanied by a convoy of commando filled cars to face charges of cheating, forgery and bribery".

Mr. V N Narayanan seems to have forgotten that the real beneficiary of the VIP protection, at large, is the society only. The state usually provides VIP protection to those who handles power. Power is a double-edged sword.

It can kill itself if not handled carefully. The dangers of power and its proper handling were prophetically envisioned by our great political master, Chanakya in his ever relevant masterpiece, 'Arthashastra'. He employed spies even in the palace coterie to nip the security threats to the king in the bud. He gave high priority to the safety of the ruler and emphasized that the weakness or insecurity of the king always lead to anarchy. His utmost priority had always been the safety and thus the prosperity of the kingdom he administered.

A peep into the recent Indian history will make us understand how farsighted Chanakya was. Did not we almost plunge into anarchy after the assassination of Mahatma Gandhi? Did not the former Prime Minister Indira Gandhi's assassination incite large scale violence throughout the country? More recently, even the natural death of former Chief Minister of Assam, Mr. Hiteshwar Saikia revived the insurgent activities of ULFA.

Perhaps, it is in the democratic set up that the protection of people who handle power is of utmost importance. In democracy, the rulers are elected by the people and any threat to them form any quarter is an indirect threat to the democratic rights of people. Thus the very objective of the VIP protection cannot be different from that of the protection of the democratic rights. Thus if we see the things in larger perspective, the hassles created by the VIP security measures is too petty to cry hoarse over.

In democracy, as is known to everybody, the power is not concentrated in an individual or a group of individuals as is in autocracy, but is decentralised. However, decentralisation does not mean dilution. The security threat, to an extent, is directly proportional to the power one wields. The more is the intensity of the power you wield the more is the number and complexity of the security threats. New Delhi being the hub of political power, Delhites should willingly adjust themselves to live in tune with the VIP security measures around.

Mr. Narayanan' dig at the security industry is groundless. In which industry he will substantiate if the same yardstick is applied to other industries as well? Then why this tirade against security industry.

Mr. V N Narayanan's concern over the large amount of money spent for the protection of VIPs is unwarranted. Their concern goes a long way to show the least level of awareness in mass psyche about security, safety and loss prevention. If we take into account the loss of life and property as a result of the violent death of a leader, the money spent for their security is minimal only. The first exercise we need to master for public administration and maintenance is that a penny spent for the security, safety and loss prevention is a penny, sometime much more than a penny, saved.

The principles of our leaders are all fine. No doubt, the security measures of VIPs curb their freedom. But their insecurity may curb the freedom of the public. When one accomplishes the status of a VIP, one cannot expect to retain the freedom of an ordinary citizen. At their level, VIPs should restrain themselves from being exposed to security threats. In Indian conditions, the society or a section of society expect them to change their lives and always look up to them. In other words, they symbolise their hopes and desires. History says it that the Indian mass had always reacted hysterically to the abuse of their hopes and desires. The VIPs must realize this fact and should not let their senses prevail over their wisdom under any circumstances. Unfortunately, many of our VIPs are short tempered and sometime react violently in disdain. Our leaders' inability to control themselves causes many a trouble to their security arrangements. The hauling up of media persons in New Delhi allegedly by Bahujan Samajwadi Party leader, Mr. Kanshi Ram and the pulling up of a public servant at the time of last Parliamentary elections by Samajwadi Party candidate and film star, Mr. Raj Babbar testify how susceptible our leaders are. Mr. Cho Ramaswamy should have insight into these facts.

Multi-diverse is our society. We have multi-party political system prevailing. No leader in the contemporary India is considered unquestionable leader of Indian masses. A leader of a particular group is alien to another group. Thus we have numerous leaders of different status and stature. The safety of all such leaders is very important to the society as a whole and it cannot be taken for chance, as it could prove fatal for the very reason of multi-diversity and multi-party system. Thus the omnipresence of security arrangements is due to the evolution of numerous groups and leaders. So, do not blame the security, blame, if you want, the changing systems.

It is the obdurate VIP security measures that cause annoyance to many. In the Lakhubhai Pathak cheating case, the shifting of court proceedings to Vigyan Bhawan from Tiz Hazari courts on security grounds has raised many eyebrows. Many' intellectuals' felt that the unusual shifting of court proceedings is absurd. Those who questioned the shifting of court proceedings have apparently no sense of security. They were predetermined and went all around shouting political intrigue behind shifting of court proceedings.

Judiciary has never been as active in our country as it is today. Many politicians and bureaucrats are facing serious criminal charges in the courts. Are the political parties and their loyalists mature enough to take the judicial activism in their stride?

It was not the security of Mr. P V Naramsimha Rao alone that mattered. What about the security of judges and the safety of important records kept in the Tiz Hazari courtrooms? It was not long time back that an accused criminal was shot dead in Patiala House courts in Delhi by organised criminals. Those who lash out at the massive security measures are actually carried away by the political intricacies and intrigue.

Yes, crores of rupees are being spent for security of VIPs in a country where majority of populace live below poverty line. Ironical it is, indeed. Look closer at social, political and economical changes blowing in the country. Crores of rupees are being spent for few hours of entertainment in a country where majority of populace live below poverty line. The real giant irony is this gross imbalance. No irony cease to exit without taking its toll. Clearly the VIPs need better security consciousness and arrangements than ever before to minimize the larger than life irony of present day Indian conditions.

If one read between the lines, it is easily understood that the annoyance over the VIP security measures is really the annoyance over the deterioration of morality in public life. If a former prime minister face the charges of bribery, cheating and forgery, surrounded by a convoy of commandoes, the editor's dismay, really, is over the degradation of morality, but his first salvo is at security of VIPs! It is quite unfortunate that security is seen as a symbol of evil. It should rather be viewed as a healthy sign of awareness and alertness. Mr. V N Narayanan should have concluded his column with 'what a state of morality to be in!, not "what a state of security to be in."

It is high time that the security agencies concerned immediately take up image building exercises. No amount of good work by security agencies will yield desired results if they do not enjoy a noble image among the public.

The question of amending the Special Protection Group (SPG) did not arise during 1990 as the people engaged in discussing the alternative security arrangements for late Prime Minister Rajiv Gandhi thought that it would be as effective as the SPG itself, the Finance Minister told the Jain Commission in his maiden appearance, reported the Hindu. Mr. Chidambaram was liasioning with the then Government on security for Rajiv Gandhi.

Mr. Chidambaram is reported to have said that the quality of security cover was not the same as that of SPG. More than the senior level of Police personnel attached with the SPG, it was the training imparted to SPG personnel which was crucial, the Finance Minister deposed before the commission. Mr. Chidambaram further said that the force had also several special features like advanced security liaison, separate official earmarked for each functions of the Prime Minister, and also the overriding powers to oversee other security agencies. According to Mr. Chidambaram, under the Delhi Police cover provided to Rajiv Gandhi, some of these features were not there.

WOMEN SECURITY

Being given that half of our population - whether in Mauritius or Rodrigues is women, we have to consider certain situations that may render our society a secure place. It is the duty of women more or less to place the family into a secure society as we always say "behind successful man, there must be a women".

Security for women, her children, the family and specially the elderliness under her control. When it comes to the security of the young child, the mother has to watch every step—whether it is the crawling child or the one who can walk. The security in the kitchen is very important. Certain basic things such as never to switch on light or an electrical fitting with bare feet or with wet hands must be considered. Boiling kettle or other kitchen utensils is must be kept away from the child's reach. The housekeeping around the yard and in the house must be of a sound nature. If you find certain wires that can be walked upon, it is your duty to make your house a safe place. Avoid keeping medicines from the reach of children especially colorful ones. Avoid keeping bottled medicines into soft drinks bottles because children know the kind of bottles not the contents. Concerning the elderlies you have to give them the same type of security that you give to your children. I think that women have the will power and the power to fight for a place under the sun in whatever field they choose, specially in the context where the law is in favour and facilities are available to them. However, to the development of women, I consider that some specific measures of positive discrimination could be taken like special loan schemes or training programmes.

The more important obstacle is the mentality and the attitude which can only be changed through education. So it is high time that women start to bring this contribution to the creation of a better world. I think that in the case of Mauritius the evolution of women occurred side by side with that of society, for example, changes the legal structure that the right to vote, the advent of women, candidates in elections, free access to education and legal reforms in the early eighties, specialty that of the civil code. These bring to the empowerment of women. However, the main factor that prepared the ground and opened minds was when family planning campaign was launched, it was education. It was only then that women accepted to control the fertility rate.

The development of women has modified the scales in the balance of power. Men find that they are no longer the sole bread winners and no longer the sole decision makers at home. Not only because women has economic power of her own, but also because she is now well read, has an opinion of her own where the management of the home and the education of children are concerned. So often, with his power eroded, the man loses his balance and is confronted with an identity problem that he is not able to solve. That can create very sad situations and this is where unhappiness and violence prevail.

I think that the empowerment of women must not be a threat to men. It is rather an opportunity to enrich the relationship, to uplift the status of both women and men. The relationship should be more that of "equal partners in greatness" than that of master and subordinate. With this new relationship of mutual respect there would be more sharing of responsibilities and decisions. Each partner would bear an equal share of the burden and each one would be a source of confidant and encouragement for the other and to the society. This type of relationship could be the basis of sound sharing and caring families where children would grow and develop in peace, harmony and in a secured society.

Free compulsory education has today made women rediscover their own academic and technical abilities and is bringing in a sense of independence, self-reliance and freedom in thinking. Furthermore, women have participated fully in the economic development of the country with growing industrialization and greater access to capital and entrepreneurial skills. There is a need to implement the National Plan of Action for children in a systematic manner and evolve a monitoring mechanism. She sees that environmental issues such as abortion and other explosive issues be contained and redressed.

Furthermore, there must be that freshness of a river from which, we create the attitude of IEC (Information, education, communication). Gender development and mainstreaming of gender issues is a must in the global plan of the development of the policies and projects of our women, our society and our security. Sometimes ago there were strong reactions from men and even women who feared that equal rights for men and women would lead to a total breakdown of the social order with drastic repercussions for family life. In a male dominated society, it was inevitable that any change in gender relations would provoke resistance and even hostility. There have been cases where women have been wearing trousers and men wearing skirts. Men have now realized the labour market of women not only for the benefit of national economies but for their own household budgets. They are having to adapt to this situation in their own interest without tension or strain. Women understanding the ways of safety and security, may become new rulers.

Men however, seem to be able to rely on considerable support from networks which they build up outside the family but this is difficult for women because of their family and the double burden, as they try to combine professional and household responsibilities. Many women and men suffer from dangerous social isolation. This is becoming a problem for our society. All I can say like a famous Indian Sociologist 'Professor Andre Beteille' that "EQUALITY IS LIKE A DJIN, ONCE RELEASED FROM A BOTTLE IT CANNOT BE PUT BACK INTO IT AGAIN".

While Mauritian Women have given to take up the opportunities and challenges created through socioeconomic development, they have had little support from the interacting systems — in the family or work or in the community at large. Mauritian society is in the danger of becoming rampant with too much materialism and self-indulgence. Such circumstances create conditions, which affect family stability thus we are seeing an increase on substance abuse and dependence, family disharmony and breakdown. It is very difficult to solve these problems once they develop. The best solution is prevention by creating the Awareness of Security and thus protect our gender society.

EVENT SECURITY

Organising security measures in respect of big events like international exhibitions, entertainment shows, fairs and festivals and places of large congregations is truly challenging. It becomes more important when exhibitions of jewellery and rare iconic artifacts are displayed in organised events. It requires sharp intelligence of a scientist and through perfection of an artist. Since a large amount of money and valuable are involved in these types of exhibitions and shows, the threat could come from any part. Therefore, a private security organisation, manning the security should look into the following matters and take security measures accordingly.

(*a*) The security head must consult, co-ordinate and colluse with the local police organisation.

(*b*) It first assess the security threats

(*c*) The classification of security threats may be

  (*i*) shop lifting,

  (*ii*) steading from the buyers,

  (*iii*) robbery at the exit points.

(*d*) It cannot rule out organised attacks, since it's happenings are being reported daily in print media in large numbers.

(*e*) To see visitors do not carry arms, while visiting the exhibition/site. This can be checked at the entry point by the security personnel.

(*f*) Physical checking of visitors need to be done if it warrants with metal detectors.

(*g*) It is incumbent on the part of security organisation to throw a net work of intelligence to thwart the attempt of organised criminals to loot the wealth displayed in public exhibition.

(*h*) In the event of terrorist sponsored kind of attack/threat, it is necessary to arrange for heavy surveillance at the outer perimeters and inside the exhibition halls. At the outer perimeter, patrol vans may do the rounds clock wise and anticlockwise round the clock.

(*i*) Number of towers (preferably two) should be erected specially for the event at the height of 30 feet each. The security personnel, at the top of the tower should be equipped with binoculars for better surveillance.

(*j*) The patrolling team should have search light for night surveillance.

(*k*) Surveillance inside the exhibition hall should be manned by CCTVs, cameras, PIRs, motion detectors, infrared detectors, magnetic conductors etc.

(*i*) Spotters, both male and female are to be deployed in the hall to detect the shop lifters.

(*m*) Although entry and exit points are strategically two vital points as far as event security is concerned, but in most of events like cricket matches, public functions, private functions, religious functions, the crowded entry and exit points are generally manned by a handful of security guards and chaos is first created at these points. Therefore, security arrangements at entry and exit points should be devised, taking into view the highly undisciplined nature and large number of Indian crowds.

Intelligent Buildings

Security systems for perimeter protection, intruder detection, metal detection, bomb detection etc., can be integrated with I.B.M.S. to provide continuous monitoring and reporting with appropriate alarm functions, such as turning on hooters, sirens, visually highlighting area of security violation on a console or mini panel etc.

Access Control

IBMS provides computerised access control and access recording facilities. Depending on requirements of a building, person identification could be alone by the following devices.

(*i*) Bar Code Readers

(*ii*) Magnetic Card Readers

(*iii*) Finger Print Scanners

These are must for most buildings. However for high security areas, fingerprint scanners provide fool poof access control.

Miscellaneous Facilities

Water level monitoring and control water softener plant monitoring, monitoring lift operations, recording usage times of various equipments in a building etc., can be provided with the help of IBMS.

System Architecture

CMC Ltd. has designed and developed intelligent building management systems indigenously with a view to provide the advantages of an intelligent building in India.

Physical Protection of Intelligent Building

Protection of personnel, data, assets and equipments, has to be ensured. Fire, flood intrusions and explosion need be ripped in bad. Prevention of catastrophic conditions is mainly dependent on the design of the building, location of equipment, building materials and occupant/operator discipline. However, IBMS can minimise damage due to fire by early detection. Smoke/heat detectors connected to IBMS enable early and round the clock fire detection zonal fire pannels connected to IBMS enable identification of fire zone. Cutting off air flow by switching off air handling units, closing fire dampers of the affected areas. Operation of water sprinklers, carbon dioxide extinguishers etc., in the affected areas. It is necessary to dial to important persons as well as nearest fire station and deliver pre- recorded messages, operate sirens and hooters to attract attention of personnel. The exit ways should be lighted for speedy evacuation.

SECURITY IN OPEN ENVIRONMENTS

Security functioning is primarily a management function. Responsibility towards its function is ever increasing with the enlarged threats by terrorist, act of sabotage, espionage and subversion in addition to the traditional threats of pilferage/theft of materials, threats on person etc. This threat multiply in a particular environment when the set up is not provided with, not aided by infrastructure complying the principle of physical security. The mischief mongers spread vandalism, play mischief, adds assault, resort to strike and intimidation. Organised crime with the help of newly found weaponry, it become a real problem for the security personnel to extend their good services to the customers. Customers may be the insiders and outsiders both.

For a captive security set up, it's own establishment is a customer whereas for a private set up, it's client is the customer, in this age of customerisation, one is the customer of another and the vice-versa. How to insulate the customer from the turbulent environment? Particularly when own's environment is not under one's control, the area cannot be brought under fence to control the situation because of many factors—the larger geographical area, the multilingual, multifaceted social and economic strata, always pose challenges to be managed. Then in such an inhospitable aggressive situation, what is desired from a security managers? The dimension of problem itself has to be used as a catalyst to find out as it's solutions. In the following, we shall try to analyse these factors attributing to it's solution:

1. **Security consciousness:** The general awareness amongst the stalwarts of the industry to down the line to the labourer is proved boon to the concept of security. Security means "the state of mind of being cared". This care come forth automatically if everybody in the establishment are aware about the need of being cared. The notion thus only Uniformed Members of the Industry can only mind the need of providing care to the establishment, fortunately diminishing. Now it is the baby of all members of the industry to form a part of uniform security by becoming conscious about this particular needs of being secured. To manage the affairs, we find the four '*Mantras*'.

   The moral courage: Story goes that in erstwhile Travancore State, a single constable was detailed to quell a riot between the villagers. The solitary representative of law out of his conviction was able to restore law and order and came back to his H.Q peacefully after ensuring the lasting peace. How was he able to sustain so. He could be able to achieve his goal because people were happened to be God-fearing and having absolute faith in the law and order agency. The people among whom the constable appeared, co-operated him in maintaining law and order.

The situation you may say in the present context is changing. It is no more possible to handle the situation single handedly. We accept but not able to venture to come out of the principle of establishing commitment and the sincerity of the purposes. This always has helped in the past and will be aiding in the future. Absolute integrity, resourcefulness, calm determination, sense of fairplay, sympathetic understanding and sincerity, particularly in the face of mounting misunderstandings and resentment, still pay the dividends. In handling assignment in the open environment, when you are not utilising the principle of physical security, you have to depend upon the above virtues in achieving your goals.

2. **Communication:** Crucial need is for ranking communication as a priority where you fail to provide security fencing around the assigned areas.
3. **Creation of sources:** Security of assets is warranting the services of everybody voluntarily not by the security personnel alone. This may go to a long way. Therefore, all persons are to be involved to participate freely and they should be motivated to share the burden and thus increase the numbers of security conscious persons of the area. They occasionally would be favoured in some way or other to keep the relation continued. For any untoward incidents these volunteers may be serving the "eyes" of the security.

The final list of such participants would be drawn up their antecedents screened and then engaged for this job. The undesirables have to be weeded out of the list. Credentials is important. Since list of persons are to be trained a bit amalgamated into the parent force, declaring them to be the second line of defense in the security set-up. Frequent meetings and discussions about the workings of security to channelise their thought process and mould them in the right direction into becoming more security conscious would be the next obvious step. A few of these volunteers who have a background of the armed forces and who show more promises and shape up well could then be selected as a 'Group' Head at various levels. A befitting nomenclature to distinguish them from others in the organisation would go far in cementing them to the security organisation.

The usefulness of such organised volunteers participating whole heartedly in the field of Industrial Security is a boon of no mean order as has been experienced in actual practice. They can be utilised to the best advantage as a source of intelligence and also as a second line in case of crisis management. Stray bits of information about errant workers, malicious propaganda of any kind, malafide intentions of contractors and their labour union activities, stirrings of labour unrest, common shop floor gossip, can then find tits way to the security set up easily. Such information when collected and collated would then give the true picture of the very pulse of the work force in the entire premises. It has been practically experienced that such feedback from the shop floors/departments prove to be of invaluable assistance while determining the course of action to be taken in various spheres of industrial life . As a second line of defence in crisis, these volunteers can be deployed sporadically, interspersed with security guards in less sensitive pockets. Such a force could be kept as reserve, which can act as a veritable ace in the hole to be used as and when the situation so warrants. If it is handled with a master hand at the helm, this volunteer force can be made to render all assistance in multifarious ways.

Workers participation is a new concept, but Tata Steel is practicing since a long years and it has been found to be very encouraging. The principles, what have been widely used in the Tata Steel as a solution to industrial peace and harmony, may proved to be guiding factors in security setup also. The greatest benefit could be derived with a lesser expenditure and the greater forces would be available all time including in crisis.

4. **Contingency plan:** Should also be taken into consideration all possibilities of unusual happenings such as civil disturbances, weather phenomenon, natural calamity etc., depending upon the circumstances, location and people's attitude.

USE OF DOGS IN SECURITY MANAGEMENT

When is a Hebrew word which, when used on a daily basis, refers to basic Kennel Management at the "ground floor" level of cleaning and feeding dogs.

When used in a global sense, it becomes a general heading for all the subjects which comprise the complex profession of calbanut. There is no equivalent single word in english which covers all the different aspects of the profession.

If we analyse the subject *calbanut* into its component parts, we get:

1. Kennels for various purposes,
2. Administrate/Accountancy/Computers,
3. Staff for different purposes,
4. Diets for different purposes,
5. Transport, internal/external to kennels, factories and military bases, etc.
6. Equipment for various tasks (Kennels, Training, Working, Care etc.),
7. Care: Daily, Clipping, Basic Veterinary,
8. Dog breeds for various duties,
9. Genetics, Breeding, Physical/Psychological Make-up,
10. Training for various functions/purposes,
11. Security/Guard Dogs (Patrol, Compound, Cable Dogs),
12. Patrol/Working Techniques and General Security,
13. Canine and Security/Guard Related Laws (According to Country),
14. Animal Welfare,
15. Miscellaneous topics not already covered above.

Security/Guard Dogs

Basically, there is no difference between the terms—*Security* and *Guard*. It's purely a matter of local usages-one country may use the term security when referring to national security and guard when referring to local matters *e.g.* We may talk of Security Forces, Guard Companies in one country and yet Security Forces and Security Companies in another one. (although we will refer to private staff as Guards and Security Officers) Security/Guard dogs may be considered under 3 classifications:

(*i*) Patrol Dogs

(*ii*) Compound Dogs

(*iii*) Cable Dogs

We'll start with Compound dogs.

Compound Dogs

Most of you will be familiar with the idea of leaving a dog loose in the grounds of a private house or the surrounding yards of a factory. As the name suggests, such dogs are turned loose in compounds to prevent access to strangers/prevent escape of intruders. Such an application is ideal when needing extra security on a building or otherwise on-secured compound (To prevent the dogs from escaping). Their Full training requires them to:

1. Be dangerous to all persons except their handler/kennel man.
2. Be able to protect themselves from intruders, which also includes their being trained to keep back from perimeter walls/boundaries thereby avoiding danger of being injured through the fence and also giving them the advantage of attacking an intruder on their own ground once he is inside and unable to "run for it" so easily.
3. Refuse food from all except the Handler/Kennel man so as to prevent the possibility of poisoning.
4. To be an alarm system in applications with nearby personnel.

   Note: In some countries, such dogs are only legally permissible If there is a Handler/Kennel man in vicinity during the dog's work. Further to this, there may also be a legal requirement of *earning notices* outside

Cable Dogs

As their name suggests, the dogs are chained to run along cables in applications requiring. For example:

1. Long protective runs.
2. Temporary protection of assets without permanent fencing.
3. Specific protection of assets within an outer perimeter.

Their training/function may be to act:

1. Purely as *alarm dogs* to inform nearby guards of intruders.
2. Both alarm and "attack" dogs are trained as per compound dogs.

Compound Dogs

The use of compound dogs and/or cable dogs is a question of cost effectiveness vs. Assets value vs. Other possible/available security solutions.

In the area surrounded by cable dogs, one dog's barking can set up a chain reaction around all of the dogs, which can be very disconcerting to handlers/patrol dogs and other searching staff. Apart from making it difficult or impossible to know which dogs originated the alarm.

In both the cases of *cable dogs* and *compound dogs*, (usually in the case of Cable Dogs), it should be possible to remove or restrict them in none-working periods to rest them and there should also be strict discipline exercised on personnel working near to the duty dogs to prevent unofficial interference with them. Lastly, there should be no authority for workers to be in the dogs' working area unless they (the dogs) have first been removed, dog cannot differentiate between an authorised worker or an unauthorised thief and repeated intrusions will eventually make it ignore all entries.

There is also a problem that, in some systems, the *canine staff* and *general security* staff come from completely separate sources with neither side understanding the other's work. This can lead to a situation in which, for example, there are guards who are unable to differentiate between the different ways of barking. This varies according to circumstances. Now we come to the most important of all the security/guard dogs.

Patrol Dogs

No matter how many active/passive (incl/electronic) defence systems are put into a security system, when these devices fail to keep out or do warn you of an intruder within your perimeter/ buildings, none of them can finally find, arrest, and escort the intruders, let alone protect the patrolling/searching guards from ambush/physical assault.

By virtue of it's extensive training, a patrol dog can:

1. quickly search buildings/land for unfamiliar objects/persons,
2. warn the handler of the intruder/s presence,
3. help to prevent escape of intruder,
4. prevent an attack on the handler/assist in his protection,
5. help Guard and Escort a detained person,
6. help to Track down an intruder who succeeds in escaping,
7. carry out various other functions according to extent/type of training (To be discussed later).

Furthermore, a dog handler will be less likely to change his job and "working partner" than a normal guard thereby reducing staff turnover/related costs.

To come down to the very basics of having a dog accompany you on your rounds, it may be pointed out the advantages of having a companion (even an untrained one) especially on night patrols and one who'll be your eyes, ears and nose to warn you of the unusual.

With further reference to the sense of *sight*, it may be pointed out, that whilst a dog is sensitive to the slightest movement, it's often possible to stand perfectly still and be unnoticed by a dog unless it hears you or smells you. There are certain hunting breeds (classified as gaze-hounds) which do rely on their sight rather than other senses, but for most dogs, the sense of sight is less developed than those of smell, and hearing. These being the ones that are highly developed and precisely the ones, we rely on for security/guard work-especially but not only at night-time.

To go one stage further, a dog trained to speak on command accompanying a uniformed guard will often be a psychological deterrent to a would be wrongdoer, even more so when there are warning notices around your site. If the dog barks on command at an intruder, (especially to a command such as "watch him"), then that intruder has no way of knowing what other training it may have and will often be reluctant to find out. Furthermore, the presence of an "apparent" patrol dog will often discourage a criminal from trying to make a run from his/her escort. However, don't rely on such untrained dogs, one day some tough-guy will "call you bluff" and put you in an embarrassing/dangerous situation.

At this point, as the main type of security/guard dog is a patrol dog, then we should expand somewhat on the various functions for which these dogs are trained and discuss, where applicable, various training alternatives, according to the dog's sphere of action *e.g.*, whether civilian or military sphere and whether there is forbidden, restricted or un-restricted access/presence of other persons.

It may be pointed out that for those of you who have seen demonstrations or competitions with trained dogs, especially demonstrations of man work/attack exercises, these are often highly dramatized and divorced from working reality and as you will see from later examples, there is often very little necessity for dogs to use their man work/attack training. Furthermore, only an idiot trains a true working dog to stand up to being hit by a stick as in competitions. It needs to point out that whilst we may give a dog extensive training to cover as many eventualities as possible, the basic aim is to work a dog in a manner which makes the most of the dog's own initiative and the minimum use of commands and reliance on it's handler.

Basic Training

There are three reasons for basic training.

Apart from-teaching the dog certain commands, you should have control over it. Basic training when done by MODERN training techniques is also the period for building up the correct relationship/partnership between the handler and dog.

Lastly, whilst you may be able to form a certain opinion of a dog in various ways before taking it into training, now is the period for really obtaining an intimate relationship/knowledge of your future partner-both positive and negative aspects.

During basic training what the dog is taught:

Positions: stand, sit, down, and to Stay in any of these positions.

Heelwork: To walk to Heel (normally on the left) of the handler, It can also be taught let's go which means to move off in the same general direction but not necessarily close at heel.

Recall: To return to handler on command come/to me hand etc.

The meaning of the word No/Leave

The release from orders command of free to do as you wish.

To go to it's Kennel/Bed/Corner.

Advanced Training

We'll now discuss a full curriculum of advanced training. There will be many aspects, which will be way above the requirements for a basic Security/Guard Dog (especially on a daily/nightly basis) and more suited to full police dogs. However, it is prudent to remember that:

Regardless of your dog's usual sphere of action, there may come the day of the unusual, so the more training you have covered, the better you are insured for that day. More advanced training includes:

Negotiation of various obstacles in various ways:

Up (clear jumping)/Over (scrambling)/ under/through/crawl and including swimming).

Directional training (Send Away and Re-direction) Go/Left/Right which are essential for searching/ quartering areas.

Retrieving: Fetch various objects to handler and either giving to hand or depositing on ground according to choice.

Note: If objects are not retrievable for some reasons *e.g.*, a body or some heavy/bulky object, the dog should speak/bark/give tongue to advise it's handler and remain with object (or return and take handler to it. In case of rescue dogs don't train normal patrol dogs this way. There's a danger/probability of it's leaving a criminal to escape).

Animal proofing: Teaching the dog to ignore other dogs (including bitches in season) and other animals/livestock.

Weapon proofing: The dog has to be familiarised to weapon fire and to be trained to deal with criminals armed with sticks etc., without getting hit/ injured/killed if possible.

Transport: A dog has to be familiarised with various means of transport (according to spheres of action). Such transport can include:

Cars, vans, trucks, cycles etc.

Various types of boats.

Various types of aircrafts.

Unusual "transport" such as fire snorkels/cranes and drug harness, parachutes (more likely to travel by helicopter today).

Crowd/noise familiarisation: A dog required to work in public places need much familiarisation training for this including prevention of indiscriminate attempts to bite legitimate passers by.

Speaking/barking: The dog should be trained to speak on command and also automatically under certain circumstances.

There are two methods of training the dog, the first of which leaves him passive, whilst the second is associated with aggression. In the security situation, teaching the dog to bark on the command "watch him" is obviously psychologically preferable to teaching the dog to "speak" to a criminal. The conflict to this is that you don't want a dog to show aggression to a lost child it has found.

So you should teach the two commands and eventually the situation and experience will teach the dog what is required of it. Also remember, where another person is involved, the dog can sense that person's feelings as part of it's experience and "trigger of reaction".

Man work/attack training: A dog may be trained on command to either arrest someone by holding his arm or to make a flat out attack, even with the intention of killing which method of training, depends on the circumstances and theatre of action.

He can also be trained to do this automatically on detecting an intruder or to prevent escape by circling and barking for his handler. However; the 2nd method is limited and can allow escape of intruder.

Note: Teaching a dog to only go for persons running away from you is negative-what do you do the day, someone brazenly and quietly, without any signs of aggression, walks up to you and then attacks you?

Lastly, if the dog's handler should be attacked, the dog should again automatically arrest/attack the culprit.

Additional to the actual training above, the dog should be given training in the guarding and escorting of arrested/detained persons. Dogs working in a "public" atmosphere will also need training in "crowd control".

For certain situations, a dog should be trained to "ware" (get out behind) like a sheep dog, for dealing with a criminal armed with a stick/knife. It's much more difficult for a man to protect himself from a dog and handler who are each approaching him from opposite sides. Equally, it's easier for the dog to catch the man's weapon arm or the handler to disarm him with a good clout from a night stick, if the criminal has a double sided problem of self defense.

Nosework: A dog uses his nose in two basic ways:

Airscenting/Groundscenting.

As the words suggest, one requires him to sniff the air whilst the 2nd requires him to sniff the ground (track). In actual working situations, it is usually necessary for the dog to detect using both methods.

Building Searches

In this, a dog usually relies on air scenting to search for hidden persons. Points to remember are that: Local drafts can cause a redirection of scents.

Dogs should not be released to work free in areas containing dangers such as electric cables, chemicals, building nails etc.

Land Searches

In this, dogs are also using mainly air scenting, especially when zigzagged into the wind—a process called quartering.

Tracking

Without going into specialized details/theories/ reasons, this basically requires the dog to follow a ground scent. There are dogs which will do this "exactly" that almost publicises the probability of their locating articles, discarded to the side of the track by the layer/criminal, whilst, there are others who will follow generally. This in turn, has other problems. In either method, there is always the possibility, that at a certain point, the wind will send the scent of the tracked person directly to the dog in which case it then finishes the job by air scenting.

Scent Discrimination/Identification

There are "exercises" in which a dog may take the scent of a person and then pick out that person's article from a group of articles. Conversely, the dog may take a scent from an article and then pick out the owner from a "line up" of persons.

I think that there may be a question in both cases as to whether this is admissible evidence in a court of law.

Canine Patrols/Security

When we are using a patrol dog, we should be familiar with specific aspects and techniques of canine security. Surround area with warning notices in all appropriate languages. This may be a legal obligation in certain countries. If armed, use short weapon whenever possible to avoid injuring dog by banging him on the head etc.

Use of radio should be restricted can disturb dog and also be heard by intruders from a long distance at night. Make sure dog is familiar with night working, this being different today in many ways, (Sounds/Shadows/etc.) Stop often to let dog hear/sniff the surroundings. Don't whistle/talk to dog unnecessarily.

Watch dog's head/reactions especially in very dark nights and be familiar with your dog's body language for different circumstances.

Now you are remember some points. Every profession has component parts which depend on many other professions (and their advances). You should keep abreast of as many other advances as possible. Very often the solution to a problem maybe found in another profession/industry and evernmore, so far more effectively and cheaper than a solution from your own industry, *e.g.*, advances in materials may enable you build kennels from stronger/lighter/cleaner etc., materials.

Even within your own spheres of action, you maybe forced to compromise/substitute for certain equipment or share certain other equipment with other departments *e.g.*, vehicles. Be as flexible/ open-minded as possible with everything that may be available/possible to use.

Your dog is your working partner like you. He can be injured/killed (in which event he can't help you further).

Don't put him into unnecessary danger and work in partnership. When you are trying to keep warm/dry/ cool, etc., so is your dog.

Just as you can feel off-colour and unable to work well or even not at all, then so can your dog.

Once trained, a dog needs regular refresher/reminder training, especially on exercises it doesn't use regularly.

Trained patrol dog is a weapon which is a substitute for or additional to another weapon such as a firearm. You cannot use it at will. Make sure you are familiar with your local/national laws and needless to say, those laws affecting you regardless of your use of a dog.

Your dog is not a neutral piece of security equipment, but a living and feeling creature whose psychology and emotions are no different to yours. Like you, it needs love, attention and respect from you. Given this, you'll have a partner more devoted to you than any other-even to the extent of being prepared to protect you unto death.

Lastly, whilst this is not usually the function of a patrol dog, and without going into any detail whatsoever, we it relevant to the general security field to mention the existence of dogs trained to detect:

Drugs, mines/explosives/weapons, sources of arson, live/dead bodies, and a variety of other unspecified items.

DEVELOPING A POSITIVE ATTITUDE TOWARDS SECURITY MANAGEMENT

A number of issues go into the developing of a positive attitude.

1. **Regard for the self and regard for others:**

   Self-regard comes when one has the ability to recognize one's real, genuine strengths, the capacity to develop these strengths, and to use them for one's own good and the good of others.

   Self-regard also means the ability to recognise one's weaknesses and the effect of these on one's own growth and on one's relationships with others, and the ability to develop and implement strategies to reduce, if not eliminate, these shortcomings.

Regard for other is determined mainly by the way we relate to others. This regard is positive when we are able to

(*a*) accept people as they are, not as we would like them to be;

(*b*) deal with relationships and problems in terms of here and now, and we do not get obsessed with the past;

(*c*) separate issues from personalities;

(*d*) treat people close to us, subordinate to us, and less important than us with the same respect and courtesy that we extend to strangers, to superiors and to VIPs.

True professionalism recognises that wisdom is not bound by organisational levels. The best decisions come when one involves others, in the belief that expertise and good ideas can come regardless of rank, level, discipline and qualifications.

Professionalism is further enhanced when we are ready to submit our plans, ideas and suggestions to scrutiny, challenge, questioning and perhaps even to contradiction by others at any level.

2. **Failure is seen as an opportunity to do better:**

For people with a negative attitude, failure is the last step of one's life a 'finality' from which there is no 'inwardness.' Those with a positive attitude, however, regard failure as the beginning of one mere learning, from where one can only go onward and upward.

That way failure is treated, thus adds to one's level of professionalism or takes away from it. Professionals treat failure as a wave on which they can ride shore-wards. Those not so professional, treat failures as the wave that will take them out into the sea and ultimately drown them.

3. **Realsticoptimism rules over eternal pessimism:**

Two executives are rushing to catch a flight. The pessimist, with the negative attitude, is more likely to repeat over and over again: We will miss this flight. I know it. We are bound to hit only red lights. I just know it. The optimist is likely to say: We may just make it. Even if we get a few red lights, I think we will make it.

Realistic optimism makes professionals action-oriented. Pessimism, on the other hand, often paralyses both thought and action.

4. **Begin by trusting others, but also use judgement to confirm if trust is to be continued:**

People with a negative attitude start on a note of distrust, and test continuously to check whether other can be trusted. Those with a positive attitude begin by trusting, but when presented with evidence of misplaced trust, are not afraid to exercise caution. "

Professionals who do not trust cannot motivate. In the ultimate analysis, it is not lots of money or the latest in technology that makes for a winning organisation. Rather, it is the motivated man, supported by technology and funds, that determines the fortunes of an organisation.

5. **Criticism is a responsibility that must be exercised carefully and appropriately:**

When people do not criticise at all, even when their experience and training tell them that there is need to correct, guide or control, they display as much of the negative attitude as when they over-criticise, control or dominate, particularly because they have the power and authority, physical and/or positional.

People with the right attitude use their experience to correct deviations, but manage this in the most appropriate way. In other words, even when they do loss their temper, they express themselves in proportion to the issue. They will not, in other words, use a hammer to break an egg.

6. **Supportive and nurturing behaviors are in evidence:**

   These are positive traits, and therefore desirable. Yet, those with a positive attitude are aware of the pitfalls of over-nurturing. Killing with over-kindness, crippling by being a constant crutch and verbalising praise where no cause for such encouragement exists, do not add to the positiveness of one's personality.

   At work or in social situations, professionals learn to give just the right amount and kind of support, that is really needed.

7. **Logic controls emotion:**

   Emotions result in spontaneous expressions, some of which may have very beautiful consequences, some very ugly. If, however, we want to ensure that the consequences are right, or correct, then emotion must give way to logic, to data collection, data analysis, and data verification, before decisions are made. A positive attitude calls for recognising that an emotion exists, for determining if the emotion can result in negative actions, and then, using the thinking process to give the right response.

   Professionals learn to manage their thoughts so that they do not interfere in objective thinking.

8. **Creativity and the readiness to try out change:**

   Change makes people feel insecure. But those with a positive attitude will give change a try, before declaring 'unworkable', "impossible", 'un-necessary', etc. While they will not seek change continuously, for that is often indicative of an uneasy restlessness, they will suggest it or be willing to consider it if it appears constructive.

   Such an attitude goes a long way in the making of a true professional. It leads to the encouragement of calculated risks, to acceptance of failure, and the determination to try new routes.

9. **A sense of humour:**

   People with a positive attitude make a conscious effort to develop a sense of humour. They develop the ability to see the funny side even of so called disasters, to laugh with others, and most importantly, at themselves.

   They take themselves lightly, and this helps them to take critical comments in their stride. It helps them to face the truth about themselves and to recognise that they are not as perfect as they had believed they were.

10. **Happiness is all, ultimately:**

    Life is full of happy surprises. All we have to do is to open our eyes, ears, minds and hearts to sense these surprises, enjoy them, and share them with others, even with strangers. And yet, scores of people go through life, grow old and die, with sour expressions, bitter thoughts, and angry words. To me, it is almost an act of ingratitude to God in heaven, and our fellow beings on earth, who have in one form or the other, blessed us with so much. Not only do such people live their own lives unhappily, they make the lives of others around then full of stress and misery.

There is no denying that life has its bad, unpleasant and sad moments. There is no denying that in everybody's life, there are deprivations of one kind or the other — of good health, of economic security, of love, affection and respect of chosen ones, and in some cases, of even the very basics of existence.

But there is also no denying that there are, along with these deprivations, a whole lot of gifts and blessings, just waiting to be acknowledged.

A positive attitude calls for living this life so that we accept the deprivations we try to manage them so that their ill effect is minimised and most importantly, we constantly keep in sight, all the beautiful things we have been blessed with.

In recognition of this, professionals do not depend only on created systems for their happiness. They also make themselves responsible for creating happiness- inducing situations. They create the happiness environment whether they are at work, with their family or friends, or with passing acquaintances. They seek outlets to life's frustrations enjoy or peace, giving recreations and hobbies.

If professionals develop the right attitude, they learn to take care of themselves, of their health, their family and friends, their assets and fortunes. In the process, they perform better and ensure that they are seen to be true professionals.

To sum up

A positive attitude is the difference between existing in this world and actually living in it, being happy and giving happiness.

A positive attitude is the vital difference between being merely a manager and being a leader.

PRIVATE SECURITY VRS. GOVT. SECURITY

As per Bayley's definition of the police as "people authorised by a group to regulate inter personal relations within the group through the application of physical force." So far policing is concerned, this definition is applicable in general sense but not restricted to the police as an organisation. Three elements are involved in this definition.

(*a*) Authorisation for the police functioning by the state.

(*b*) Regulatory nature of their services.

(*c*) Application of physical force in due performance of their duties.

The private security functioning also rests on these three pillars. The police operates strictly under the laws of the land enacted by the legislatures and parliament. On the other hand, the private security draws authority from the terms of agreement between service provider and service user. The law enforcement agency *i.e.*, police is essentially a crime prevention and investigation agency,strictly adhering to the prescribed laws and procedures. Private security is responsible for loss prevention and detection,which goes beyond the realm of handling crime and criminals. All the activities of private security organisation is in sense, supplimentary in nature. In a broader sense, it is a way that assist the police organisation, so far apprehending the criminals, their prosecution and trial. It has no legal sanction except the right of private defense, which is liberally used by any individual. However, the private security is more flexible, innovative and cost-effective. In case of police, the regulatory functions are more in the nature of prohibitory, intrusive, or restraining action. Non conformance with police direction, may invite a penal action and police charge in a competent caurt of law. In case of private security, their functions are designed to be less prohibitory, non-intrusive, least restraining and more persuasive in nature. Use of force by police is legalized in certain specific situations. But no such legal power devolves on private security. It's activities, if do not violate the Law of land, it is ok. As told earlier, it can act under the right or private defence in relation to application of force to defend the person or property entrusted to their charge and apprehend a person committing a crime in their presence, as liberally available to all citizens.

The Indian subcontinent, as a whole is generally beset with rising crime graph and unprecedented turbulence on the public order front. Mid air hijacking, serial killings, international swindling and other forms of organised crime are causing concern through out, the reason on the public order front, terrorism and extremism of different varieties, civil strife, communal and racial violence have dotted the region. Religious fundamentalism has clearly emerged in our horizon. As per the latest U.N. report, "At this huge rate the crime is rapidly outstripping the capacity of many govts, to keep pace with it and has already outpaced their capacity to reduce it".

Though this crime situation has overburdened all the components of criminal justice system, the police has come in for severe criticism,. who has failed to cope with the mounting figure and practically lost confidence of the general public. People simply donot trust the govt, any more. So they are buying guns, cordoning their streets and hiring private security.

Role and Status of Private Security

1. It started as a watch and ward service provider.
2. Due to increase of crime figure in relation to kidnappings of business and corporate heads and their family members, the private security was hired to extend protection to their lives as it is not possible for the police force to afford general protection to the individual public. The private security has been born to provide direct security cover to an individual or an organisation against any specific or all possible threats to the person and property of that individual.
3. Police and private security have both supporting roles. It is true that if private security services are drastically reduced or eliminated, crime, fear of crime and prices of retail merchandise would rapidly rise and the security of vulnerable section of people in the society will be in jeopardy.
4. A closer look at the functioning of the police and private security would show that these agencies are exclusive, complementary and supplementary to each other. An illustrative but not exhaustive list of these three spheres is given below.

A. Exclusive Areas of Functioning

Police	Private security
1. Patrolling public places	1. Patrolling private areas
2. Arrest on reasonable suspicion with or without warrants.	2. Regulate entry in private premises
3. Enter house for arrest, search and seizure	3. Frisk incoming and outgoing persons at private places.
4. Conduct searches and seizures	4. Protect private goods, godowns, car parks and other property in its charge.
5. Keeps criminals in custody for interrogation	5. Prevent non criminal, intentional, unintentional or negligent losses of resources under it's charge.
6. Take finger prints, photographs and physical measurements of arrested or convicted persons with use of force, if necessary	6. Check documentation of incoming or outgoing goods at the entry —exit points or supply gates, private industry etc.
7. Prosecute or associate with prosecution	7. Stock taking of materials for detection of losses.
8. Service of court summons and warrants.	8. Check against misuse or abuse of human or material resources
9. Summon witness and record statements	9. Regulate vehicular traffic on private roads
10. Maintain personal and criminal history of persons under certain conditions	
11. Regulate entry to public places	
12. Regulate vehicular traffic on public roads	
13. Protect public property, car parks, public entertainment places etc.	

B. Supplementary Areas of Functioning

Police	Private security
1. Prevention and detention of crime and maintenance of peace.	1. Prevent and detect security breaches, crime, violence and drug abuse at work place
2. Surveillance over known and suspected criminals.	2. Surveillance over known and suspected persons posing threat to security and safety in their area of responsibility
3. Keeping records of crimes and incidents.	3. Keeping records of crimes and incidents
4. Conduct enquires, investigations and collect evidence.	4. Conduct in house enquires and investigate security breaches.
5. Interrogate arrested persons in custody or otherwise	5. Arrest if caught red handed.
6. Inspection of scene of crime/Incident at the earliest opportunity.	6. Hands over the arrested person to police.
7. Help and assist fire fighting operations and control crowd at the site.	7. Help and assist in fire fighting and crowd control at the site.
8. Launch public education campaigns on crime prevention, road safety, drug abuse etc.	8. Preserve the scene of crime for police inspection.
9. Educate the general public on individual security, crime prevention co-operation in investigation watch and ward etc.	9. Educate the security users and management regarding modern security system technologies and approaches like total loss prevention concept.

In U.K. and U.S.A., some prisons in the country have been placed under private management without any interference by govt. Based on their performance-appraisal, the educational and reformatory services in some prisons have been contracted out to the voluntary sector, which is now know in U.K. as the "Third sector". Securitas, a private firm has taken over the management of Wolds prison in Humberside and secured two contracts–one to escort prisoners to and from court and the other run the immigration centre near Heathrow airport, thus affording relief to the police and enabling them to devote more time and energy for maintenance of law and order and prevention of crime, which are their primary duties and responsibilities. Further privatization of police has been achieved by having civilians to maintain police transport and to take care of stolen property during investigation. India could take its cue from British and American venture of privatization of police and prison services.

First, certain peripheral services rendered by the police in our country need to be privatized so that they can devote more time and energy to their basic duties by prevention and detection, specially when the country is under the grip of both internal and external terrorism. Imagine, the amount of time and resources are wasted by the police to escort prisoners to and from the courts on the day to day basis. So also the traffic duties and security cover given to V.I.P.s, V.V.I.P.s and threatened persons of various categories. Similarly, serving of summons, warrants (bailable etc.) on behalf of judiciary could be transferred to private security which could save considerable police man power and resources, which could otherwise be utilised for effective policing.

Apart from notorious delay in dispensation of justice in judicial courts, one important reason for the higher rate of acquitals lies in faulty and slipshod investigation by police, which in its turn, is largely due to inadequate staff for investigation work and the inability of the police officers to peruse investigation with a keen sense of commitment and determination. A sample survey conducted under the auspices of the national police commission in six states covering different parts of the country, revealed that an average investigating officer was able to devote only 37 percent of his time for investigation work, while the rest of his time was taken by V.I.P. arrangement, petition enquiries, preventive petrol and surveillance, court attendance, collection of intelligence, dispensing security cover to the designated individuals and other administrative work. On the contrary, due to increasing threats of violence from the terrorists, militants, naxalities, Maoists in large parts of the country. The police officials, in general are now more preoccupied with VIP and VVIP security and other odd jobs, when their own personal protection is in jeopardy, being in direct target of terrorists and Naxalities. It is therefore necessary, to contract out these services to private security agencies. The problem of policing particularly in urban and metropolitan cities are becoming increasingly complex, which could grow in the next decades due to growing unemployment, unplanned growth of cities, the population explosion, consumerism, displacement of tribal for major industrial projects, increase of economic inequalities, exploitation of landed gentry, corruption at high places, large scale misappropriation of public fund by the groups having political and bureaucratic nexus and several other factors. To meet these challenges effectively, the police must have a more innovative approach to crime control and its mechanism. The following steps may be taken.

(*a*) As part of this, one could consider farming out peripheral services to private security agencies of repute.

(*b*) The design, construction, if not management, of all new prisons could be contracted out to private agencies without any interference by govt.

(*c*) Based on performance appraisals, educational and reformatory services in prisons, could either be privatized or contracted out to the voluntary sector.

(*d*) The management of remand/reformatory home for children and women could also be privatized.

(*e*) Once, it is started in a planned and positive manner, as in U.K. and U.S.A., our bureaucracy may undergo a sea change, which I am in great doubt. The recent happenings in Indian prisons have started each and every intelligible citizen. The prisoners are found in possession of drugs, mobile phones, arms etc., and the dreaded under trial prisoners operate from their prison cells in relation to commission of crimes, like kidnapping their opponents, political bigwigs, business tycoons, for ransom, extortion, murder, through their hench men outside. This is being done in connivance of prison staff. The corrupt Govt. security personnel are heavily bribed and the mafia elements housed in prison cells lead luxurious life. Therefore the very purpose of curtailment of their liberty for a temporary period is defeated when Govt, staff in-charge of prison security are adding and abeting commission of crime in exchange of personal gain. Time would say, whether private security would be able to remain unpolluted and resist temptation of their sudden upliftment into a world of opulence. Some honest prison personnel indicate all about their complicity with the criminals inside, is a compulsion under duress. They are forced to toe their dictation under grave threat to their lives and lives of their family members. In fact, one jail warden on duty in Bhubaneswar prison was shot dead on the broad day light, due to his honesty and fair play. Being private employees of a security establishment, will they dare to withstand the threat perception of hardcore criminals ? These are some of the burning issues, the administration would face, when taking a decision on privatisation. U.K. and U.S.A. scenario is entirely different. They do not have corrupt bureaucracy as in India, who is in the ninth position in the list of most corrupt states in the globe.

Besides, without framing the appropriate laws regulating the activities of private security organisations, it would be perhaps immature to go ahead with this programme. A large number of private security agencies already exist in the country and are manned by retired police and defence personnel. Hence, there is enormous scope for competition, professionalism and cost effectiveness. It is indeed necessary to ensure that provide security agencies conform to minimum standards of structural organisation and supervision with a view to protect not only their own interest but also of those, whom they employ. It is heartening that Govt, of India is taking steps to bring out a legislation in order to regulate the functioning of private security agencies.

Security laws

1. Central Industrial Security Act
2. CBI Act
3. Prevention of Corruption Act
4. VIP Security Act
5. Preventive Sections of Indian Penal Code, Criminal Procedure Code.
6. Vigilance Act
7. Official Secrecy Act

Security defence systems

1. Metal Detectors/Mine Detector.
2. Mine Sweeper
3. Search Light
4. Spot Light
5. Inspection Mirrors
6. Binoculars
7. Turnstiles
8. Perimeter Surveillance and Control System
9. Barbed Intrusion Alarms
10. Bank Token Displays
11. Distress Alarms
12. Guard Post Alarms and other high-tech Electronics Equipment
13. Explosive Detectors
14. Bomb Detection/Disposal Equipment
15. Mobile Phone
16. Dry Shell Torch
17. Emergency Big Lighting Equipment
18. EAS (Electronic Article Surveillance)
19. CCTV (Close Circuit Television)
20. Reusable Hand Stock Tags
21. Disposal Stock Tags

22. Sensing Screens
23. Intruder Alarm
24. CCD Camera (Various types)
25. Door Vision
26. Automatic Alarms (Various types)
27. Bullet Proof Jackets
28. 24hours Emergency Vehicles fitted with bullet proof glasses
29. First Aid Box
30. Public Address System
31. Video Door Phone
32. Guard Vigil
33. Access Control – AP-530
34. Passive infrared (motion sensor light)
35. Door frame metal detector
36. Ultraviolet scanner
37. Auto gate system
38. Intelligent wireless security system
39. Fire Detector
40. LPG Detector
41. Motion Detector
42. Glass Break Detector
43. Roller Shutter Contact
44. Video Entry System
45. Bomb Blankets
46. Anti shop lifting equipment
47. Commercial lock system
48. Anti Bomb blast Glass
49. Skin testing camera
50. Image Processor.

10

SECURITY ORGANISATION

SECURITY

After World War II, Commerce and Industry began to suffer financial losses, due to theft and other illegal activities, which resulted in companies deciding that they needed some form of protection for their assets and property in about 1947, the age of the security officer was born. At that time, it was thought that, having a man present on the site during silent hours would be sufficient to prevent the types of losses that were occurring. This was the advent of the security officer but he was no more than a night watchman.

In almost all cases, the security officer was an elderly often, a retired person, who reported for duty when the factory or office closed for the day. He had no instructions, either written or verbal, other than that, his job was to prevent theft and losses. The average security officer sat in a hut in-front of blazing fire, listened to the radio. He did no patrols, had no communication with any outside agency and no idea about what he should do in the event of an incident. Infact, the general feeling of the client was that the mere presence of the individual would deter the potential thief. Cirme at this time was comparatively small and the person was a novice in comparison to today's sophisticated animal. No training was given nor it was felt necessary for the security officer. As the criminals became more professional, it was realized that, in order to combat crime, a trained person was required. The image of the modern security officer was born and training programmes were introduced to give the person some guidance on the duties that were required of him.

It was later felt that the security officer should be intelligent and imitative. He needs to be polite yet firm, when the necessity demands. It was found that the elderly retired man who might be incapacitated in some way, was not the right person for the job. In fact the security officer was no longer the person, who is the misfit within the organisation and who could not be used in any other function. As a result the security officer of the 1970 was born.

The companies realized that to combat the ever increasing levels of crime at both national and international level, they must have well trained and managed security team. This resulted in the appointment of security managers within the organisations who are charged with the task of building up effective security teams, who were given clear and concise training and written instructions. The increase in violence in security and the professionalism of the criminal necessitated, able bodied persons to counter their activities. Although the probability was that the security officer would never have to defend himself from attack, the chances of this happening were considerably reduced if the individual was of good physique, was well turned out and was personable and able to command respect.

RECRUITMENT OF SECURITY TEAM

Personal Standard

(a) *Age range:* The security managers are also required to be conscious of the age range of their staff and in order to be able to provide a reasonable career structure, the balance between the younger and more experienced individual had to be struck. Although maturity is a distinct advantage with security officers, it has to be born in mind that a mass exit of mature staff would result in manning problems, other points that have to be considered are the domestic circumstances of the individual and his/her past employers and clean record from criminal offence.

(b) *Health:* Physical fitness and health are factors that have to be considered. The long hours of work and the requirement to patrol continuously dictate that the security officer must not have back or foot problem. His eye sight should be fairly good and he should not suffer from bronchitis or similar chest complaints, because this would result in spells of sickness. The above yardstick is applicable to both contract and in house security.

(c) *The selection process:* The selectors/Interviewers should keep these following points in their mind.

   (*i*) Is the applicant stable enough to settle down in a job, where promotion prospects are limited?

   (*ii*) The job requires shift work. Is he capable for it?

   (*iii*) Is he prepared for over time engagement and unexpected shift charge?

   (*iv*) Is he confident of his timely and regular attendance, failing which there will be total dislocation as it affects the time off of another member of the team?

   (*v*) Is he capable to work alone and often at night?

   (*vi*) Does he realize that he will be subject to sever provocation on occasions and he must be able to control his temper and emotion?

   Through checking of the candidates' past record, qualifications etc., is a precondition for one's selection. All references must be taken in writing.

Training

After selection the person should be thoroughly trained with regard to training of security officer/guard, the matter has been exhaustibly dealt in the next chapter.

Discipline

Having outlined in broad terms of selection process, vetting and training, it is necessary to ensure that they are subject to discipline which lays down the parameters in which they are required to operate. No person, who is required to work in uniform should be allowed to operate without a set of regulations concerning his/her dress and conduct. This is the discipline guide and should be provided in the form of a security officers handbook.

Duties and Responsibility of Security Officers

In the corporate environment, the role of security is recognised as one of protecting the bottom line of corporate activities. Clearly, how the role of security is seen in the organisation is often determined by how far it is placed from the level of policy making and of the impact security can have on the operations and consequently on the profits. The security function must be so organised and placed in the organisation hierarchy as to allow effective implementation policy, responsiveness to business requirements and efficient reactions to local emergency and routine security needs. Security cuts across organisational lines. The chief security needs have the authority to inspect performances and evaluate security risk throughout the organisation. The security chief operates best, when his functional authority flows from highest echelons. Most important however, security master planning facilities the formation of a security programme that provides a balanced approach and channel resources to the organisations most pressing concerns.

In brief some important tips in relation to duties of security manager/officer are given below:

(*a*) He must first prepare security master plan in the lines elaborately discussed earlier, in the beginning.

(*b*) Prepare an exhaust security hand book for the entire security staff, highlighting their duties and responsibilities and prevention of security hazards.

(*c*) Direct the assistant security officer/supervisor to prepare daily roster for guards including night shift.

(*d*) Ensure security drill in the morning hours.

(*e*) Prescribe formats for registers to be maintained in gates, interview slips etc.

(*f*) He is required to design and print the identity cards, passes for the employees and get the punch card machines installed in the main gate.

(*g*) He is required to make round visits atleast three times a day and twice in the night to ensure alertness, performance of duties earnestly by the guards. In case of default, absence, illegal actions, indecent dealings of the members of the staff with visitors , he may suggest disciplinary action. He can issue warnings, word of caution and his displeasure to correct the erring staff.

(*h*) He is responsible in framing the security code of the corporate house.

(*i*) He is required to submit weekly, monthly and annual reports in the prescribed formats devised by him to the security committee, which is constituted to review the performance of security personnel. He is to work as the convenor of the committee with members chosen by the management. Generally, the head of the company or his representative, the factory manager, the general managers of each department like mechanical, electrical, instrumental, electronics, safety and fire prevention, finance , HRD, admin, law and workshop are the ex-officio members of the committee. A secretary may be appointed to assist the chief security officer in this regard.

(*j*) He is required to go round all the departments, branches, sections, duly inspect security aspects, and report events of any theft, wall breaking, trespass, intrusion etc., report to the management with action taken report (ATR). He is responsible for filing F.I.R. with local police and handover the accused to the police and assist in investigation work. He should maintain excellent rapport with the police admn., general admn, and other law enforcing authorities.

(*k*) He is to handle all cases of negligence, fraud, embezzlement of company fund, theft, violence at the work places, cyber offences, theft of trade secrets, cases of sabotage, espionage, and handover the offenders to the law enforcing authorities. He is required to act as the principal co-ordinator inside the factory premises as well as outside.

(*l*) He is to provide escort guards for drawal/deposit of heavy cash in the bank.

(*m*) Procure arms and ammunitions and licenses for the arms from the local licensing authorities.

(*n*) He is to procure uniforms, security appliances for the entire security personnel and the guards and distribute them.

(*o*) He is to provide protection to the head of the organisation factory, his family members, top ranking functionaries, and even management level officials in case of threat.

(*p*) He is responsible for training and education of security personnel , prepare security related posters, hoardings, pamphlets, for wide exhibition in the factory premises.

(*q*) Conduct mock security drill at intervals.

(*r*) He is to work as the co-ordinator for fire-prevention and fire fighting operation as his guards are the grassroot officers who handle fire prevention equipments/machineries. Along with the plant safety officer/engineer he is to train his own guards in fire fighting.

(*s*) He is to act as the event security officer, when ever events are organised in the plant premises or plant sponsored events outside.

(*t*) He is to deal with security related confidential matters, classified documents and remain in-charge of very sensitive legal documents, formula related papers and other research related paper and records.

(*u*) He is to maintain datas and records of glaring security lapses for monthly and annual review and take remedial measures.

(*v*) He is to work as the civil defence officer and train his staff on this score. He is to keep liaison with the district civil defence officer.

(*w*) He is required to act as relief and rehabilitation incharge of the employees of the company and even general public at times of natural calamities like flood, draught, fire havoc, earth-quake, tornado, storm and even man made big hazards, like communal riots etc.

(*x*) He is to look after the whole gamut of patrolling operation in the working area. The tidbits of this operation have been enumerated in detail under the heading "General Security"and elaborately discussed earlier.

(*y*) With regard to canine management in the area of patrolling, investigating and apprehending criminals etc., he is the chief planner. He is required to monitor all types of activities, where the dog's role is involved. He is also to look after the basic training of the dog, advanced training, man work/attack training, land searches, training, scent discrimination/identification etc. He is squarely responsible for dog squad's management from the stage of procurement, food and equipment, daily care and maintenance, veterinary care, accommodation, trainings, maintenance of special kennel staff/handlers, special insurance and legal provisions etc.

(*z*) He is required to frame security budget for the plant, conduct security audit at intervals and manage the security department as its team leader and strictly follow Ergonomics formulas/ procedures for efficient administration.

Duties and Responsibilities of Security Supervisor

(*a*) It is necessary that all security guards should be visited al least once in every shift (day or night) by the supervisor.

(*b*) He should complete a site visit inspection report which covers all aspects of the Guards' operating performances. The guards should be shown the report and asked to sign it. Any administrative problems, such as uniform requirements, or wages queries, should be included in this inspection report. These inspection reports need be presented to the operations manager/assistant security officer/security officer/manager including his recommendation or any disciplinary action.

Guard Duties

This has been elaborately discussed and duties enumerated in the preceding chapter under "General Security" heading, which may please be gone through.

SECURITY ACCOMMODATION ON SITE

Ideally, a security office should be purpose built but more often not, it will be conversion of an existing building or even a porta cabin. The security office is the first contact point for visitors to the firm and must convey an appearance of tidiness and efficiency. Dingy and poor working conditions will be reflected in the attitude of the staff to the manner in which they carry out their duties. The essential points in a security lodge may be obvious but some general points worth considering are:

(*a*) Must have a full and uninterrupted view of the entrance and surrounding area.

(*b*) Should be located as near as possible to the entrance gates so that no vehicle could pass without being challenged. It is also preferable, if car parks are located near by.

(*c*) The entrance area and gate house should be flood lit and there should be an emergency generator, provided in case of power failure.

(*d*) Electrically operated barriers installed at the entrance would make access and exit of vehicles, much easily controllable.

(*e*) In case of necessity for a weigh bridge, it should be installed in close proximity of the gate lodge and make part of security guards' duties.

(*f*) The interior lay out of the gate lodge should be such that visitors are segregated from security staff by means of a counter. This also ensures privacy of office and its contents and also prevents the gate lodge from being used as a point at which employees meet and waste time.

(*g*) Seating accommodation should be available for visitors.

(*h*) Toilet facilities should be available.

(*i*) Telephone and other equipments should be built into a console, which gives security of the equipment from both theft and inquisitive eyes.

(*j*) A rest area with facilities to warm food and make tea/coffee should be made available.

(*k*) Lockers for storage of personal equipments should be provided.

(*l*) A first aid box and facilities should be provided.

(*m*) A case safe to store items of found property, cash or other delivered items should be provided.

(*n*) A personal attack button should be installed with a high pitch siren on the exterior of the building.

(*o*) The controls for all external lighting should be located in the gate lodge.

Security as a Profession

With the advance in technology throughout industrial sector in the modern world and the ever increasing opportunities that this brings to the security industry, it is important that a career in the security industry be regarded by people as both worth while and lucrative. To achieve this, the industry throughout the industrial world must concentrate on enhancing the image of the security officer. The days of the night watch man or the untrained security guard are gone forever. The industry must develop a professional status with suitable accreditation, performance and training. Training programme must be devised to incorporate the ever advancing technology and continuous updating should be normal practice.

There is currently a distinct lack of quality and professional leadership in the industry throughout the world. This is because of salaries not being competitive in relation to other professions. The industry has to demonstrate that it has credibility and is, professionally trained, if it is to remain self regulated. Failure to do this will result in Govt. regulation, which security organizations throughout the world do not want to happen. Without these three important ingredients, the Industry would never gain the confidence of its Govt. and its customers. To achieve this confidence, the industry must strive to employ a better educated individual, who comes from a good background and is capable of absorbing the technology required to keep up with the advances being made in security equipment. The ongoing training and updating must become a part of the profession. The benefits of doing this for out weigh the obvious costs, that will be incurred. The training will improve the performance of individuals and the continuity in the labour force which will result providing experience to perfect their individual skills. In order that the security industry achieve recognition in the career opportunities section, they must produce a qualification system, which is evaluated and recognized by its association. Once this is achieved, it will result in level playing field in relation to standards. The training of managers and supervisors should form part in this programme.

SECURITY MANAGEMENT FOR 21ST CENTURY

The modern approach to combating the loss of assets to develop a security team that is trained professionally to operate technology so as to combat crime and protect property and people.

The points to be addressed, when setting up a security team, should cover the following

(*a*) The threat to the company assets and people.

(*b*) The risk assessment to the assets and how the company can continue trading without disruption.

(*c*) The preparation of security policy to protect these assets and implementation of this policy.

In the past, the tendency of security management was to concentrate maximum effort on personal physical security. Although protection from external sources is always present and therefore essential. At present, there is ever increasing threat from internal activities within the organization. There are blacksheeps who can, through inciting labour disputes, industrial espionage, theft of trade secrets and sabotaging development and growth, very quickly ruin the business. The security manager, of the 21st century will have the ability to gather information and intelligence from within the work force in order to keep the management up to date with this information. He needs to have exercise to understand the financial systems within the organization and the operating environment as it affects the work force. The security manager must always remain aware of the possibility of industrial espionage and effect policies to prevent this happening. He should strive to reduce internal losses by the effective use of information and intelligence, gained through sources within the work force. In nut shell, he is to provide total risk management by putting in place a strategy to identify, whether or not, his policies, he has implemented actually works. By identifying internal crime, he will be greatly assisting the company to control its costs and this in turn, will keep the cost of its merchandise competitive within the market place.

Security Training and Education

1. **Basic and mandatory training:** It must be informative, participative and must be a fun to gain new skills and improve attitudes.

   The following are some of the subjects of training given below:

   (*a*) Report writing
   (*b*) Personal hygiene
   (*c*) Public relations
   (*d*) Safety habits
   (*e*) Fire prevention
   (*f*) Communications
   (*g*) Patrolling
   (*h*) Reporting, logging
   (*i*) Time sheets (pay roll and client result)
   (*j*) Criminal law
   (*k*) Civil responsibility
   (*l*) Use of force
   (*m*) First aid
   (*n*) Equipment use and care
   (*o*) General liability
   (*p*) Company security manual.

Continued Education

After basic training many other job related subjects should be covered. Training can be given by class room instruction lecture, audio cassette, video presentation, class discussion and text books. The training should cover not only security matters but safety related subjects, health betterments, good driving habits etc. The training programmes must be informative and fun. Extra incentive-maintain skills, gather knowledge, improve quality and performance should form part of the curricula. The second phase of training should include the following subjects.

(*a*) Armed subjects
(*b*) Legal use of fire arms
(*c*) Ethical and moral considerations
(*d*) Liability for acts
(*e*) Use of deadly force
(*f*) Search seizure and arrest procedures
(*g*) Fire arm safety and maintenance
(*h*) Fundamentals of weapon use
(*i*) Stance, grip, sighting picture and trigger control etc.

Additional subjects to be covered under training programme.

(*a*) Advanced First Aid
(*b*) Fire equipment
(*c*) Advance report writing
(*d*) Vehicle patrol
(*e*) Hazardous material handling
(*f*) Telephone / Radio use and care
(*g*) Handling unruly persons
(*h*) Drug identification
(*i*) Supervision
(*j*) Deadly weapons
(*k*) Loss prevention
(*l*) Alarm systems
(*m*) Crime prevention
(*n*) Security surveys
(*o*) Principles of leadership
(*p*) Inspection procedure
(*q*) Equipment use
(*r*) Accident investigation
(*s*) Alarm equipment
(*t*) Typing/Computer skills
(*u*) Theft investigation
(*v*) Self defence

(*w*) Work place violence
(*x*) Employee theft
(*y*) Courtesy in security
(*z*) Handling emergency situations
(*aa*) Bomb threats
(*ab*) Fire systems
(*ac*) Time management
(*ad*) Ergonomics.

Training evaluation (ongoing process)

1. Willingness to learn
2. Speed of comprehension
3. During probation:
   (*a*) Ability to learn
   (*b*) What effect to put forward
   (*c*) Performance expected.
4. Quarterly/Semi/yearly – promotion self motivation to achieve success and career development.
5. Exit interview = likes and dislikes of company. How to improve system, what to change for better most of all.

Training calendars of SITO (Security Industry Training Organisation) (UK)

(*a*) Intruder alarm engineering
(*b*) Close circuit television operation
(*c*) Access control
(*d*) System surveying
(*e*) Project management
(*f*) Security guarding
(*g*) Retail security
(*h*) Leisure security
(*i*) Investigation
(*j*) Risk management
(*k*) Transport property under guard
(*l*) Supervisory management
(*m*) Industry wise security as enumerated in the preceding chapter
(*n*) Holding mock security drill at intervals
(*o*) How to become a successful security officer.

Guard Training

In addition to the above disciplines, the security guards need some additional tips as detailed below.

How training is accomplished.

(*a*) Written material – Reading
(*b*) Personal Lecture – Instructor speaking
(*c*) Audio and/or video programmes
(*d*) Printed hand outs and news letters

(*e*) On the job training
(*f*) Correspondence studies
(*g*) Conferences and seminars
(*h*) College and University programmes

Parts of Training

(*a*) *Basic general knowledge appearance:* Personal hygiene (Fresh and clean, nobody odour of offensive breath), hair in place, hands clean, good posture, awake and alert.

Dress properly – Clean and neat looking uniform, well identified or marked dress for the job. Indoor/Outdoor in weather/day or night/selected post or in public view, traffic, high visibility etc., command respect – Don't project a wrong or bad image (dirty, slept in look, overtired no energy). Be a professional.

(*b*) *Job knowledge:* Laws – Use of force, detention, public relation and proper etiquette.

Proper skills – Read, write, spelling and form, speak proper language.

Code of conduct — Illness and absenteeism. Tardiness standard – Procedures for company (Do's and Don'ts)

Specific duties – Post instructions (must be written) for each location or job position.

(*c*) *Communication:* Security officer is the eyes and ears of management for majority of time. Management/owner only work 40/60 hours per week minus vacation, personal days, away on business etc.

Security officer is their replacement and must communicate what takes place when they are gone and not around. He is able to write reports on who/what/when/whether/why and how things happened in such a way that reader can get a mind picture of what went on. A good report should become a written log of all facts, collected at the time of the investigation of the incident. Remember 5 "Ws" and one "H".

Who – was involved

What – happened

When – time of occurrence

Where – it took place

Why – what led to incident

How – tell the story.

(*d*) *How to react or handle certain happenings:*

1. How to use equipment: Two way radio, telephones, recording devices, vehicles, emergency and safety devices.
2. How to handle people and their many attitudes or ways: Be polite but firm, be fair and informative. Don't loose your temper. Remind them what you have job to do and take proper action. Remember "Courtesy is Courteous".
3. First aid procedure: Help to relieve the pain and suffering while beginning the healing process. Fast proper action may save a life.
4. Fire protection procedures and how to use equipment: Warnings – Hazards areas, know what to do and what not to do. Alarm – Positive notification to fire department. Life safety first – Notify subjects in building to get out, help if necessary. Direct – Fire fighting persons to correct building or area. Attempts to contain if possible.

5. Deter crime by being visible yet don't set a routine rounds, breaks etc., check doors/ windows, any access to location over 8" × 10" opening. Inquire about all suspicious persons, actions or packages. Be alert and ask if you don't know.
6. Never take granted or assume, check it out – Don't be afraid to ask questions, Be polite Be alert – Be courteous – But be firm.
7. Always document – Written logs and reports are a must. They help us remember, they communicate information to others , they do what we can't do personally. They protect our reliability and honesty. They may even help save our jobs.

Specific information for each location

(*a*) Access control, ID systems how they operate.
(*b*) Alarm system's and CCTV consols what they protect.
(*c*) Emergency procedures, fire, weather, disaster.
(*d*) Patrols when and where.
(*e*) Where to find/locate things – Emergency shut off switches or valves, fire extinguishers or hoses, alarm panels, hazard materials, special items or persons (know your job location) especially all entrances/exits, lights, visitors and stairs.
(*f*) Information (contacts) visitors or emergencies, names, phone numbers, times, places etc.
(*g*) How to do/act/react/perform.

Training can be done in person by trainer, in a school type atmosphere or on the job. Supervisors must continue to train and enforce programmes. Officers can also improve by self taught methods *i.e.*, books, films, videos, audiotape programmes, seminars and magazines etc.

Each training item must be documented and possibly tested for validity of knowledge. Permanent records are to be kept in personal files for promotion and liability support. Self motivated and better educated officers are usually easier to sell as an asset for the security departments/agency and they usually produce a much higher quality of performance.

List of subjects for training

The following are some of the subjects for training.

(*a*) Stress analysis and motivation
(*b*) Emergency and disaster planning
(*c*) Weapon, how and when to use
(*d*) Fire prevention and equipment
(*e*) Access control
(*f*) Vehicle and foot patrol
(*g*) Report writing and note taking
(*h*) Substance abuse
(*i*) Accountability
(*j*) Conflict resolution
(*k*) Future of the private security industry
(*l*) Local ordinances
(*m*) Personal appearance
(*n*) Self defense
(*o*) Investigations
(*p*) Fraud and theft prevention
(*q*) Public relation
(*r*) Plant and building protection
(*s*) Leadership skills
(*t*) Executive protection
(*u*) Crime scene containment
(*v*) Field contracts/Interview techniques
(*w*) Effective communications
(*x*) Officer safety considerations
(*y*) Arrest control techniques
(*z*) State laws and crimes

(*aa*) Code of ethics
(*ab*) Telephone manners
(*ac*) Searching
(*ad*) Care of uniforms and equipments.

This list is an indication. It may be broadened or shortened.

CODE OF ETHICS FOR SECURITY GUARD

As a security guard, my fundamental duty is to serve mankind, to safeguard life and property of my employer, to protect the innocent against deception, the weak against oppression or intimidation and the peaceful against violence or disorder, and to respect the constitutional rights of all men to liberty, equality and justice.

I will keep my private life unsullied as an example to all, maintain courageous calm in the face of danger, scorn or ridicule, develop self restraint and be constantly mindful of the welfare of others. Honest in thought and deed in both my personal and professional life. I will be examplary in obeying the laws of the land and the regulation of my employer. Whatever I see or hear of a confidential in nature or that is confined to me in my official capacity, will be kept ever secret unless revelation is necessary in the performances of my duty.

I will never act officiously or permit personal feelings, prejudices, animosities or friendships to influence my decisions. With no compromise for crime, I will perform my duty courteously and appropriately without fear, favour, malice or ill will, never utilize unnecessary force or violence.

I recognize the badge, I wear as a symbol of public faith and accept it as a trust to be held so along as I am true to the ethics of the security profession. I will constantly strive to achieve these objectives and ideas, dedicating myself before God to my chosen profession.

Model format of incident report

Facility – Write store name

Address – Number, street, city or country

Owner/Manager – Give name

Date – Month, date, year, day, time

Type of investigation – Give all pertaining information as to who, what, when, how and how much.

Subject:

Who – Give name address, DOB (Date of barth), social security no., sex, height, weight, hair, eye, complexion, clothing, mannerism, identification marks if any.

When – Approximate time

Where — Location

What – Items taken for instance if it is theft

How – How the event took place

Any other relevant information and Investigation made by (Signature).

Guard rules and regulations

1. Report for duty in full uniform, clean and neat in appearance, ready for work.
2. Report for duty on time and don't leave before relieved or assigned time, if directed.
3. If for any reason, you are unable to report for assigned duty or become ill while in duty, call supervisor in charge.

4. Read and follow all instructions for position assigned.
5. Report all emergency situations immediately.
6. Fill out "Guard daily report" completely and legible before leaving at the end of your duty.
7. No eating or smoking other than the designated place.
8. Be sharp and well informed. Be firm (when needed) but courteous at all times. Remember, good public relation is very important.
9. Friends and/or relations are not permitted on post with you. Don't congregate with other employees.
10. Clean up your area before leaving post. Do your job properly and stay alert.

What every guard should know?

1. The exact layout of the area, he is to protect, including all buildings and occupations, yard areas, entrances, road ways, stair ways, elevator shifts and other features of the premises.
2. How to call the local force and police department and other sources of help.
3. Location of all doors and fire exits.
4. Location of fire alarms boxes and telephones inside and outside the plant and how to use them.
5. Whether alarm are only internal or are connected directly to the fire department.
6. The proper pressure on the regular sprinkler system valve.
7. Location of various controlling sprinkler systems, inside riser valves, post indicator valves, sectional valves and drain valves.
8. Location of hydrants, stand pipes, small hose and fire extinguishers.
9. Locations and care of hazardous manufacturing processes, specially those continuing during the night, holidays and weekends.
10. Location of dangerous or extra hazardous machinery or material.

Hints on patrolling

(*a*) The first patrol made after joining the duty is very much important. The guard should check for hazards, which departing employee may have created in his haste, at the end of day.

(*b*) He must be prepared to handle any emergency, he would likely to face during his tour.

(*c*) An observant guard will note many conditions on his first round, then check them on his later tours. New dangers may develop at any time during the night and a good guard will be alert for them. He should follow these rules.

1. Never leave his post except for good and sufficient cause.
2. Never allow unknown or unauthorized persons to visit the property.
3. Never permit personal friends or acquaintances to visit him during his tour.
4. Be sure to visit each station on time and in the order, prescribed by the supervisor. Never take shortcuts or omit visiting any station nor change the sequence in which, he is to visit them.
5. Cultivate a "nose for trouble". Investigate strange odours specially smoke or gas. If the odour persists, but he cannot find the source, immediately he should call the proper authorities.

6. Never smoke or light matches while on tour, except in authorized areas.
7. If for good cause, you are forced to deviate from your route, make a report to the supervisor.
8. Remember a mere "clock - puncher" is not a good guard. He should keep his eyes, ears, nose and common sense open at all times. A guard owes to his employer his fellow employees, his community and himself to perform his job as it is his sacred duty.

Model Format for Report to Police Incase of Robbery

Date ____ Time ____ Store No. ____

(Fill the blanks. Give the top copy to the police officer on the scene and send copy to security Deptt.)

Sex ____ Age ____ Height ____ Weight ____

Hair ____ Hat (Colour, Type) ____

Eyes ____ Glass type ____

Tattoos ____ Scars/Marks ____

Complexion ____ Shirt worn ____

Trouser ____ Shoes ____

Description of weapon used ____

Automobile description ____

(Registration number, make, colour)

What Robber said ____

Details of talks among

The members of Gang ____

(Signature)

(Stay on the phone, if you are calling the police. Don't hang up.)

CCTV OPERATION TRAINING

An operator requires a wide range of knowledge for proper operation and maintenance of CCTV. The training topics should cover the following.

1. Operation of installed monitoring equipment.
2. Maintenance of video tapes and preservation of evidence.
3. Observations including identification, suspicious behaviour and descriptions.
4. Communication in the control room.
5. Topography of the area observed by day and by night.
6. Code of practice for the operation of the system.
7. Court room procedures and witness statements.

Motivating Security Personnel

The various motivating factors are out lined below. Man's needs are arranged in a hierarchy of importance like-

1. *Physiological needs:* The basic life sustaining needs are air, food, water, sleep and sex etc.
2. *Safety needs:* Desires for protection against danger, threat and deprivation.
3. *Social needs:* Giving and receiving love and friendship, belonging association, acceptance by peers.
4. *Esteem needs:* It includes both self esteem and esteem of others and include the need for status, appreciation, importance and prestige. Satisfaction of these needs leads to self confidence, self respect, competence, achievement, independence, freedom, a feeling of worth, capability, strength and being useful and necessary in the world.
5. *Self actualization needs:* These include realization of one's potentialities, self fulfillment and continued self development. These needs could be fulfilled through one or any combination of athletics, politics academics, religion or business. While the first two needs are satisfied externally, the balance two are satisfied internally. Research has proved that achievers are motivated by jobs that offer personal responsibility, feed back and moderate risks. This theory is called as Abraham Maslow's theory.

   As per Frederick Herzberg's motivation hygiene theory, dissatisfaction occurs, when following are not present in the job; good money, status, job security, congenial people to work with, high quality supervision, company policies, administration and excellent working conditions. Presence of these factors bring to a satisfactory neutral state. But they themselves are not strong motivating factors. So the satisfying motivating factors are:

   (*a*) Feelings of achievement
   (*b*) Professional growth
   (*c*) Recognition
   (*d*) Challenge
   (*e*) Scope in the job.

These are all intrinsic factors. The company employing, should provide the factors that prevent dissatisfaction. But the leader should provide the factors that both satisfy and motivate.

Douglas Mc Gregor's 'X' & 'Y' theory speaks that an employee is to be coerced to perform, as they dislike work and are generally lazy. 'Y' theory says, just the opposite way, employees like work, seek responsibility, are creative. These are major theories on motivation but there are other host of theories like goal setting, reinforcement, equity and expectancy etc.

Characteristic of the high achiever

1. He has desire to take personal responsibility.
2. He likes to take calculated risks.
3. He is keen to know the results of his efforts.
4. He tends to persist in the face of set backs.
5. He is likely to be innovative.
6. He develops interpersonal competence.
7. He is future oriented.
8. He regards time as valuable.
9. He is restless and likes to take initiative.
10. He is never completely content.

Creating the Achieving Environment

It is a fact that people are seldom motivated directly by others. They basically motivate themselves. However, even if they have high potential for self motivation, it is only the creation of an achieving atmosphere, that will spark the individual himself to greater effort and hence achievement. Here are some suggestions on how to develop such an atmosphere.

1. Know your staff.
2. Tell them where they fit in the organization.
3. Tell them why they and what they do are important.
4. Establish clear cut objectives and standards.
5. Evaluate progress against set objectives and standards.
6. Delegate and invite participation in decision making.
7. Give credit and support.
8. Take prompt corrective action.
9. Communicate
10. Develop talent continuously and rigorously.

The real motivators

1. Money
2. Responsibility, challenge and achievement
3. Recognition

What the security head to do to motivate his subordinates?

1. He must reward the behaviour or action, that the wants to be repeated.
2. He must not reward activities that are not desired.
3. He must not punish or ignore the right behaviours and activities.
4. He must set the right example.

What do we really want?

1. We want superior quality work.
2. We want costs to be cut. We seek financial discipline.
3. We want time schedules to be met.
4. We want people to be creative, to think better ways of doing things.
5. We want people to take decisions quickly instead of wasting time on meetings, voluminous report preparations, multiple analysis figures etc., in the name of planning "that stayed played" but not acted upon.
6. We want reliable people, who do what needs to be done, without making too much of noise.
7. We seek loyalty from our staff.
8. We look for team work.
9. We look for responsibility in behaviour and attitude.
10. We want demonstration of leadership behaviour.

11. We want to communicate to our subordinates clearly the penalties accrued on contrary to good behaviour demonstrated.
12. We need to be sure that we set the right example always.
13. We want to apply penalty and reward procedure quickly, whenever required.

To sum up, to really motivate subordinates, you need to ensure that, they have a job that challenges them, a pay packet that is little more than just right for the job. The security heads must create conditions, which enable subordinates to reach their personal goals. A business establishment cannot run profitably, unless its manpower is motivated to contribute its best effort to the organization. Creating motivation is the prime duty of security manager, who has even only one person, reporting to him.

RETENTION

A better trained security person is more knowledgeable of his responsibility and becomes a highly quality person. Shows initiative and self discipline. Therefore, security officer is more valuable to both the guard agency and the client. This makes it easier to sell additional service at even higher rate, which helps compensate for any bonus and/or higher wages paid. Historical problems of the private security industry has been in the hiring of poorly qualified persons and then no proper training given for those persons put into positions of responsibility and 'account ability'. Today, clients are forcing the private security industry to introduce approved training programmes and even demanding proof of their training completion. The market, forces this action because, it fears for any legal liability action that might go against their company. This could be disastrous.

Today's security officer must handle functions that are increasing in complexity and scope. These expanded duties require a high caliber candidate, who already possesses the necessary skills or one, who is capable of acquiring them quickly. Looking into the 21st century, security management needs technical advances to lower costs for education and training of private security personnel.

INDEX

S

T

U

V

W

Y